An Introduction to Integral Transforms

An Introduction to
Integral Transforms

Baidyanath Patra, *Ph. D.*

Ex-Professor, Bengal Engineering Science University
(Recently named as IIEST, Shibpur)

CRC Press
Taylor & Francis Group
Boca Raton London New York

CRC Press Is an imprint of the
Taylor & Francis Group, an **informa** business

CRC Press
Taylor & Francis Group
6000 Broken Sound Parkway NW, Suite 300
Boca Raton, FL 33487-2742

First issued in paperback 2023

© 2018 by Baidyanath Patra and Levant Books
CRC Press is an imprint of the Taylor & Francis Group, an informa business

No claim to original U.S. Government works

ISBN-13: 978-1-138-58803-5 (hbk)
ISBN-13: 978-1-03-265335-8 (pbk)
ISBN-13: 978-0-429-50358-0 (ebk)

DOI: 10.1201/9780429503580

Print edition not for sale in South Asia (India, Sri Lanka, Nepal, Bangladesh, Pakistan or Bhutan)

Library of Congress Cataloging in Publication Data
A catalog record has been requested

Visit the Taylor & Francis Web site at
http://www.taylorandfrancis.com

and the CRC Press Web site at
http://www.crcpress.com

LEVANT

Dedication

I dedicate this book to my beloved wife, Mrs. Minu Patra, for standing beside me throughout my career and during the writing of this book. She has been my inspiration and motivation to expand my knowledge and without her selfless support and encouragement this book would not have been possible.

Preface

Integral transform is one of the powerful tools in Applied Mathematics. This book provides an introduction to this subject for solving boundary value and initial value problems in mathematical physics, Engineering Science and related areas.

During my fifty years teaching and research experiences at the Indian Institute of Engineering Science and Technology, Shibpur, formerly known as Bengal Engineering College and Bengal Engineering & Science University, I have been introduced to the subject of Integral Transforms.

Here an attempt has been made to cover the basic theorems of commonly used various integral transform techniques with their applications.

The transforms that have been covered in this book in detail are Fourier Transfoms, Laplace Transforms, Hilbert and Stieltjes Transforms, Mellin Transforms, Hankel Transforms, Kontorovich-Lebedev Transforms, Legendre Transforms, Mehler-Fock Transforms, Jacobi-Gegenbauer-Laguerre- Hermite Transforms, and the Z Transform.

This textbook targets the graduate and some advanced undergraduate students. My endeavor will be considered fruitful if it serves the need of at least one reader.

Date : January, 2016
                                                                    B. Patra

Acknowledgement

I would like to take this opportunity to express my gratitude to my respected teacher, Professor S C Dasgupta, D. Sc.. He introduced me to this subject and developed my career in this field of research with special emphasis to application of Integral Transforms.

I am indebted to my many research students who worked under my supervision on this subject. I also acknowledge the authors of all the books that I read and results of which have been used in preparation of this book.

I would also like to thank my sons and their families for their constant encouragement during the writing of this book. Special thanks goes to my grandsons for their support for this work despite all the time it took me away from them.

During the process of writing, I have been greatly inspired by the textbooks of I. N. Sneddon and L. K. Debnath on this subject. Many other books and research papers, all of which are not possible to be mentioned individually, have also been consulted during this process.

I am thankful to the publishers, Levant Books for their unfailing co-operation in bringing this book to light within a reasonable amount of time.

Contents

Chapter 1

FOURIER TRANSFORM

1.1 Introduction.

The method of integral transforms is one of the most easy and effective methods for solving problems arising in Mathematical Physics, Applied Mathematics and Engineering Science which are defined by differential equations, difference equations and integral equations. The main idea in the application of the method is to transform the unknown function, say, $f(t)$ of some variable t to a different function, say, $F(p)$ of a complex variable p. Then the associated differential equation can be directly reduced to either a differential equation of lower dimension or an algebraic equation in the variable p. There are several forms of integral transforms and one form may be obtained from the other by a transformation of the coordinates and the functions. The choice of integral transform depends on the structure of the equation and on the geometry of the domain under consideration. This method of integral transform simplifies the computational techniques considerably.

Suppose that there exists a known function $K(p,t)$ of two variables p and t and that

$$\int_a^b K(p,t)\ f(t)\ dt = F(p) \tag{1.1}$$

is convergent. Then $F(p)$ is called the integral transform of the function $f(t)$ by the function $K(p,t)$, which is called the **kernel** of the transform. Here, the variable p is a parameter, which may be real or complex. Different forms of kernels will generate different form of transforms. Below we discuss a variety of integral transforms with their applications after using different forms of kernels in variety of domains. To begin with, we consider Fourier Transform.

1.2   Classes of functions

A singlevalued function $f(x)$ of the independent variable x which is continuous in an interval [a,b], is said to belong to a class denoted by $f \in C\ [a, b]$.

A function $f(x)$ is said to be piecewise continuous in an interval (a, b) if the interval can be partitioned into finite number of non-intersecting subintervals $(a, a_1), (a_1, a_2), ..., (a_{n-1}, b)$, in each of which the function is continuous and has finite limits as x approaches the end points of each of the sub-intervals. Such a function is said to belong to a class denoted by $f \in P(a, b)$.

A piecewise continuous function f in (a, b), whose first order derivative is also a piecewise continuous function in (a, b) and belongs to a class denoted by $f \in P^1(a, b)$.

The set of functions $f(x)$ is said to be absolutely integrable over Ω, if $\int_\Omega |f(x)| dx$ is finite. Then we say that this function belongs to a class denoted by $f(x) \in A_1(\Omega)$. Similarly, the statement $f \in A_m(\Omega)$ implies $\int_\Omega |f(x)|^m dx$ is finite.

Finally, we introduce a class of functions $f(x)$, which satisfies the following conditions

   (i) $f(x)$ is defined in $c < x < c + 2\ l$
   (ii) $f(x)$ is periodic function of period $2\ l$ and
   (iii) $f(x)$ and $f'(x)$ are piecewise continuous in $c < x < c + 2l$ denoting by $f(x) \in P^1(c, c + 2l)$.

This class of function $f(x)$ is said to satisfy Dirichlet's conditions.

A function $f(x)$ is said to be of exponential order σ as $x \to \infty$, if constants $\sigma, m(> 0)$ can be so found that $|e^{-\sigma x} f(x)| < m$ implying $|f(x)| < m\ e^{m\sigma}$ for $x > x_0$. Equivalently, we also write $f(x) = 0\ (e^{m\sigma})$ as $x \to \infty$.

1.3   Fourier Series and Fourier Integral Formula

Suppose that a function $f(x)$ satisfies Dirichlet's conditions over the interval $(-l, l)$ and it belongs to the class $P^1(R)$ and also to the class

$A_1(R)$. Let the principal period of $f(x)$ be $2\,l$. Then $f(x)$ admits of the Fourier series

$$f(x) = a_0 + \sum_{n=1}^{\infty}\left[a_n \cos\frac{n\pi x}{l} + b_n \sin\frac{n\pi x}{l}\right]$$

where

$$a_0 = \frac{1}{2l}\int_{-l}^{l} f(x)\,dx$$

$$(a_n, b_n) = \frac{1}{l}\int_{-l}^{l} f(x)\left[\cos\frac{n\pi x}{l}, \sin\frac{n\pi x}{l}\right]dx$$

so that

$$a_n - i\,b_n = \frac{1}{l}\int_{-l}^{l} f(x)\,e^{\frac{-in\pi x}{l}}\,dx$$

Let us now set

$$a_0 = c_0$$

$$a_n = c_n + c_{-n}$$

$$i\,b_n = c_n - c_{-n}$$

Then,

$$f(x) = c_0 + \sum_{n=1}^{\infty}\left[c_n e^{\frac{-in\pi x}{l}} + c_{-n}e^{\frac{in\pi x}{l}}\right]$$

$$= \sum_{-\infty}^{+\infty} c_n e^{\frac{-in\pi x}{l}} \tag{1.2}$$

where,

$$c_n = \frac{1}{2l}\int_{-l}^{l} f(x)\,e^{\frac{in\pi x}{l}}\,dx = \frac{1}{2l}\int_{-l}^{l} f(t)\,e^{\frac{in\pi t}{l}}\,dt \tag{1.3}$$

This form of Fourier series is called the complex form of Fourier series in $(-l, l)$. Thus, from (1.2) and (1.3), one gets

$$f(x) = \sum_{-\infty}^{+\infty}\left[\frac{1}{2l}\int_{-l}^{l} f(t)\,e^{\frac{in\pi t}{l}}\,dt\right]e^{-\frac{in\pi x}{l}} \tag{1.4}$$

Let us now put $\frac{\pi}{l} = \delta\xi$. We note that $\delta\xi \to 0$ as $l \to \infty$ such that $l.\delta\xi = \pi = a$ finite number. Then the series (1.4), before taking the limit as $\delta\xi \to 0$, becomes

$$f(x) = \frac{1}{2\pi}\sum_{-\infty}^{+\infty}\delta\xi\left[\int_{\frac{-\pi}{\delta\xi}}^{\frac{\pi}{\delta\xi}} f(t)\,e^{int\,\delta\xi}\,dt\right]e^{-inx\,\delta\xi}$$

$$= \frac{1}{2\pi}\int_{\frac{-\pi}{\delta\xi}}^{\frac{\pi}{\delta\xi}} f(t)\left[\sum_{-\infty}^{+\infty}\delta\xi.c^{in(t-x)\delta\xi}\right]dl$$

after interchanging formally the summation and integration signs. Since, by assumption $f(x) \in P^1(R)$ and also $f(x) \in A_1(R)$, making $\delta\xi \to 0$ and using the definition of Riemann definite integral as limit of a sum we get

$$f(x) = \frac{1}{2\pi} \int_{-\infty}^{+\infty} f(t) \left[\int_{-\infty}^{+\infty} e^{i(t-x)y} dy \right] dt \qquad (1.5)$$

$$\Rightarrow f(x) = \frac{1}{\sqrt{2\pi}} \int_{-\infty}^{+\infty} e^{-iyx} [F(y)] \, dy, \qquad (1.6)$$

where
$$F(y) = \frac{1}{\sqrt{2\pi}} \int_{-\infty}^{+\infty} f(t) e^{iyt} \, dt \qquad (1.7)$$

Again, from eqn. (1.5) one gets

$$f(x) = \frac{1}{\pi} \int_0^\infty dy \int_{-\infty}^{+\infty} f(t) \, \cos \, y(-t + x) \, dt \qquad (1.8)$$

Eqn. (1.8) is known as the Fourier Integral formula. At a point of finite discontinuity of $f(x)$, the left hand side of (1.8) is replaced by $\frac{1}{2}[f(x+0) + f(x-0)]$ in the sense of limiting values.

A detailed proof of this statement is included in the following corollaries.

Coro 1.1   Fourier's Integral Theorem.

If $f(t) \in P^1(R)$ and also $f(t) \in A_1(R)$, then for all $x \in R$,

$$\frac{1}{\pi} \int_0^\infty dy \int_{-\infty}^{+\infty} f(t) \, \cos[y(x-t)] dt = \frac{1}{2}[f(x+0) + f(x-0)]$$

Proof.   If $k > 0$,

$$\int_0^\infty f(t) dt \int_0^\lambda \cos[y(t-x)] dy - \int_0^\lambda dy \int_0^\infty f(t) \cos[y(t-x)] dt$$

$$= \int_0^k f(t) dt \int_0^\lambda \cos[y(t-x)] dy + \int_k^\infty f(t) dt \int_0^\lambda \cos[y(t-x)] dy$$

$$- \int_0^\lambda dy \int_0^k f(t) \cos[y(t-x)] dt - \int_0^\lambda dy \int_k^\infty f(t) \cos[y(t-x)] dt$$

$$= \int_k^\infty f(t) dt \int_0^\lambda \cos[y(t-x)] dy - \int_0^\lambda dy \int_k^\infty f(t) \cos[y(t-x)] dt$$

Since $f(t) \in A_1(R)$, for every arbitrary positive number $\in$, we can find a number K_1 such that

$$\int_k^\infty |f(t)| \, dt < \frac{\in}{2\lambda}, \quad k > K_1$$

Thus,

$$\left| \int_0^\lambda dy \int_k^\infty f(t)\cos[y(t-x)]dt \right| < \int_0^\lambda dy \int_k^\infty |f(t)|dt < \frac{\in}{2}.$$

Also, since

$$\int_0^\lambda \cos[y(t-x)]dy = \frac{\sin[\lambda(t-x)]}{t-x}$$

it follows that

$$\left| \int_k^\infty f(t)dt \int_0^\infty \cos[y(t-x)]dy \right| = \left| \int_{k-x}^\infty f(t+x)\frac{\sin\lambda t}{t}dt \right|$$

Thus, there exists a number K_2 such that if $k > K_2 + x$,

$$\left| \int_{k-x}^\infty f(t+x)[\sin\lambda t/t]dt \right| < \frac{\in}{2}.$$

$$\left| \int_{k-x}^\infty f(t)dt \int_0^\lambda \cos[y(t-x)]dy - \int_0^\lambda dy \int_k^\infty f(t)\cos[y(t-x)]dt \right| < \in,$$

for $k > K = max(K_1, K_2)$ even for large values of λ

In other words,

$$\lim_{\lambda\to\infty} \int_0^\infty f(t)dt \int_0^\lambda \cos[y(t-x)]dy$$

$$= \lim_{\lambda\to\infty} \int_0^\lambda dy \int_0^\infty f(t)\cos[y(t-x)]dt \qquad (1.9)$$

In an exactly similar way we can show that

$$\lim_{\lambda\to\infty} \int_{-\infty}^0 f(t)dt \int_0^\lambda \cos[y(t-x)]dy$$

$$= \lim_{\lambda\to\infty} \int_0^\lambda dy \int_{-\infty}^0 f(t)\cos[y(t-x)]dt \qquad (1.10)$$

Adding (1.9) and (1.10) we get

$$\lim_{\lambda\to\infty} \int_{-\infty}^\infty f(t)dt \int_0^\lambda \cos[y(t-x)]dy$$

$$= \lim_{\lambda\to\infty} \int_0^\lambda dy \int_{-\infty}^{+\infty} f(t)\cos[y(t-x)]dt$$

Again since, $\int_{-\infty}^0 f(x+t)\frac{\sin\lambda t}{t}dt = \int_0^\infty f(x-t)\frac{\sin\lambda t}{t}dt$

and for all $x \in R$, $\int_{-\infty}^{+\infty} f(x+t)\frac{\sin\lambda t}{t}dt \to \frac{\pi}{2}[f(x+0) + f(x-0)].$

We have,

$$
\begin{aligned}
\frac{1}{2}[f(x+0)+f(x-0)] &= \lim_{\lambda \to \infty} \frac{1}{\pi} \int_{-\infty}^{+\infty} f(x+u) \frac{\sin \lambda\, u}{u}\, du \\
&= \lim_{\lambda \to \infty} \frac{1}{\pi} \int_{-\infty}^{+\infty} f(t) \frac{\sin[\lambda(t-x)]}{t-x}\, dt \\
&= \lim_{\lambda \to \infty} \frac{1}{\pi} \int_{-\infty}^{+\infty} f(t) dt \int_0^{\lambda} \cos[y(t-x)]\, dy \\
&= \lim_{\lambda \to \infty} \frac{1}{\pi} \int_0^{\lambda} dy \int_{-\infty}^{+\infty} f(t) \cos[y(t-x)]\, dt
\end{aligned}
$$

Thus, the result of the Fourier Integral Theorem follows immediately. In addition, if $f(t)$ is continuous at the point $t = x$, we have

$$
\frac{1}{2\pi} \int_{-\infty}^{+\infty} e^{-iyx} dy \int_{-\infty}^{+\infty} f(t)\, e^{iyt} dt = f(x) \tag{1.11}
$$

From eqn. (1.11) the result in eqn. (1.8) of article 1.3 will follow.

1.4   Fourier Transforms

We are now ready to define Fourier transform of a piecewise continuous function $f(x) \in P^1(R)$ and $f(x) \in A_1(R)$. It is defined by

$$
F[f(x); x \to \xi] = F[f(x)] = F(\xi) = \bar{f}(\xi) = \frac{1}{\sqrt{2\pi}} \int_{-\infty}^{+\infty} f(x) e^{i\xi x} dx \tag{1.12}
$$

and $\ f(x) = F^{-1}[F(\xi)] = \dfrac{1}{\sqrt{2\pi}} \displaystyle\int_{-\infty}^{+\infty} F(\xi)\, e^{-i\xi x}\, d\xi$ $\hspace{2cm}$ (1.13)

at a point of continuity of $f(x)$.

$F(\xi)$ in (1.12) is called Fourier transform of $f(x)$ and $f(x)$ in (1.13) is called the inverse Fourier transform of $F(\xi)$.

Some authors define Fourier transform (1.12) of $f(x)$ and inversion formula in (1.13) by

$$
F[f(x)\ ;\ x \to \xi] = F(\xi) = \int_{-\infty}^{+\infty} f(x)\, e^{i\xi x} dx \tag{1.14}
$$

and $\hspace{3cm} f(x) = \dfrac{1}{2\pi} \displaystyle\int_{-\infty}^{+\infty} F(\xi)\, e^{-i\xi x} d\xi$ $\hspace{2cm}$ (1.15)

respectively.

At a point of discontinuity of $x \in R$, the left hand sides of eqns. (1.13) and (1.15) take the form

$$\frac{1}{2} \left[f(x+0) + f(x-0) \right]$$

instead of $f(x)$.

From the definitions of Fourier transform and Inverse Fourier transform we can, therefore, write from (1.12)

$$F(\xi) = F[f(x)] = F[F^{-1}(F(\xi))] \Rightarrow F \, F^{-1}[F(\xi)] \equiv I[F(\xi)]$$

$\Rightarrow F \, F^{-1} = I$, an identity operator, F and F^{-1} being Fourier and its inverse operators respectively and from (1.13)

$$f(x) = F^{-1}[F(\xi)] = F^{-1}[F(f(x))] = F^{-1}F[f(x)] = I[f(x)]$$
$$\Rightarrow F^{-1}F = I \tag{1.16}$$

Thus in operator notation $F F^{-1} = F^{-1} \, F = I$. It shows that these operators F and F^{-1} are commutative.

1.4.1 Fourier sine and cosine Transforms.

Let in addition belonging to the classes $P^1(R)$ and $A_1(R)$, the function $f(x)$ be an odd function of $x \in R$. Then, clearly $f(-x) = -f(x)$, Now, by (1.12)

$$\begin{aligned}
F[f(x); x \to \xi] &= F(\xi) = \frac{1}{\sqrt{2\pi}} \int_0^\infty f(x) \left[e^{i\xi x} - e^{-i\xi x} \right] dx \\
&= i \sqrt{\frac{2}{\pi}} \int_0^\infty f(x) \sin \xi x \, dx \\
&\equiv i \, F_s(\xi), \quad \text{say} \tag{1.17}
\end{aligned}$$

where, $$F_s(\xi) = \sqrt{\frac{2}{\pi}} \int_0^\infty f(x) \sin \, \xi x \, dx \tag{1.18}$$

$F_s(\xi)$ in (1.18) is defined to be **Fourier sine transform** of the function $f(x)$. Its connection with the Fourier transform is given by (1.17) provided $f(x)$ in an odd function of $x \in R$. The inversion formula of (1.18) is then defined by

$$f(x) = \sqrt{\frac{2}{\pi}} \int_0^\infty F_s(\xi) \, \sin \, \xi x \, d\xi \tag{1.19}$$

Let $f(x)$ be an even function of $x \in R$ in addition belonging to classes $P^1(R)$ and $A_1(R)$. Then, $f(-x) = f(x)$ and therefore, by (1.12) above

$$F[f(x); x \to \xi] = F(\xi) = \frac{1}{\sqrt{2\pi}} \int_0^\infty f(x) \left[e^{i\xi x} + e^{-i\xi x} \right] dx$$

$$= \sqrt{\frac{2}{\pi}} \int_0^\infty f(x) \cos \, \xi x \, dx$$

$$\equiv F_c(\xi), \quad \text{say,} \tag{1.20}$$

where
$$F_c(\xi) = \sqrt{\frac{2}{\pi}} \int_0^\infty f(x) \cos \, \xi x \, dx \tag{1.21}$$

$F_c(\xi)$ in (1.21) is defined to be **Fourier cosine transform** of the function $f(x)$. Its connection with the Fourier transform is given by eqn. (1.20), provided $f(x)$ is an even function of $x \in R$. The inversion formula of Fourier cosine transform defined by (1.21) is then defined as

$$f(x) = \sqrt{\frac{2}{\pi}} \int_0^\infty F_c(\xi) \cos \xi x \, d\xi \tag{1.22}$$

1.5   Linearity property of Fourier Transforms.

If c_1 and c_2 are constants, then

(i) $F[c_1 \, f_1(x) + c_2 \, f_2(x)] = c_1 \, F[f_1(x)] + c_2 \, F[f_2(x)]$

(ii) $F_s[c_1 \, f_1(x) + c_2 \, f_2(x)] = c_1 \, F_s[f_1(x)] + c_2 \, F_s[f_2(x)]$

(iii) $F_c[c_1 \, f_1(x) + c_2 \, f_2(x)] = c_1 \, F_c[f_1(x)] + c_2 \, F_c[f_2(x)]$

Proof.

(i) By definition of Fourier Transform in eqn.(1.12) of article 1.4, we have

$$F[c_1 \, f_1(x) + c_2 \, f_2(x); x \to \xi]$$

$$= \frac{1}{\sqrt{2\pi}} \int_{-\infty}^{+\infty} [c_1 \, f_1(x) + c_2 \, f_2(x)] \cdot e^{i\xi x} \, dx$$

$$= c_1 \cdot \frac{1}{\sqrt{2\pi}} \int_{-\infty}^{+\infty} f_1(x) e^{i\xi x} \, dx + c_2 \cdot \frac{1}{\sqrt{2\pi}} \int_{-\infty}^{+\infty} f_2(x) e^{i\xi x} \, dx$$

$$= c_1 \, F[f_1(x); x \to \xi] + c_2 \, F[f_2(x); x \to \xi]$$

(ii) By definition of Fourier sine transform in eqn. (1.18) of article 1.4.1 we get

$$F_s[c_1 \; f_1(x) + c_2 \; f_2(x); x \to \xi] = \sqrt{\frac{2}{\pi}} \int_0^\infty c_1 \; f_1(x) \sin \xi x \; dx$$

$$+ \sqrt{\frac{2}{\pi}} \int_0^\infty c_2 \; f_2(x) \sin \xi x \; dx$$

$$= c_1 \; F_s[f_1(x); x \to \xi] + c_2 \; F_s[f_2(x); x \to \xi].$$

(iii) By definition of Fourier cosine transform in eqn.(1.21) of article 1.4.1 proceeding similarly as above, we have

$$F_c[c_1 \; f_1(x) + c_2 \; f_2(x)] = c_1 \; F_c[f_1(x)] + c_2 \; F_c[f_2(x)] \; .$$

1.6 Change of Scale property.

(i) If $F[f(x); x \to \xi] = F(\xi)$, then $F[f(ax) \; ; \; x \to \xi] = \frac{1}{a} F\left(\frac{\xi}{a}\right)$

Proof. Since $F[f(x); x \to \xi] = \dfrac{1}{\sqrt{2\pi}} \int_{-\infty}^{+\infty} f(x) e^{i\xi x} \; dx = F(\xi)$,

we have $F[f(ax); x \to \xi] = \dfrac{1}{\sqrt{2\pi}} \int_{-\infty}^{+\infty} f(ax) e^{i\xi x} \; dx$

$$= \frac{1}{\sqrt{2\pi}} \int_{-\infty}^{+\infty} f(t) e^{i\frac{\xi}{a}t} \frac{1}{a} \; dt$$

$$= \frac{1}{a} \cdot \frac{1}{\sqrt{2\pi}} \int_{-\infty}^{+\infty} f(t) e^{i\left(\frac{\xi}{a}\right)t} \; dt = \frac{1}{a} F\left(\frac{\xi}{a}\right).$$

(ii) If $F_s[f(x)] = F_s(\xi)$, then $F_s[f(ax)] = \frac{1}{a} F_s\left(\frac{\xi}{a}\right)$

Proof. By definition

$$F_s[f(x); x \to \xi] \quad = \quad \sqrt{\frac{2}{\pi}} \int_0^\infty f(x) \sin \xi x \; dx = F_s(\xi).$$

Therefore,

$$F_s[f(ax); x \to \xi] \quad = \quad \sqrt{\frac{2}{\pi}} \int_0^\infty f(ax) \sin \xi x \; dx$$

$$= \quad \frac{1}{a} \sqrt{\frac{2}{\pi}} \int_0^\infty f(t) \sin \left(\frac{\xi}{a} t\right) dt$$

$$= \quad \frac{1}{a} F_s \left(\frac{\xi}{a}\right)$$

(iii) Similarly, it is easy to prove that if

$$F_c[f(x); x \to \xi] = F_c(\xi), \text{ then } F_c[f(ax); x \to \xi] = \frac{1}{a} F_c\left(\frac{\xi}{a}\right).$$

1.6   The shifting property of Fourier Transform.

If $F(\xi)$ is Fourier transform of $f(x)$ in R, then Fourier transform of $f(x - a)$ is $e^{i\xi a} F(\xi)$.

Proof.   By definition, we have

$$F(\xi) = \frac{1}{\sqrt{2\pi}} \int_{-\infty}^{+\infty} f(x) \, e^{i\xi x} \, dx = F[f(x); x \to \xi]$$

Therefore,

$$F[f(x - a); x \to \xi] = \frac{1}{\sqrt{2\pi}} \int_{-\infty}^{+\infty} f(x - a) \, e^{i\xi x} \, dx$$

$$= \frac{1}{\sqrt{2\pi}} \int_{-\infty}^{+\infty} f(t) \, e^{i\xi(t+a)} \, dt$$

$$= e^{i\xi a} \cdot \frac{1}{\sqrt{2\pi}} \int_{-\infty}^{+\infty} f(t) \, e^{i\xi t} \, dt$$

$$= e^{i\xi a} F(\xi).$$

1.7   The Modulation theorem.

Theorem :   If $F[f(x); x \to \xi] = F(\xi)$, then $F[f(x) \cos ax \; ; x \to \xi]$ $= \frac{1}{2}[F(\xi + a) + F(\xi - a)]$.

Proof.   Since it is given that

$$F[f(x); x \to \xi] \;\; = \;\; F(\xi) = \frac{1}{\sqrt{2\pi}} \int_{-\infty}^{+\infty} f(x) e^{i\xi x} \, dx$$

we have

$$F[f(x) \cos ax \; ; x \to \xi] \;\; = \;\; \frac{1}{\sqrt{2\pi}} \int_{-\infty}^{+\infty} f(x) \cdot \frac{e^{iax} + e^{-iax}}{2} \cdot e^{i\xi x} \, dx$$

$$= \;\; \frac{1}{2}\left[\frac{1}{\sqrt{2\pi}} \int_{-\infty}^{+\infty} f(x) \, e^{i(\xi+a)x} \, dx \right.$$

$$\left. + \frac{1}{\sqrt{2\pi}} \int_{-\infty}^{+\infty} f(x) \, e^{i(\xi-a)x} \, dx\right]$$

$$= \;\; [F(\xi + a) + F(\xi - a)].$$

Other three parts of the theorem are

(a)   if $F_s[f(x); x \to \xi] = F_s(\xi)$ then

$$F_s[f(x)\cos ax\ ;x \to \xi] = \frac{1}{2}[F_s(\xi + a) + F_s(\xi - a)]$$

(b) if $F_s[f(x); x \to \xi] = F_s(\xi)$ then

$$F_c[f(x)\sin ax\ ;x \to \xi] = \frac{1}{2}[F_s(\xi + a) - F_s(\xi - a)]$$

and (c) if $F_c[f(x); x \to \xi] = F_c(\xi)$ then

$$F_s[f(x)\sin ax\ ;x \to \xi] = \frac{1}{2}[F_c(\xi - a) - F_c(\xi + a)]$$

These results can similarly be proved as was done in the first case on replacing $\cos ax$ and $\sin ax$ by their exponential forms in the definitions of the corresponding transforms of the left hand side.

1.8 Evaluation of integrals by means of inversion theorems.

By definitions of Fourier transform and its inversion formula in eqns. (1.12), (1.13), (1.17), (1.19), (1.21) and (1.22) of article 1.4, we may employ these results to evaluate certain integrals involving trigonometric functions. For example, first of all we consider

$$I_1 = \int_0^\infty e^{-bx} \cos ax\ dx \quad , \quad I_2 = \int_0^\infty e^{-bx} \sin ax\ dx$$

Integrating by parts we can now evaluate I_1 and I_2 as

$$I_1 = \frac{b}{a^2 + b^2} \quad , \quad I_2 = \frac{a}{a^2 + b^2}$$

These results means that if we set $f(x) = e^{-bx}$, then its cosine and sine transforms are given by

$$F_c\left[e^{-bx}; x \to \xi\right] = \sqrt{\frac{2}{\pi}}\frac{b}{\xi^2 + b^2}$$

$$F_s\left[e^{-bx}; x \to \zeta\right] = \sqrt{\frac{2}{\pi}}\frac{\xi}{\xi^2 + b^2}$$

Substituting these expressions in eqns. (1.19) and (1.22) of article 1.4.1, we get

$$\int_0^\infty \frac{\cos \xi x}{\xi^2 + b^2}\ d\xi = \frac{\pi}{2b}\ e^{-bx}$$

and $$\int_0^\infty \frac{\xi \sin \xi x}{\xi^2 + b^2}\ d\xi = \frac{\pi}{2}\ e^{-bx}$$

As a second example, if we take $\quad f(x) = \begin{cases} 1, & 0 < x < a \\ 0, & x > a, \end{cases}$

then

$$F_c(\xi) = \sqrt{\frac{2}{\pi}} \int_0^a 1 \cos \, \xi x \, dx = \sqrt{\frac{2}{\pi}} \, \frac{\sin \xi a}{\xi}$$

and so it gives

$$\frac{2}{\pi} \int_0^\infty \frac{\sin \xi a}{\xi} \cos \xi x \, d\xi = \begin{cases} 1, & \text{if } 0 < x < a \\ 0, & \text{if } x \geqslant a \end{cases}$$

Let us consider a third example when $f(x) = \begin{cases} 0, & 0 < x < a \\ x, & a \leq x \leq b \\ 0, & x > b \end{cases}$

Then,

$$\begin{aligned} F_s(\xi) &= \sqrt{\frac{2}{\pi}} \int_a^b x \sin \, \xi x \, dx \\ &= \sqrt{\frac{2}{\pi}} \left[\frac{a \cos \xi \, a - b \cos \xi \, b}{\xi} + \frac{\sin \xi \, b - \sin \xi \, a}{\xi^2} \right] \end{aligned}$$

and this result gives

$$\frac{2}{\pi} \int_0^\infty \sin \, \xi \, x \, [F_s(\xi)] \, d\xi = \begin{cases} 0, & 0 < x < a \\ x, & a \leq x \leq b \\ 0, & x > b \end{cases}$$

In the fourth case if we choose $\quad f(x) = \begin{cases} 1, & 0 < x < a \\ 0, & x > a \end{cases}$

Then

$$F_s(\xi) = \frac{1 - \cos \xi \, a}{\xi}$$

and it gives

$$\frac{2}{\pi} \int_0^\infty \sin \xi \, x \, \left[\frac{1 - \cos \xi \, a}{\xi} \right] d\xi = f(x) = \begin{cases} 1, & 0 < x < a \\ 0, & x > a \end{cases}$$

Similarly taking $\quad f(x) = \begin{cases} 1, & a < x < b \\ 0, & \text{otherwise} \end{cases}$

we get

$$\frac{2}{\pi} \int_0^\infty \sin \xi \, x \left[\frac{\cos \, a \, \xi - \cos \, b \, \xi}{\xi} \right] \, d\xi = \begin{cases} 0 \, , \, 0 < x < a \\ 1 \, , \, a < x < b \\ 0 \, , \, x > b \end{cases}$$

Combining these above results, we get

$$\frac{2}{\pi} \int_0^\infty \sin \, \xi \, x \left[\frac{a - b}{\xi} + \frac{\cos \, a \, \xi - \cos \, b \, \xi}{\xi^2} \right] \, d\xi = \begin{cases} a - b \, , \, 0 < x < a \\ x - b \, , \, a < x < b \\ 0 \, , \, x > b \end{cases}$$

1.9   Fourier Transform of some particular functions.

We consider the following step function $f(x)$ defined by

$$f(x) = \begin{cases} 0 \, , \, x < 0 \\ 1 \, , \, x > 0 \end{cases}$$

This function is called **Heaviside** unit step function and it is denoted by the symbol $H(x)$. Thus, we have

$$H(x) = \begin{cases} 0 \, , \, x < 0 \quad (\text{or sometimes for } x \leq 0) \\ 1 \, , \, x > 0 \end{cases}$$

Almost all step functions can be expressed as a combination of **Heaviside** unit step functions. For example,

$$f(x) = \begin{cases} 1 \, , \, |x| < a \\ 0 \, , \, \text{otherwise} \end{cases}$$

is equivalent to $f(x) = H(x + a) - H(x - a) \equiv H(a - |x|)$

and the step function $f(x) = \begin{cases} k \, , \, x > a \\ 0 \, , \, \text{otherwise} \end{cases}$

is equivalent to $f(x) = k \, H(x - a),$ for all x

Example 1.1. Evaluate Fourier transform of $H(x + a) - H(x - a)$

Solution. By definition,

$$F[H(x + a) - H(x - a); x \to \xi] = \frac{1}{\sqrt{2\pi}} \int_0^\infty e^{i\xi x}[H(x + a) - H(x - a)]dx$$

$$= \frac{1}{\sqrt{2\pi}} \int_{-a}^{a} e^{i\xi x} \, dx$$

$$= \frac{1}{\sqrt{2\pi}} \left[\frac{e^{i\xi x}}{i\xi} \right]_{-a}^{a}$$

$$= \frac{1}{\sqrt{2\pi} i\xi} \left[e^{i\xi a} - e^{-i\xi a} \right]$$

$$= \sqrt{\frac{2}{\pi}} \frac{\sin a \, \xi}{\xi}$$

Therefore, using the formula for inverse $F\,T$ we get

$$\frac{1}{\sqrt{2\pi}} \int_{-\infty}^{+\infty} \sqrt{\frac{2}{\pi}} \frac{\sin a \, \xi}{\xi} e^{-i\xi x} \, d\xi = H(x+a) - H(x-a)$$

This result implies,

$$\frac{1}{\pi} \int_{-\infty}^{+\infty} \frac{\sin a \, \xi}{\xi} \cdot [\cos \xi \, x - i \, \sin \xi \, x] \, d\xi = H(x+a) - H(x-a)$$

i.e $\displaystyle \int_{-\infty}^{+\infty} \frac{\sin a \, \xi}{\xi} \cos \xi \, x \, d\xi = \pi[H(x+a) - H(x-a)]$

$$= \begin{cases} \pi \,, & \text{for} \quad |x| < a \\ 0 \,, & \text{for} \quad |x| > a \end{cases}$$

Putting $x = 0$ and $a = 1$, in the above we get $\displaystyle \int_{-\infty}^{+\infty} \frac{\sin \xi}{\xi} \, d\xi = \pi$

implying, $\displaystyle \int_{0}^{\infty} \frac{\sin \xi}{\xi} \, d\xi = \frac{\pi}{2}$

and $\displaystyle \int_{-\infty}^{+\infty} \frac{\sin a \, \xi \sin \, \xi \, x}{\xi} \, d\xi = 0$

Example 1.2.   Evalute FT of

$$f(x) = \begin{cases} x \,, & |x| < a \\ 0 \,, & |x| > a \end{cases}$$

Solution. By definition,

$$F[f(x) \,;\, x \to \xi] = \frac{1}{\sqrt{2\pi}} \int_{-a}^{a} x \, e^{i\xi x} \, dx$$

$$= \frac{1}{\sqrt{2\pi}} \left\{ \left[x \frac{e^{i\xi x}}{i\xi} \right]_{-a}^{a} - \frac{1}{i\xi} \int_{-a}^{a} e^{i\xi x} \, dx \right\}$$

$$= \frac{1}{\sqrt{2\pi}} \left[\frac{a \, e^{i\xi a} + a \, e^{-i\xi a}}{i\xi} + \frac{1}{\xi^2} \left\{ e^{ia\xi} - e^{-ia\xi} \right\} \right]$$

$$= \frac{1}{\sqrt{2\pi}} \left[\frac{2 \, a \, \cos \, a \, \xi}{i\xi} + \frac{2 \, i \, \sin \, a \, \xi}{\xi^2} \right]$$

$$= \sqrt{\frac{2}{\pi}} \, i \, \left[\frac{\sin \, a \, \xi - a \, \xi \cos \, a \, \xi}{\xi^2} \right].$$

Example 1.3. If $F \, T$ of $f(x)$ is $F(\xi)$, calculate $F \, T$ of
(i) $e^{i\lambda x} f(x)$ and (ii) $f(-x)$ in terms of $F(\xi)$.

Solution. We know that,

$$F[f(x) \, ; \, x \to \xi] = \frac{1}{\sqrt{2\pi}} \int_{-\infty}^{+\infty} f(x) \, e^{i\xi x} \, dx = F(\xi).$$

Therefore,

(i) $F\left[e^{i\lambda x} f(x) \, ; \, x \to \xi \right] = \dfrac{1}{\sqrt{2\pi}} \displaystyle\int_{-\infty}^{+\infty} f(x) \cdot e^{i(\xi+\lambda)x} \, dx = F(\xi + \lambda).$

(ii) $\begin{aligned}[t] F[f(-x) \, ; \, x \to \xi] &= \frac{1}{\sqrt{2\pi}} \int_{-\infty}^{+\infty} f(-x) \, e^{i\xi x} \, dx \\ &= \frac{1}{\sqrt{2\pi}} \int_{-\infty}^{+\infty} f(y) \, e^{-i\xi y} \, dy = F(-\xi) = F^{-1}(\xi) \end{aligned}$

Example 1.4. Find the function whose cosine transform is $\sqrt{\dfrac{2}{\pi}} \dfrac{\sin \, a \, \xi}{\xi}$.

Solution. Let $f(x)$ be the required function such that

$$F_c[f(x) \, ; \, x \to \xi] = \sqrt{\frac{2}{\pi}} \, \frac{\sin \, a \, \xi}{\xi}$$

$$f(x) = \frac{2}{\pi} \int_0^\infty \frac{\sin \, a \, \xi}{\xi} \cdot \cos \, \xi \, x \, d \, \xi$$

$$= \frac{1}{\pi} \int_0^\infty \frac{1}{\xi} [\sin \xi(a + x) + \sin \, \xi(a - x)] \, d\xi$$

$$= \frac{1}{\pi} \left[\int_0^\infty \frac{\sin \, \xi(a + x)}{\xi} \, d\xi + \int_0^\infty \frac{\sin \, \xi(a - x)}{\xi} \, d\xi \right]$$

$$= \frac{1}{\pi} \left[\frac{\pi}{2} + \frac{\pi}{2} \right] , \text{ when } x < a \text{ and } \frac{1}{\pi} \left[\frac{\pi}{2} - \frac{\pi}{2} \right], \text{ when } x > a$$

So, $f(x) = \begin{cases} 1 \, , \text{ when } \, x < a \\ 0 \, , \text{ when } \, x > a \end{cases}$

Example 1.5. Find the $\quad F\,T\quad$ of $\quad f(x) = \begin{cases} 1 - x^2 & , \quad |x| \le 1 \\ 0 & , \quad |x| > 1 \end{cases}$

and hence evaluate $\displaystyle\int_0^\infty \frac{x\,\cos\,x - \sin\,x}{x^3}\,\cos\,\frac{x}{2}\,dx$

Solution. From definition, we have

$$F(\xi) \;=\; F[f(x)\,;\,x \to \xi] = \frac{1}{\sqrt{2\pi}}\int_{-1}^1 (1 - x^2)\,e^{i\xi x}\,dx$$

$$= \;\left[-\frac{2}{\xi^2}\,(e^{i\xi} + e^{-i\xi}) + \frac{2}{i\xi^3}\,(e^{i\xi} - e^{-i\xi})\right]\cdot\frac{1}{\sqrt{2\pi}}$$

Therefore,

$$F(\xi) = \frac{-4}{\sqrt{2\pi}}\,[(\xi\,\cos\,\xi - \sin\,\xi)\,/\xi^3]$$

Now, the inversion formula gives

$$\frac{1}{\sqrt{2\pi}}\int_{-\infty}^{+\infty}\frac{-4}{\sqrt{2\pi}}\left[\frac{\xi\,\cos\,\xi - \sin\,\xi}{\xi^3}\right]\cdot e^{-i\xi x}\,dx = f(x)$$

Hence, $\displaystyle\int_{-\infty}^{+\infty}\left[\frac{-4(\xi\,\cos\,\xi - \sin\,\xi)}{\xi^3}\right][\cos\,\xi\,x - i\,\sin\,\xi\,x]\,d\xi = 2\pi f(x)$

Equating real parts, we have

$$\int_{-\infty}^\infty \frac{\sin\,\xi - \xi\,\cos\,\xi}{\xi^3}\,\cos\,\xi\,x\,d\xi = \frac{\pi}{2}\,f(x)$$

Or, $\displaystyle 2\int_0^\infty \frac{\sin\xi - \xi\,\cos\,\xi}{\xi^3}\,\cos\,\xi\,x\,d\xi = \begin{cases} \frac{\pi}{2}(1 - x^2) & , \quad |x| \le 1 \\ 0 & , \quad |x| > 1 \end{cases}$

Now, putting $x = \frac{1}{2}$ in above we get

$$\int_0^\infty \frac{\sin\xi - \xi\,\cos\,\xi}{\xi^3}\,\cos\,\frac{\xi}{2}\,d\xi = \frac{1}{2}\cdot\frac{\pi}{2}\cdot\frac{3}{4} = \frac{3\pi}{16}$$

This gives, $\displaystyle\int_0^\infty \frac{x\,\cos\,x - \sin\,x}{x^3}\,\cos\,\frac{x}{2}\,dx = -\frac{3\pi}{16}$

Example 1.6. Find FT of $f(x) = [1 - |x|]H[1 - |x|]$

Solution.

Here $\quad f(x) = \begin{cases} 1 - |x| , & \text{when} \quad -1 < x < 1 \\ 0 & , \quad \text{otherwise} \end{cases}$

Here, $f(x)$ is an even function of x and so

$$
\begin{aligned}
F[f(x) ; x \to \xi] &= F_c[f(x) ; x \to \xi] \\
&= \sqrt{\frac{2}{\pi}} \int_0^1 (1 - x) \cos \xi \, x \, dx \\
&= \sqrt{\frac{2}{\pi}} \frac{1 - \cos \xi}{\xi^2}
\end{aligned}
$$

So, $F_c^{-1}[F_c\{f(x) ; x \to \xi\} ; \xi \to x] = [1 - |x|] \, H \, [1 - |x|]$

Or, $\quad \dfrac{2}{\pi} \displaystyle\int_0^\infty \dfrac{1 - \cos \xi}{\xi^2} \cos \xi \, x \, d\xi = \begin{cases} 1 - x , & 0 < x < 1 \\ 0 , & x > 1 \end{cases}$ (i)

Some Deductions :

(a) Putting $x = 0$ in the final result (i) of the example 1.6 we get

$$
\int_0^\infty \frac{1 - \cos \xi}{\xi^2} \, d\xi = \frac{\pi}{2}
$$

Or, $\quad \displaystyle\int_0^\infty \dfrac{\sin^2 x}{x^2} \, dx = \dfrac{\pi}{2}$

(b) Integrating the result (i) in the example 1.6 with respect to x from 0 to X, we get

$$
\int_0^\infty \frac{1 - \cos \xi}{\xi^3} \frac{\sin \xi \, X}{\xi} \, d\xi = \frac{\pi}{2} \left[X - \frac{X^2}{2} \right]
$$ (ii)

Now, putting $X = \frac{1}{2}$, one gets

$$
\int_0^\infty \frac{2 \sin^3 \frac{\xi}{2}}{8 \left(\frac{\xi}{2} \right)^3} \cdot 2d \left(\frac{\xi}{2} \right) = \frac{\pi}{2} \left[\frac{3}{8} \right]
$$

$$
\Rightarrow \quad \int_0^\infty \frac{\sin^3 x}{x^3} \, dx - \frac{3\pi}{8}
$$

(c) We have seen in Example 1.1 that

$$
\int_0^\infty \frac{\sin x}{x} \, dx = \frac{\pi}{2}
$$

Again $\quad \displaystyle\int_0^\infty \dfrac{\sin^2 x}{x^2} \, dx = \dfrac{\pi}{2},$ from (a)

Also in (b), $\quad \displaystyle\int_0^\infty \dfrac{\sin^3 x}{x^3} \, dx = \dfrac{3\pi}{8} < \dfrac{\pi}{2}$

These results show that

$$\frac{\sin x}{x} \geqslant \left[\frac{\sin x}{x}\right]^2 \geqslant \left[\frac{\sin x}{x}\right]^3$$

(d) Again integrating the result (ii) in (b) above with respect to X from $X = 0$ to $X = 1$, we get

$$\int_0^\infty \frac{1 - \cos \xi}{\xi^2} \left[\frac{-\cos \xi X}{\xi^2}\right]_0^1 d\xi = \frac{\pi}{2} \left[\frac{X^2}{2} - \frac{X^3}{6}\right]_0^1$$

Or, $\displaystyle\int_0^\infty \frac{\sin^4 x}{x^4} dx = \frac{\pi}{3}$

Example 1.7. Find the $F\,T$ of (i) $f(x) = e^{-a|x|}$, $a > 0$ and (ii) $f(x) = |x| e^{-a|x|}$, $a > 0$.

Solutions (i) $\quad F[f(x)\,;\ x \to \xi] = \dfrac{1}{\sqrt{2\pi}} \displaystyle\int_{-\infty}^{+\infty} e^{-a|x|} \cdot e^{i\xi x} \, dx$

$$= \frac{1}{\sqrt{2\pi}} \left[\int_{-\infty}^0 e^{ax} \cdot e^{i\xi x} \, dx + \int_0^\infty e^{-ax} \cdot e^{i\xi x} \, dx\right]$$

$$= \frac{1}{\sqrt{2\pi}} \left[\frac{1}{a + i\xi} + \frac{1}{a - i\xi}\right]$$

$$= \sqrt{\frac{2}{\pi}} \frac{a}{a^2 + \xi^2}$$

(ii) $\quad F\left[\,|x| e^{-a|x|}\,;\ x \to \xi\,\right]$

$$= \frac{1}{\sqrt{2\pi}} \int_{-\infty}^{+\infty} |x| \, e^{-a|x|} \cdot e^{i\xi x} \, dx$$

$$= \frac{1}{\sqrt{2\pi}} \left[-\frac{d}{da} \int_{-\infty}^{+\infty} e^{-a|x|} \, e^{i\xi x} \, dx\right]$$

$$= -\frac{d}{da} \left[\sqrt{\frac{2}{\pi}} \frac{a}{a^2 + \xi^2}\right]$$

$$= \sqrt{\frac{2}{\pi}} \left[\frac{a^2 - \xi^2}{(a^2 + \xi^2)^2}\right]$$

Example 1.8. (a) Find the $F\,T$ of $\quad f(x) = \begin{cases} e^{-ax}, & x > 0,\ a > 0 \\ -e^{ax}, & x < 0,\ a > 0 \end{cases}$

Solution. Clearly, the given function is an odd function of x in R.

Therefore

$$F[f(x) ; x \to \xi] = i \sqrt{\frac{2}{\pi}} \int_0^\infty e^{-ax} \sin \xi x \, dx$$

$$= i \sqrt{\frac{2}{\pi}} \cdot \frac{\xi}{\xi^2 + a^2}$$

(b)   Find $F T$ of $f(x) = x \, e^{-a|x|}$, $a > 0$.

Solution. The given function $f(x)$ is an odd function of $x \in R$. Hence,

$$F[f(x) ; x \to \xi] = i \sqrt{\frac{2}{\pi}} \int_0^\infty x \, e^{-ax} \cdot \sin \xi x \, dx$$

$$= i \sqrt{\frac{2}{\pi}} \left(-\frac{d}{da}\right) \int_0^\infty e^{-ax} \sin \xi x \, dx = -i \sqrt{\frac{2}{\pi}} \frac{d}{da} \left[\frac{\xi}{a^2 + \xi^2}\right]$$

$$= i \sqrt{\frac{2}{\pi}} \frac{2 \, a \, \xi}{(\xi^2 + a^2)^2}$$

Example 1.9. Find the $F T$ of $f(x) = e^{-\alpha x^2}$, $\alpha > 0$.

Solution. We know from definition that

$$F[f(x) ; x \to \xi] = \frac{1}{\sqrt{2\pi}} \int_{-\infty}^{+\infty} e^{-\alpha x^2} e^{i\xi x} \, dx$$

$$= \frac{1}{\sqrt{2\pi}} \int_{-\infty}^\infty e^{-\left[\sqrt{\alpha}x - \frac{i\xi}{2\sqrt{\alpha}}\right]^2 - \frac{\xi^2}{4\alpha}} \, dx$$

$$= \frac{e^{-\frac{\xi^2}{4\alpha}}}{\sqrt{2\pi}} \int_{-\infty}^{+\infty} e^{-z^2} \cdot \frac{dz}{\sqrt{\alpha}}$$

$$= \frac{2 \cdot e^{-\frac{\xi^2}{4\alpha}}}{\sqrt{2\pi\alpha}} \int_0^\infty e^{-z^2} dz$$

$$F[e^{-\alpha x^2}] = \frac{2 \cdot e^{-\frac{\xi^2}{4\alpha}}}{\sqrt{2\pi\alpha}} \cdot \frac{\sqrt{\pi}}{2} = \frac{1}{\sqrt{2\alpha}} e^{\frac{-\xi^2}{4\alpha}} \qquad \text{(i)}$$

Some important deductions.

(i) Let $\alpha = 1$. Then $F[e^{-x^2} ; x \to \xi] = \frac{1}{\sqrt{2}} e^{\frac{-\xi^2}{4}}$

(ii) Let $\alpha = \frac{1}{2}$. Then $F[e^{\frac{-x^2}{2}} ; x \to \xi] = e^{\frac{-\xi^2}{2}}$

This result shows that $F^{-1}[e^{\frac{-\xi^2}{2}} ; \xi \to x] = e^{\frac{-x^2}{2}}$

Such a function $f(x) = e^{\frac{-x^2}{2}}$ having the property that

$$F[f(x) \; ; \; x \to \xi] = f(\xi)$$

is called self reciprocal.

(iii) Differentiating (i) with respect to ξ we get

$$\frac{d}{d\xi}\left[\sqrt{\frac{2}{\pi}} \int_0^\infty e^{-\alpha x^2} \cos \xi x \; dx\right] = \frac{d}{d\xi} \frac{1}{\sqrt{2\alpha}} \cdot e^{\frac{-\xi^2}{4\alpha}}$$

$$\Rightarrow F_s[xe^{-\alpha x^2} \; ; \; x \to \xi] = \frac{1}{\sqrt{8\alpha^3}} \xi \; e^{\frac{-\xi^2}{4\alpha}}$$

Putting , $\quad \alpha = \dfrac{1}{2} \quad$ we get $\quad F_s\left[x \; e^{\frac{-x^2}{2}} \; ; \; x \to \xi\right] = \xi \; e^{\frac{-\xi^2}{2}}$

Thus $f(x) = x \; e^{\frac{-x^2}{2}}$ is self reciprocal with regard to Fourier sine transform.

Putting, $\quad \alpha = \dfrac{i}{2} \quad$ we get $\quad F_c\left[e^{\frac{ix^2}{2}} \; ; \; x \to \xi\right] = \dfrac{1-i}{\sqrt{2}} \; e^{\frac{i\xi^2}{2}}$

Therefore, $\quad F_c\left[\cos\dfrac{x^2}{2} \; ; \; x \to \xi\right] = \dfrac{1}{\sqrt{2}} \; Re\left[(1-i)e^{\frac{i\xi^2}{2}}\right]$

$$= \cos\left(\frac{\xi^2}{2} - \frac{\pi}{4}\right) \tag{ii}$$

Also, $\quad F_c\left[\sin\dfrac{x^2}{2} \; ; \; x \to \xi\right] = \dfrac{1}{\sqrt{2}}\left[\cos\dfrac{\xi^2}{2} - \sin\dfrac{\xi^2}{2}\right]$

$$= -\sin\left(\frac{\xi^2}{2} - \frac{\pi}{4}\right) \tag{iii}$$

after equating real and imaginary parts of both sides. Combining results in (ii) and (iii) above one gets

$$F_c\left[\cos\left(\frac{x^2}{2} - \frac{\pi}{8}\right) \; ; \; x \to \xi\right] = \cos\left[\frac{\xi^2}{2} - \frac{\pi}{8}\right]$$

1.10   Convolution or Faltung of two integrable functions.

The convolution or Falting or Faltung of two integrable functions $f(x)$ and $g(x)$, where $-\infty < x < \infty$ is denoted and defined as

$$f * g = \frac{1}{\sqrt{2\pi}} \int_{-\infty}^{+\infty} f(x-u) \; g(u) \; du \tag{1.23}$$

It possesses many formal properties such as

$$f * (\lambda g) \;=\; (\lambda f) * g = \lambda(f * g), \lambda = \text{constant}$$
$$f * g \;=\; g * f \quad, \text{ (the commutative property)}$$
$$f * (g + h) \;=\; f * g + f * h ,$$

which can be verified easily directly from definition (1.23). Further, if both $f(x)$ and $g(x)$ belong to $C^1(R)$ and $A_1(R)$ classes, then so does their convolution $h(x) = f * g$, since

$$\sqrt{2\pi} \int_{-\infty}^{\infty} |h(x)| dx \;=\; \int_{-\infty}^{+\infty} dx \left| \int_{-\infty}^{+\infty} f(u)g(x-u)du \right|$$
$$<\; \int_{-\infty}^{+\infty} dx \int_{-\infty}^{+\infty} |f(u)g(x-u)| \; du$$
$$=\; \int_{-\infty}^{+\infty} |f(u)| du \int_{-\infty}^{+\infty} |g(x-u)| \; dx$$
$$\Rightarrow \int_{-\infty}^{+\infty} |h(x)| dx \;<\; \frac{1}{\sqrt{2\pi}} \int_{-\infty}^{+\infty} |f(u)| du \int_{-\infty}^{+\infty} |g(v)| dv$$

and the result follows from the fact that both $f(x)$ and $g(x)$ belong to the class $A_1(R)$.

Again, $(f*g)*h = f*(g*h)$ is true by direct verification. Therefore, the convolution property is also associative.

We shall now discuss Fourier transform of the convolution of a pair of functions as detailed below, by the name Convolution or Falting or Faltung theorem for $F\ T$.

1.11   Convolution or Falting or Faltung Theorem for FT.

Theorem.

The FT of the convolution of $f(x)$ and $g(x)$, both belonging to the classes $C^1(R)$ and $A_1(R)$, is the product of the $F\ T$ of $f(x)$ and $g(x)$. That means that,

$$F\ [f * g; x \rightarrow \xi] = F(\xi)\ G(\xi)$$

where $\qquad F(\xi) = \dfrac{1}{\sqrt{2\pi}} \displaystyle\int_{-\infty}^{+\infty} f(x)\ e^{i\xi x}\ dx \quad$ and

$$G(\xi) = \frac{1}{\sqrt{2\pi}} \int_{-\infty}^{+\infty} g(x)\ e^{i\xi x} dx$$

Proof.   Since $f(x)$ and $g(x)$ both belong to the classes $C^1(R)$ and $A_1(R)$, $f * g$ also belongs to the classes $C^1(R)$ and $A_1(R)$ and hence FT of $f * g$ also exists. Therefore,

$$
\begin{aligned}
F[f * g \; ; \; x \to \xi] &= \frac{1}{\sqrt{2\pi}} \int_{-\infty}^{+\infty} e^{i\xi x} \frac{dx}{\sqrt{2\pi}} \int_{-\infty}^{+\infty} f(x-u)g(u)du \\
&= \frac{1}{2\pi} \int_{-\infty}^{+\infty} \int_{-\infty}^{+\infty} f(x-u)g(u)e^{i\xi x} \, dx \, du \\
&= \frac{1}{2\pi} \int_{-\infty}^{+\infty} g(u) \left[\int_{-\infty}^{+\infty} e^{i\xi x} f(x-u)dx \right] du \\
&= \frac{1}{\sqrt{2\pi}} \int_{-\infty}^{\infty} g(u)e^{i\xi u} \left[\frac{1}{\sqrt{2\pi}} \int_{-\infty}^{+\infty} e^{i\xi v} f(v) \, dv \right] du \\
&= \frac{1}{\sqrt{2\pi}} \int_{-\infty}^{+\infty} g(u)e^{i\xi u} \cdot F(\xi) \, du
\end{aligned}
$$

$$
F[f * g \; ; \; x \to \xi] \;\; = \;\; F(\xi) \cdot G(\xi) \tag{1.24}
$$

This proves the theorem.

Coro 1.2   Another interpretation of convolution theorem from eqn. (1.24) is that

$$
F^{-1} \left[F(\xi)G(\xi) \; ; \; \xi \to x \right] \;\; = \;\; \frac{1}{\sqrt{2\pi}} \int_{-\infty}^{+\infty} f(x-u)g(u)du
$$

Or, $\quad \dfrac{1}{\sqrt{2\pi}} \displaystyle\int_{-\infty}^{+\infty} F(\xi)G(\xi)e^{-i\xi x}d\xi \;\; = \;\; \dfrac{1}{\sqrt{2\pi}} \displaystyle\int_{-\infty}^{+\infty} f(v) \, g(x-v)dv$

$$
= \;\; f * g \tag{1.25}
$$

Coro 1.3   From the result (1.25) above, we have

$$
\frac{1}{\sqrt{2\pi}} \int_{-\infty}^{+\infty} F(\xi) \, G(\xi) \, e^{-i\xi x}d\xi = f * g = \frac{1}{\sqrt{2\pi}} \int_{-\infty}^{+\infty} f(v)g(x-v)dv
$$

Putting $x = 0$ in the above equation, we get

$$
\int_{-\infty}^{+\infty} F(\xi) \, G(\xi) \, d\xi = \int_{-\infty}^{+\infty} f(v) \, g(-v) \, dv \tag{1.26}
$$

Coro 1.4   Let $f(x) \in R$. Then $F(\xi) = F[f(x) \; ; \; x \to \xi]$ is a complex valued function and its conjugate, denoted by $\bar{F}(\xi)$ is then given by

$$
\bar{F}(\xi) = \frac{1}{\sqrt{2\pi}} \int_{-\infty}^{+\infty} f(x)e^{-i\xi x}dx = F^{-1}[f(x); x \to \xi]
$$

i.e, $\bar{F}\{f(x); x \rightarrow \xi\} = F^{-1}[f(x) ; x \rightarrow \xi]$

Therefore, $\bar{F} \equiv F^{-1}$ (1.27)

Coro 1.5 Similar results for sine and cosine transforms of the functions $f(x)$ and $g(x) \in R$ may also be established in regard to their convolution. For example, if $f(x), g(x)$ are even functions and

$$F_c \left[f(x) ; x \rightarrow \xi \right] = \sqrt{\frac{2}{\pi}} \int_0^\infty f(x) \cos \xi x \, dx$$

$$F_c \left[g(x) ; x \rightarrow \xi \right] = \sqrt{\frac{2}{\pi}} \int_0^\infty g(x) \cos \xi x \, d\xi \, ,$$

then $F_c \left[f * g ; x \rightarrow \xi \right] = F_c(\xi) \, G_c(\xi)$ (1.28)

Also, if $f(x), g(x)$ are odd functions of x in R and if

$$F_s[f(x) ; x \rightarrow \xi] = F_s(\xi) \text{ and } F_s[g(x) ; x \rightarrow \xi] = G_s(\xi)$$

then $F_s \left[f * g ; x \rightarrow \xi \right] = F_s(\xi) \, G_s(\xi)$ (1.29)

It may be noted that the convolution had already been defined in article 1.10 and keeping this together with evenness and oddness character of the associated functions one can very easily deduce the results in equations (1.28) and (1.29) respectively.

1.12 Parseval's relations for Fourier Transforms.

Theorem.

If $F(\xi)$ and $G(\xi)$ are complex F.T of $f(x)$ and $g(x)$ respectively, then

(i) $\dfrac{1}{\sqrt{2\pi}} \displaystyle\int_{-\infty}^{+\infty} F(\xi)\overline{G(\xi)} \, d\xi = \dfrac{1}{\sqrt{2\pi}} \int_{\infty}^{+\infty} f(x)\overline{g(x)} \, dx$ (1.30)

(ii) $\dfrac{1}{\sqrt{2\pi}} \displaystyle\int_{-\infty}^{+\infty} |F(\xi)|^2 d\xi = \dfrac{1}{\sqrt{2\pi}} \int_{-\infty}^{+\infty} |f(x)|^2 dx$ (1.31)

where bar sign over function signifies complex conjugate of the complex functions or absolute value for real functions

Proof (i) By Fourier Inverse transform formula, we have

$$g(x) = \frac{1}{\sqrt{2\pi}} \int_{-\infty}^{+\infty} G(\xi)e^{-i\xi x} d\xi$$

Taking complex conjugates on both sides, the above equation gives

$$\overline{g(x)} = \frac{1}{\sqrt{2\pi}} \int_{-\infty}^{+\infty} \overline{G}(\xi) \, e^{i\xi x} \, d\xi$$

Therefore, $\dfrac{1}{\sqrt{2\pi}} \displaystyle\int_{-\infty}^{+\infty} f(x) \, \overline{g}\,(x) dx$

$$= \frac{1}{\sqrt{2\pi}} \int_{-\infty}^{+\infty} f(x) \left[\frac{1}{\sqrt{2\pi}} \int_{-\infty}^{+\infty} \overline{G}(\xi) \, e^{i\xi x} d\xi \right] dx$$

$$= \frac{1}{\sqrt{2\pi}} \int_{-\infty}^{+\infty} \overline{G}(\xi) \left[\frac{1}{\sqrt{2\pi}} \int_{-\infty}^{+\infty} f(x) e^{i\xi x} dx \right] d\xi$$

$$= \frac{1}{\sqrt{2\pi}} \int_{-\infty}^{+\infty} \overline{G}(\xi) F(\xi) \, d\xi \, , \quad \text{which proves the result (i)}.$$

(ii) In the result (i) if we take $g(x) = f(x)$, we get

$$\frac{1}{\sqrt{2\pi}} \int_{-\infty}^{+\infty} f(x) \, \overline{f(x)} \, dx = \frac{1}{\sqrt{2\pi}} \int_{-\infty}^{+\infty} F(\xi) \, \overline{F(\xi)} \, d\xi$$

i.e, $\dfrac{1}{\sqrt{2\pi}} \displaystyle\int_{-\infty}^{+\infty} |f(x)|^2 dx = \dfrac{1}{\sqrt{2\pi}} \int_{-\infty}^{+\infty} |F(\xi)|^2 \, d\xi$

This proves the part (ii) of the Parseval's relations.

Note : In the above parseval's relations or identities one may drop the factors $\frac{1}{\sqrt{2\pi}}$ from either sides of (1.30) and (1.31)

There exists other four Parseval's identities given by

$$\text{(iii)} \quad \sqrt{\frac{2}{\pi}} \int_0^\infty F_c(\xi) G_c(\xi) d\xi = \sqrt{\frac{2}{\pi}} \int_0^\infty f(x) g(x) dx \qquad (1.32)$$

$$\text{(iv)} \quad \sqrt{\frac{2}{\pi}} \int_0^\infty F_s(\xi) G_s(\xi) d\xi = \sqrt{\frac{2}{\pi}} \int_0^\infty f(x) g(x) dx \qquad (1.33)$$

$$\text{(v)} \quad \sqrt{\frac{2}{\pi}} \int_0^\infty |F_c(\xi)|^2 d\xi = \sqrt{\frac{2}{\pi}} \int_0^\infty |f(x)|^2 dx \qquad (1.34)$$

$$\text{and (vi)} \quad \sqrt{\frac{2}{\pi}} \int_0^\infty |F_s(\xi)^2 d\xi = \sqrt{\frac{2}{\pi}} \int_0^\infty |f(x)|^2 dx \qquad (1.35)$$

connecting to Fourier cosine and sine transforms. Where again the constant terms $\sqrt{\frac{2}{\pi}}$ may be dropped from either sides in the above results. Their proofs can similarly be deduced as were done in cases (i) or (ii) above.

Example 1.10. Let $f(x) = e^{-bx}$, $g(x) = e^{-ax}$. Use parseval's relation of Fourier cosine transform to evaluate the integral to prove

$$\int_0^\infty \frac{dx}{(a^2 + x^2)(b^2 + x^2)} = \frac{\pi}{2ab(a+b)}$$

Solution.

We have $\quad F_c\left[f(x) ; x \to \xi\right] = \sqrt{\frac{2}{\pi}} \int_0^\infty e^{-bx} \cos \xi x \, dx$

$$= \sqrt{\frac{2}{\pi}} \frac{b}{b^2 + \xi^2} \equiv F_c(\xi)$$

Similarly, $\quad G_c(\xi) = \sqrt{\frac{2}{\pi}} \frac{a}{a^2 + \xi^2}$

Therefore, by the Parseval's relation (iii) above, we get

$$\sqrt{\frac{2}{\pi}} \int_0^\infty \frac{2}{\pi} \frac{ab}{(\xi^2 + a^2)(\xi^2 + b^2)(\xi^2 + b^2)} d\xi = \sqrt{\frac{2}{\pi}} \int_0^\infty e^{-(a+b)x} dx$$

i.e, $\quad \int_0^\infty \frac{d\xi}{(\xi^2 + a^2)(\xi^2 + b^2)} = \frac{\pi}{2ab(a+b)}$.

Example 1.11. Let $f(x) = \begin{cases} 1 & , & 0 < x < a \\ 0 & , & x > a \end{cases}$

Use Parseval's identity to evaluate $\int_0^\infty \frac{\sin^2 ax}{x^2} dx$.

Solution. We have from example 4 of article 1.9 that

$$F_c[f(x) ; x \to \infty] = \sqrt{\frac{2}{\pi}} \frac{\sin a \xi}{\xi} \,.$$

Then by the Parseval's identity (1.34) of the corollary above

$$\sqrt{\frac{2}{\pi}} \int_0^\infty \frac{2}{\pi} \left| \frac{\sin a \xi}{\xi} \right|^2 d\xi = \sqrt{\frac{2}{\pi}} \int_0^\infty |f(x)|^2 dx$$

i.e, $\quad \int_0^\infty \frac{\sin^2 a \xi}{\xi^2} d\xi = \frac{\pi}{2} \int_0^a dx = \frac{\pi a}{2}$

Example 1.12. Taking $g(x) = e^{ax}$ and $f(x) = \begin{cases} 1 & , & 0 < x < b \\ 0 & , & x > b \end{cases}$

and using proper Parseval's relation prove that

$$\int_0^\infty \frac{\sin at}{t(a^2 + t^2)}\, dt = \frac{\pi}{2} \frac{1 - e^{-a^2}}{a^2}$$

Solution. By Fourier cosine transform

$$G_c(\xi) = F_c[g(x); x \to \xi] = \frac{a}{\xi^2 + a^2} \quad \text{and} \quad F_c[f(x); x \to \xi] = \frac{\sin a\xi}{\xi} = F_c(\xi)$$

Then by the Parseval's relation connecting Fourier cosine transform we get

$$\int_0^\infty \frac{a \sin a\, \xi}{\xi(\xi^2 + a^2)}\, d\xi = \int_0^a e^{-ax} \cdot 1 \, dx = \frac{1 - e^{-a^2}}{a}$$

$$\Rightarrow \quad \int_0^\infty \frac{\sin at}{t(t^2 + a^2)}\, dt = \frac{1 - e^{-a^2}}{a^2}$$

1.13   Fourier Transform of the derivative of a function.

In many applications of the theory of Fourier transform to boundary value problems of Mathematical Physics it is necessary to express FT of the derivative of a function $f(x)$ in terms of FT of the function $f(x)$. To this direction we now denote FT of $\frac{d^n f(x)}{dx^n}$ by $F^{(n)}(\xi)$, say. Then, we get

$$F^{(n)}(\xi) = \frac{-1}{\sqrt{2\pi}} \int_{-\infty}^{+\infty} (i\xi) \frac{d^{n-1} f(x)}{d\, x^{n-1}}\, e^{i\xi x} dx + \left[\frac{1}{\sqrt{2\pi}} \frac{d^{n-1} f(x)}{d\, x^{n-1}} \cdot e^{i\xi x}\right]_{-\infty}^{+\infty},$$

which is obtained after integrating by parts the right hand side of the definition of

$$F^{(n)}(\xi) = \frac{1}{\sqrt{2\pi}} \int_{-\infty}^{+\infty} \frac{d^n f(x)}{dx^n}\, e^{i\xi x}\, dx \qquad (1.36)$$

If we assume that $\frac{d^{n-1} f(x)}{dx^{n-1}}$ tends to zero as $|x| \to \infty$, the above result takes the form

$$F^{(n)} = -i\xi F^{(n-1)}$$

By repeated application of this rule and by the assumption

$$\lim_{|x| \to \infty} \left[\frac{d^r f(x)}{dx^r}\right] = 0 \ , \ r = 1, 2, \cdots n - 1,$$

we have finally

$$F^{(n)} = (-i\xi)^n \ F$$

This implies that FT of the nth derivative of a function $f(x)$ is $(-i\xi)^n$ times FT of the function, provided that the first $(n-1)$ derivative of the function vanish as $|x| \to \infty$.

The corresponding results for Fourier cosine and Fourier sine transform of a function are not so simple. For discussion towards this direction, we define $F_s^{(n)}$ and $F_c^{(n)}$ by the equations

$$F_s^{(n)} = \sqrt{\frac{2}{\pi}} \int_0^\infty \frac{d^n f(x)}{dx^n} \sin \xi x \ dx \ ,$$

$$F_c^{(n)} = \sqrt{\frac{2}{\pi}} \int_0^\infty \frac{d^n f(x)}{dx^n} \cos \xi x \ dx \tag{1.37}$$

Then, integrating by parts the right hand side of the second of the above equations in (1.37) we get

$$F_c^{(n)} = \sqrt{\frac{2}{\pi}} \left[\frac{d^{n-1} f(x)}{dx^{n-1}} \cos \xi x \right]_0^\infty + \xi \sqrt{\frac{2}{\pi}} \int_0^\infty \frac{d^{n-1} f(x)}{dx^{n-1}} \sin \xi x \ dx$$

$$= -a_{n-1} + \xi \ F_s^{(n-1)} \ , \tag{1.38}$$

on the assumption that $\displaystyle\lim_{x \to \infty} \frac{d^{n-1} f(x)}{dx^{n-1}} = 0,$

$$\lim_{x \to 0} \sqrt{\frac{2}{\pi}} \frac{d^{n-1} f(x)}{dx^{n-1}} = a_{n-1}$$

Similarly, from the first of the equation in (1.37) we derive

$$F_s^{(n)} = -\xi \ F_c^{(n-1)} \tag{1.39}$$

Using the result (1.39) in (1.38) we have

$$F_c^{(n)} = -a_{n-1} - \xi^2 \ F_c^{(n-2)} \tag{1.40}$$

Thus, by repeated application of equation (1.40) one can reduce $F_c^{(n)}$ to a sum of ξ's and either $F_c^{(1)}$ of F_c according as n is odd or even integer respectively. Therefore, we obtain the following formulae :

$$F_c^{(2m)} = -\sum_{s=0}^{m-1} (-1)^s a_{2m-2s-1} \ \xi^{2s} + (-1)^m \ \xi^{2m} \ F_c \tag{1.41}$$

$$F_c^{(2m+1)} = -\sum_{s=0}^{m} (-1)^s a_{2m-2s} \ \xi^{2s} + (-1)^m \ \xi^{2m+1} \ F_s \tag{1.42}$$

Similar results hold for sine transforms too. From (1.39) and (1.40) we obtain,

$$F_s^{(n)} = \xi\, a_{n-2} - \xi^2\, F_s^{(n-2)}$$

From this formula we can derive the formulas

$$F_s^{(2n)} = -\sum_{k=1}^{n}(-1)^k \xi^{2k-1} a_{2n-2k} + (-1)^{n+1}\xi^{2n}\, F_s \tag{1.43}$$

$$F_s^{(2n+1)} = -\sum_{k=1}^{n}(-1)^k \xi^{2k-1} a_{2n-2k+1} + (-1)^{n+1}\xi^{2n+1} F_c \tag{1.44}$$

Certain special cases of the above formulae arise frequently. For example, if $\frac{df}{dx} = \frac{d^3 f}{dx^3} = 0$, when $x = 0$, then

$$\sqrt{\frac{2}{\pi}} \int_0^\infty \frac{d^2 f}{dx^2} \cos \xi\, x\, dx \;=\; -\xi^2\, F_c$$

and $\quad \displaystyle\sqrt{\frac{2}{\pi}} \int_0^\infty \frac{d^4 f}{dx^4} \cos \xi\, x\, dx \;=\; \xi^4\, F_c$

On the other hand, if $\quad f(x) = \frac{d^2 f(x)}{dx^2} = 0$, when $x = 0$, then

$$\sqrt{\frac{2}{\pi}} \int_0^\infty \frac{d^2 f}{dx^2} \sin \xi\, x\, dx = -\xi^2\, F_s$$

and $\quad \displaystyle\sqrt{\frac{2}{\pi}} \int_0^\infty \frac{d^4 f}{dx^4} \sin \xi\, x\, dx = \xi^4\, F_s$

Corollary 1.6. If a function $f(x)$ has a finite discontinuity at a single point $x = a \in R$ for simplicity, the above formula for $F\,T$ of its derivative has to be modified in the following manner :

We write

$$F[f'(x)\,;\; x \to \xi] = \frac{1}{\sqrt{2\pi}} \int_{-\infty}^{a-0} f'(x)\, e^{i\xi x} dx + \frac{1}{\sqrt{2\pi}} \int_{a+0}^{\infty} f'(x)\, e^{i\xi x} dx$$

As before, integrating the right hand side integrals by parts, one gets

$$F[f'(x)\,;\; x \to \xi] = \frac{1}{\sqrt{2\pi}} \left[f(x)\, e^{i\xi x} \right]_{-\infty}^{a-0} + \frac{1}{\sqrt{2\pi}} \left[f(x) e^{i\xi x} \right]_{a+0}^{\infty}$$

$$- \frac{i\xi}{\sqrt{2\pi}} \int_{-\infty}^{a-0} f(x) e^{i\xi x} dx - \frac{i\xi}{\sqrt{2\pi}} \int_{a+0}^{\infty} f(x) e^{i\xi x} dx,$$

which can be written in the form

$$F[f'(x) \; ; \; x \to \xi] = -i\xi \; F[f(x) \; ; \; x \to \xi] - \frac{1}{\sqrt{2\pi}} e^{i\xi a}[f(x)]_a \qquad (1.45)$$

where $[f(x)]_a = f(a+0) - f(a-0)$, called the jump of $f(x)$ at $x = a$.

As a generalisation if we now suppose that $f(x)$ has n points of finite discontinuities at points $x = a_i$, $i = 1, 2, \cdots n$, the modified form of eqn.(1.45) is

$$F[f'(x); x \to \xi] = -i\xi \; F[f(x); x \to \xi] - \frac{1}{\sqrt{2\pi}} \sum_{i=1}^{n} [f(x)]_{a_i} \; e^{i\xi a_i} \qquad (1.46)$$

Corollary 1.7.

We know that $f(x) \in C^1(R)$, then regarding the function $f(x) \; e^{i\xi x}$ as a function of x and ξ, we see that $f(x) \; e^{i\xi x} \in C^1(R \times R)$ and therefore $F(\xi) = \frac{1}{\sqrt{2\pi}} \int_{-\infty}^{+\infty} f(x) e^{i\xi x} dx$ is convergent and the integral

$$\frac{i}{\sqrt{2\pi}} \int_{-\infty}^{+\infty} x f(x) \; e^{i\xi x} dx = \frac{1}{\sqrt{2\pi}} \int_{-\infty}^{+\infty} \frac{\partial}{\partial \xi} \left[f(x) e^{i\xi x} \right] dx$$

is uniformly convergent. Then

$$F'(\xi) = iF \; [x f(x) \; ; \; x \to \xi] \qquad (1.47)$$

We can continue the above process any finite number of times to obtain

$$\frac{d^r}{d\xi^r} F(\xi) = i^r F \; [x^r f(x) \; ; \; x \to \xi], \qquad r = 0, 1, \cdots \qquad (1.48)$$

Corollary 1.8.

$$\text{If} \quad a > 0, \quad F_c \; [f(ax) \; ; \; x \to \xi] = \sqrt{\frac{2}{\pi}} \int_0^\infty f(ax) \cos \xi \, x \, dx$$

$$= a^{-1} \sqrt{\frac{2}{\pi}} \int_0^\infty f(x) \cos \left(\frac{\xi x}{a} \right) dx$$

Therefore, $\quad F_c[f(ax) \; ; \; x \to \xi] = \dfrac{1}{a} F_c[f(x) \; ; \; x \to \dfrac{\xi}{a}], \; a > 0, \qquad (1.49)$

Then, $\quad F_c[f(x) \cos(\omega x) \; ; \; x \to \xi] = \dfrac{1}{2}[F_c(\xi + \omega) + F_c(\xi - \omega)]$

$$(1.50)$$

and $\quad F_c[f(x) \sin \omega \, x \; ; \; x \to \xi] = \dfrac{1}{2}[F_s(\omega + \xi) + F_s(\omega - \xi)]$

$$(1.51)$$

Corollary 1.9.

As in corollary 1.8 above, we can easily deduce that

$$F_s[f(ax) \; ; \; x \to \xi] = a^{-1} F_s\left[f(x) \; ; \; x \to \frac{\xi}{a}\right], \; a > 0 \qquad (1.52)$$

$$F_s[f(x)\cos\omega x \; ; \; x \to \xi] = \frac{1}{2}[F_s(\xi+\omega) + F_s(\xi-\omega)]$$

and $\quad F_s[f(x)\sin\omega x \; ; \; x \to \xi] = \frac{1}{2}[F_c(\xi-\omega) - F_c(\xi+\omega)]$

Corollary 1.10.

In actual calculation of Fourier sine transform the following result may be found useful. Let us denote

$$F_s\left[f(x) \; ; \; x \to \xi\right] = \varphi(\xi) \; , \text{ for } \xi > 0$$

Then, for $\xi < 0$ we have

$$-\varphi(\eta) \equiv -\varphi(-\xi) = F_s\left[f(x) \; ; \; x \to \xi\right], \xi = -\eta \; , \; \eta > 0$$

Therefore, we have in general

$$F_s[f(x) \; ; \; x \to \xi] = \varphi(|\xi|) \cdot sgn \; \xi$$

where $\quad sgn \; \xi = \begin{cases} +1 \; , & \text{for } \xi > 0 \\ -1 \; , & \text{for } \xi < 0 \end{cases}$

1.14   Fourier Transform of some more useful functions.

(a) We require to find $F_c[e^{-ax}x^{n-1}; x \to \xi]$ and $F_s[e^{-ax} \; x^{n-1}; \; x \to \xi]$ where $a > 0$ and $n > 0$. For this purpose let us assume

$$\sqrt{\frac{2}{\pi}} \int_0^\infty e^{-ax}x^{n-1}\cos\xi \; x \; dx = C$$

and $\sqrt{\dfrac{2}{\pi}} \displaystyle\int_0^\infty e^{-ax}x^{n-1}\sin\xi \; x \; dx = S$

Then, $C - iS = \sqrt{\dfrac{2}{\pi}} \displaystyle\int_0^\infty x^{n-1} \cdot e^{-\alpha x}dx$, where $\alpha = a + i\xi = re^{i\theta}$, say

$$= \sqrt{\frac{2}{\pi}} \int_0^\infty \frac{1}{\alpha^n}e^{-z}z^{n-1}dz$$

$$= \sqrt{\frac{2}{\pi}} \frac{1}{\alpha^n} \Gamma(n) \,,$$

$$= \sqrt{\frac{2}{\pi}} \frac{\Gamma(n)}{r^n} \cdot e^{-in\theta}$$

where $r = \sqrt{a^2 + \xi^2}$ and $\theta = \tan^{-1} \frac{\xi}{a}$.

Thus, $F_c [x^{n-1} e^{-ax} ; x \to \xi] = \sqrt{\frac{2}{\pi}} \frac{\cos n\,\theta \cdot \Gamma(n)}{(a^2 + \xi^2)^{\frac{n}{2}}}$

and $F_s [x^{n-1} e^{-ax} ; x \to \xi] = \sqrt{\frac{2}{\pi}} \frac{\Gamma(n) \cdot \sin n\theta}{(a^2 + \xi^2)^{\frac{n}{2}}}$

Deductions :

Since $F_c \, F_c^{-1} = I$ and $F_s \, F_s^{-1} = I$, we get

$$x^{n-1} \, e^{-ax} \;=\; F_c \left[\sqrt{\frac{2}{\pi}} \, \Gamma(n) r^{-n} \cos n\,\theta \; ; \; \xi \to x \right]$$

$$= \; \frac{2}{\pi} \, \Gamma(n) \int_0^\infty r^{-n} \cos \, n\theta \cos \xi x \; d\xi$$

Putting $a = 1$ in the above equation we get

$$\frac{\pi}{2} \cdot \frac{x^{n-1} \, e^{-x}}{\Gamma(n)} = \int_0^\infty (1 + \xi^2)^{-\frac{n}{2}} \cos(n \tan^{-1} \, \xi) \, \cos \xi x \; d\xi$$

As $\tan \theta = \xi \Rightarrow \sec^2 \, \theta \; d\theta = d\xi$, we have

$$\frac{\pi}{2} \frac{x^{n-1} \, e^{-x}}{\Gamma(n)} = \int_0^{\frac{\pi}{2}} \cos^{n-2} \, \theta \, \cos \, n\theta \, \cos(x \tan \theta) \; d\theta \qquad (1.53)$$

Similarly,

$$x^{n-1} \, e^{-x} \;-\; F_s \left[\sqrt{\frac{2}{\pi}} \, \frac{\sin \, n\theta \, \Gamma(n)}{(1 + \xi^2)^{\frac{n}{2}}} \; ; \; \xi \to x \right]$$

$$= \; \frac{2}{\pi} \int_0^\infty \frac{\sin \, n\theta \cdot \Gamma(n)}{(1 + \xi^2)^{\frac{n}{2}}} \, \sin \xi x \; d\xi$$

This result gives,

$$\frac{\pi}{2} \cdot \frac{x^{n-1} e^{-x}}{\Gamma(n)} = \int_0^{\pi/2} \cos^{n-2} \theta \cdot \sin n\theta \cdot \sin(x \tan \theta) \; d\theta \qquad (1.54)$$

If we multiply (1.53) by $e^{-x} x^{m-1}$ and integrate the result with respect to x from 0 to ∞, we get

$$\frac{\pi}{2\Gamma(n)} \int_0^\infty e^{-2x} x^{m+n-2} dx = \int_0^{\pi/2} \cos^{n-2} \theta \cdot \cos n\theta \cdot A \, d\theta$$

where $A = \int_0^\infty e^{-x} x^{m-1} \cdot \cos(x \tan \theta) \, dx$ \hfill (1.55)

Similarly, multiplying (1.54) by $e^{-x} x^{m-1}$ and integrating the result with respect to x from 0 to ∞, we get

$$\frac{\pi}{2\Gamma(n)} \int_0^\infty e^{-2x} x^{m+n-2} \, dx = \int_0^{\pi/2} \cos^{n-2} \theta \cdot \sin n\theta \cdot B \, d\theta$$

where $B = \int_0^\infty e^{-x} x^{m-1} \sin(x \tan \theta) \, dx$ \hfill (1.56)

Therefore, from (1.55) and (1.56)

$$
\begin{aligned}
A - iB &= \int_0^\infty e^{-x} x^{m-1} e^{-ix \tan \theta} \, dx \\
&= \int_0^\infty e^{-\alpha x} x^{m-1} \, dx, \text{ where } \alpha = 1 + i \tan \theta = \frac{e^{i\theta}}{\cos \theta} \\
&= \frac{1}{\alpha^m} \Gamma(m) = \Gamma(m) \cos^m \theta \, e^{-im\theta}
\end{aligned}
$$

This result implies, $A = \Gamma(m) \cos^m \theta \cos m\theta, B = \Gamma(m) \cos^m \theta \sin m\theta$.

Again, $\quad \dfrac{\pi}{2\Gamma(n)} \displaystyle\int_0^\infty e^{-2x} x^{m+n-2} dx = \dfrac{\pi}{2^{m+n}} \dfrac{\Gamma(m+n-1)}{\Gamma(n)}$

Therefore, $\quad \dfrac{\pi \, \Gamma(m+n-1)}{2^{m+n}\Gamma(n)} = \displaystyle\int_0^{\pi/2} \cos^{n-2} \theta \cos n\theta$

$$[\Gamma(m) \cos^m \theta \cos m\theta] d\theta$$

$$\Rightarrow \frac{\pi \, \Gamma(m+n-1)}{2^{m+n}\Gamma(m)\Gamma(n)} = \int_0^{\pi/2} \cos^{m+n-2} \theta \cos m\theta \cos n\theta \, d\theta$$

Also, $\quad \dfrac{\pi \, \Gamma(m+n-1)}{2^{m+n}\Gamma(m)\Gamma(n)} = \displaystyle\int_0^{\pi/2} \cos^{m+n-2} \theta \sin n\theta \, \sin m\theta \, d\theta$

Adding these two results, we get

$$\frac{\pi \, \Gamma(m+n-1)}{2^{m+n-1} \, \Gamma(m)\Gamma(n)} = \int_0^{\pi/2} \cos^{m+n-2} \theta \cos(m-n)\theta \, d\theta$$

If we now set $m + n = 2$, then we get

$$\frac{\pi}{2\Gamma(n)(1-n)\Gamma(1-n)} = \frac{\sin n\pi}{2(1-n)}$$

Therefore, we get $\Gamma(n)\Gamma(1-n) = \frac{\pi}{\sin n\pi}$, the duplication formula for the Gamma function.

(b) We now evaluate $F_c\left[(a^2 - x^2)^{\nu - \frac{1}{2}} H(a - x) \; ; \; x \to \xi\right]$

Solution. Following the definition, we have

$$F_c[(a^2 - x^2)^{\nu - \frac{1}{2}} H(a - x) \; ; \; x \to \xi] = \sqrt{\frac{2}{\pi}} \int_0^a (a^2 - x^2)^{\nu - \frac{1}{2}} \cos \xi x \, dx$$

$$= \sqrt{\frac{2}{\pi}} \cdot I, \text{ say}$$

Now, from the series expansion of $\cos \xi x$, we can write

$$I = \int_0^a (a^2 - x^2)^{\nu - \frac{1}{2}} \sum_{r=0}^{\infty} \frac{(-1)^r (\xi x)^{2r}}{(2r)!} \, dx$$

$$= \sum_{r=0}^{\infty} \frac{(-1)^r \xi^{2r}}{(2r)!} \int_0^a (a^2 - x^2)^{\nu - \frac{1}{2}} x^{2r} \, dx$$

$$= \sum_{r=0}^{\infty} \frac{(-1)^r \xi^{2r}}{(2r)!} \cdot \frac{a^{2\nu + 2r}}{2} \int_0^1 z^{r - \frac{1}{2}} (1 - z)^{\nu - \frac{1}{2}} dz, \text{where } x^2 = a^2 z$$

$$= \sum_{r=0}^{\infty} \frac{(-1)^r \xi^{2r} a^{2\nu + 2r}}{2(2r)!} \frac{\Gamma(\nu + \frac{1}{2})\Gamma(r + \frac{1}{2})}{\Gamma(\nu + r + 1)}$$

$$= \sum_{r=0}^{\infty} \frac{(-1)^r \Gamma(\nu + \frac{1}{2})}{\Gamma(\nu + r + 1)} \cdot \left(\frac{\xi}{2}\right)^{2r} \cdot \frac{a^{2\nu + 2r}}{r!} \frac{\sqrt{\pi}}{2}$$

Thus,

$$F_c\left[(a^2 - x^2)^{\nu - \frac{1}{2}} H(a - x) \; ; \; x \to \xi\right]$$

$$= \sqrt{\frac{2}{\pi}} \frac{\sqrt{\pi}}{2} \left[\sum_{r=0}^{\infty} \frac{(-1)^r}{\Gamma(r+1)\Gamma(\nu + r + 1)}\right] \left(\frac{a\xi}{2}\right)^{2r+\nu} \left(\frac{\xi}{2}\right)^{-\nu} a^\nu \Gamma\left(\nu + \frac{1}{2}\right)$$

$$= \frac{1}{\sqrt{2}} J_\nu (a\xi) \, a^\nu \, \Gamma\left(\nu + \frac{1}{2}\right) \left(\frac{\xi}{2}\right)^{-\nu}$$

$$= \left(\frac{a}{\xi}\right)^\nu 2^{\nu - \frac{1}{2}} \Gamma\left(\nu + \frac{1}{2}\right) J_\nu(a\xi)$$

$$\Rightarrow F_c\left[\left(\frac{a}{\xi}\right)^\nu 2^{\nu-\frac{1}{2}} J_\nu(a\xi) \; ; \; \xi \to x\right] = (a^2 - x^2)^{\nu-\frac{1}{2}} H(a-x)$$

$$= \begin{cases} (a^2 - x^2)^{\nu-\frac{1}{2}} , & 0 < x < a \\ 0 & , \text{ otherwise} \end{cases}$$

Deduction. Putting $\nu = 0$, $a = 1$ in the above, we get

$$F_c\left[\sqrt{\frac{\pi}{2}} \, J_0(\xi); \xi \to x\right] = \begin{cases} \frac{1}{\sqrt{1-x^2}} , & 0 < x < 1 \\ 0 & , \text{ otherwise} . \end{cases}$$

(c) Find Fourier cosine transforms of $e^{-ax} \cos ax$ and $e^{-ax} \sin ax$ and hence evaluate that of $\frac{1}{x^4+k^4}$

Solution.

Let $\qquad C = F_c[e^{-ax} \cos ax \; ; \; x \to \xi]$

and $\qquad S = F_c[e^{-ax} \sin ax \; ; \; x \to \xi]$

Then, $\; C - iS = \sqrt{\dfrac{2}{\pi}} \displaystyle\int_0^\infty e^{-\alpha x} \cos \xi \, x \, dx$, where $\alpha = a(1+i)$

$$= \sqrt{\frac{2}{\pi}} I, \text{ say.}$$

So, $\qquad I = \dfrac{\alpha}{\xi^2} - \dfrac{\alpha}{\xi^2} I$

$\Rightarrow \qquad I = \dfrac{\alpha}{\alpha^2 + \xi^2}$

Therefore, $C - iS = \sqrt{\dfrac{2}{\pi}} \dfrac{a(1+i)}{\xi^2 + a^2(1+i)^2}$

$$= \sqrt{\frac{2}{\pi}} \cdot \frac{a(1+i)(\xi^2 - 2a^2 i)}{(\xi^2 + 2a^2 i)(\xi^2 - 2a^2 i)}$$

$$= \sqrt{\frac{2}{\pi}} \frac{a[(\xi^2 + 2a^2) + i(\xi^2 - 2a^2)]}{\xi^4 + 4a^4}$$

So, $\; F_c[e^{-ax} \cos ax; x \to \xi] = \sqrt{\dfrac{2}{\pi}} \dfrac{a(\xi^2 + 2a^2)}{\xi^4 + 4a^4}$

and $\; F_c[e^{-ax} \sin ax; x \to \xi] = \sqrt{\dfrac{2}{\pi}} \dfrac{a(-\xi^2 + 2a^2)}{\xi^4 + 4a^4}$.

Then, $\; F_c[e^{-ax}(\lambda \cos ax + \mu \sin ax) \; ; \; x \to \xi]$

$$= \sqrt{\frac{2}{\pi}} a \frac{(\lambda - \mu)\xi^2 + 2a^2(\lambda + \mu)}{\xi^4 + 4a^4}$$

Now choosing, $1 = \lambda = \mu \neq 0$, we get

$$F_c[e^{-ax}(\cos ax + \sin ax); x \to \xi] = \sqrt{\frac{2}{\pi}} \frac{4a^3}{\xi^4 + 4a^4}$$

$$\therefore \quad F_c\left[\sqrt{\frac{2}{\pi}} \frac{4a^3}{\xi^4 + 4a^4} ; \xi \to x\right] = e^{-ax}(\cos ax + \sin ax)$$

Thus $\quad F_c\left[\frac{1}{\xi^4 + 4a^4} ; \xi \to x\right] = \sqrt{\frac{\pi}{2}} \frac{e^{-ax}}{4a^3}(\cos ax + \sin ax)$

$$\therefore \quad F_c\left[\frac{1}{\xi^4 + k^4} ; \xi \to x\right] = \sqrt{\frac{\pi}{2}} \cdot \frac{e^{\frac{-kx}{\sqrt{2}}}}{\sqrt{2}k^3}\left(\cos\frac{kx}{\sqrt{2}} + \sin\frac{kx}{\sqrt{2}}\right)$$

Or, $\quad F_c\left[\frac{1}{x^4 + k^4} ; x \to \xi\right] = \sqrt{\frac{\pi}{2}} \cdot \frac{e^{\frac{-k\xi}{\sqrt{2}}}}{\sqrt{2}k^3}\left(\cos\frac{k\xi}{\sqrt{2}} + \sin\frac{k\xi}{\sqrt{2}}\right).$

(d) From $\quad P_n(x) = \dfrac{1}{2^n \, n!} \dfrac{d^n}{dx^n}(x^2 - 1)^n$ evaluate

$\quad F[P_n(x) \, H(1 - |x|) \; ; \; x \to \xi]$

Solution.

$F[P_n(x)H(1 - |x|); x \to \xi]$

$$= \frac{1}{\sqrt{2\pi}} \cdot \frac{1}{2^n \, n!} \int_{-1}^{1} \frac{d^n}{dx^n}(x^2 - 1)^n e^{i\xi x} \, dx$$

$$= \frac{(-i\xi)}{\sqrt{2\pi}2^n \, n!} \int_{-1}^{1} \frac{d^{n-1}}{dx^{n-1}}(x^2 - 1)^n \cdot e^{i\xi x} \, dx \, , \quad \text{on integration by parts}$$

$$= \frac{(-i\xi)^2}{\sqrt{2\pi}2^n \, n!} \int_{-1}^{1} \frac{d^{n-2}}{dx^{n-2}}(x^2 - 1)^n \cdot e^{i\xi x} \, dx$$

Similarly proceeding after n-th step we get

$$= \frac{1}{\sqrt{2\pi}} \frac{(-i\xi)^n}{2^n \, n!} \int_{-1}^{1} (x^2 - 1)^n e^{i\xi x} dx$$

$$= \frac{i^n \xi^n}{\sqrt{2\pi}2^n \, n!} \int_{-1}^{1} (1 - x^2)^n e^{i\xi x} dx$$

$$= \frac{i^n \xi^n}{\sqrt{2\pi}2^n n!} \cdot 2 \cdot \left[\int_{0}^{1} (1 - x^2)^n \cos\xi \, x \, dx\right]$$

But from the result of (b) above, after putting $a = 1$ and $\nu - \frac{1}{2} = n$, we get

$$F[P_n(x) \, H(1 - |x|) \; ; \; x \to \xi]$$

$$= \sqrt{\frac{2}{\pi}} \frac{i^n \xi^n}{2^n \, n!} \cdot \sqrt{\frac{\pi}{2}} \, \xi^{-n-\frac{1}{2}} \, 2^n J_{n+\frac{1}{2}} \, (\xi) \cdot \Gamma(n+1)$$

$$= \frac{i^n}{\sqrt{\xi}} \, J_{n+\frac{1}{2}} \, (\xi)$$

1.15   Fourier Transforms of Rational Functions.

Using calculus of residues *FT* of rational functions can be calculated. Let us assume that the function of a complex variable z admits of the following properties :

   (i) $f(z)$ has finite number of singularities at $z = a_1, a_2 \cdots a_n$ in the half plane $Im \ z > 0$

  (ii) $f(z)$ is analytic everywhere on real z axis except at the points $b_1, b_2, \cdots b_m$, which are simple poles and

 (iii) $\left| z f(z) \, e^{i\xi z} \right| \to 0$ as $|z| \to \infty$, $Im \ z \geqslant 0$.

Let us now integrate $e^{i\xi z} f(z)$, $\xi > 0$ round the contour shown in the adjoining figure and making use of Cauchy's residue theorem to obtain

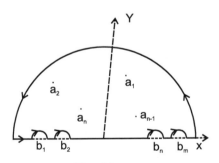

Fig. 1.1

$$\frac{1}{2\pi i} \int_{-\infty}^{+\infty} f(x) e^{i\xi x} dx = \frac{1}{2} \sum_{s=1}^{m} res \, [f(b_s)] \cdot e^{ib_s\xi} + \sum_{s=1}^{n} res \, [f(a_s)] \cdot e^{i\xi a_s}$$

This formula yeilds for $\xi > 0$

$$F[f(x); x \to \xi] = \sqrt{2\pi} i \left[\sum_{s=1}^{n} res \, \{f(a_s)\} e^{i\xi a_s} + \frac{1}{2} \sum_{s=1}^{m} res \, \{f(b_s)\} e^{i\xi b_s} \right]$$

To provide an application of the above result we first consider the simplest example when $f(z) = \frac{1}{z}$.

This function has a simple pole at $z = 0$ and it has no other singularities inside and on the contour shown above. Then, for $\xi > 0$

$$F[\frac{1}{x} \; ; \; x \to \xi] = \sqrt{2\pi}i \cdot \frac{1}{2} \, res \, [f(0)] \cdot e^{i\xi.0}$$

$$= i\sqrt{\frac{\pi}{2}}$$

Again $f(x) = \frac{1}{x}$ is an odd function of x in R. This gives

$$F_s \left[\frac{1}{x} \; ; \; x \to \xi \right] = \sqrt{\frac{\pi}{2}} \, , \; \xi > 0 \, .$$

To provide another example, consider $f(z) = \frac{1}{z(z^2+a^2)}$ where a is a real number. This function $f(z)$ has simple poles inside the contour as described above at $z = 0$ and at $z = ia$ having residues a^{-2} and $-\frac{1}{2}a^{-2}$ respectively, when $\xi > 0$. Then, we get

$$F \left[\frac{1}{x(x^2 + a^2)} \; ; \; x \to \xi \right] = \sqrt{\frac{\pi}{2}} \frac{1}{a^2} i \left(1 - e^{-\xi|a|} \right)$$

and from it we have

$$F_s \left[\frac{1}{x(x^2 + a^2)} \; ; \; x \to \xi \right] = \sqrt{\frac{\pi}{2}} \frac{1}{a^2} \left(1 - e^{-\xi|a|} \right) sgn(\xi) \, .$$

1.16 Other important examples concerning derivative of FT .

The following example will depict different methods for finding $F\,T$ of given function using derivative of the transformed function whenever necessary.

Example 1.13. Find Fourier sine and cosine transform of $f(x) = x$.

Solution.

$$\text{Let} \quad F_c(\xi) = \sqrt{\frac{2}{\pi}} \int_0^\infty x \cos \xi \, x \, dx$$

$$\text{and} \quad F_s(\xi) = \sqrt{\frac{2}{\pi}} \int_0^\infty x \sin \xi \, x \, dx$$

Thus, $\quad F_c(\xi) - i\,F_s(\xi) = \sqrt{\dfrac{2}{\pi}} \displaystyle\int_0^\infty x\,e^{-i\xi x}\,dx$

$$= \sqrt{\dfrac{2}{\pi}} \int_0^\infty \dfrac{t\,e^{-t}}{i\xi}\,\dfrac{dt}{i\xi} \quad , \text{ where } \ i\xi x = t$$

$$= -\dfrac{\Gamma(2)}{\xi^2} = -\dfrac{1}{\xi^2}$$

Equating real and imaginary parts, we get the required result.

Example 1.14. Find $f(x)$, if $F_s[f(x); x \to \xi] = \dfrac{e^{-a\xi}}{\xi}$. Hence, evaluate $F_s^{-1}\left[\dfrac{1}{\xi}\,;\,\xi \to x\right]$.

Solution. we know that

$$f(x) = F_s^{-1}\,[F_s(\xi)] = \dfrac{2}{\pi} \int_0^\infty \dfrac{e^{-a\xi}}{\xi} \sin \xi\,x\,d\xi \tag{1.57}$$

Therefore, $\quad \dfrac{df(x)}{dx} = \dfrac{2}{\pi} \displaystyle\int_0^\infty e^{-a\xi} \cos \xi\,x\,d\xi = \dfrac{2}{\pi}\,\dfrac{a}{a^2 + x^2}$

Integrating, we get

$$f(x) = \dfrac{2a}{\pi}\,\dfrac{1}{a}\,\tan^{-1}\dfrac{x}{a} + A \ , \ A \text{ is an arbitrary constant}$$

Putting $x = 0$ in the above, we get $A = f(0) = 0$, from (1.57).

So, $\qquad f(x) = \dfrac{2}{\pi}\,\tan^{-1}\dfrac{x}{a} = F_s^{-1}\,[F_s(\xi)] = F_s^{-1}\left[\dfrac{1}{\xi}\,e^{-a\xi}\right]$

Putting $a = 0$ in the above, we get

$$\dfrac{2}{\pi}\tan^{-1}\left(\dfrac{x}{0}\right) = \dfrac{2}{\pi}\cdot\dfrac{\pi}{2} = 1 = F_s^{-1}\left(\dfrac{1}{\xi}\cdot 1\right) \Rightarrow F_s^{-1}\left(\dfrac{1}{\xi}\right) = 1$$

Example 1.15. Find Fourier sine transform of $f(x) = \dfrac{1}{x(x^2+a^2)}$ [An alternative method].

Solution. Let $\quad I(\xi) = F_s\left[\dfrac{1}{x(x^2 + a^2)}\,;\ x \to \xi\right]$

$$= \sqrt{\dfrac{2}{\pi}} \int_0^\infty \dfrac{\sin \xi\,x}{x(x^2 + a^2)}\,dx \ ,$$

$$\dfrac{dI}{d\xi} = \sqrt{\dfrac{2}{\pi}} \int_0^\infty \dfrac{\cos \xi\,x}{x^2 + a^2}\,dx$$

Therefore,
$$\frac{d^2 I}{d\xi^2} = -\sqrt{\frac{2}{\pi}} \int_0^\infty \left[1 - \frac{a^2}{x^2 + a^2} \right] \frac{\sin \xi\, x}{x}\, dx$$

$$= -\sqrt{\frac{2}{\pi}} \int_0^\infty \frac{\sin \xi\, x}{x}\, dx + a^2\, I(\xi)$$

$$= -\sqrt{\frac{\pi}{2}} + a^2\, I(\xi)$$

Solving this differential equation, we get

$$I(\xi) = A\, e^{a\xi} + B\, e^{-a\xi} + \sqrt{\frac{\pi}{2}}\, \frac{1}{a^2}$$

$$\Rightarrow \quad \frac{dI(\xi)}{d\xi} = Aa\, e^{a\xi} - Ba\, e^{-a\xi}$$

Taking $\xi = 0$, $Aa - Ba = \sqrt{\dfrac{2}{\pi}}\, \dfrac{\pi}{2} \cdot \dfrac{1}{a} = \sqrt{\dfrac{\pi}{2}} \cdot \dfrac{1}{a}$

$$\Rightarrow A - B = \sqrt{\frac{\pi}{2}} \cdot \frac{1}{a^2}$$

Also,
$$A + B = -\sqrt{\frac{\pi}{2}} \cdot \frac{1}{a^2}$$

Solving we get,
$$A = 0 \quad \text{and} \quad B = -\sqrt{\frac{\pi}{2}} \cdot \frac{1}{a^2}$$

$$\therefore \quad F_s \left[\frac{1}{x(x^2 + a^2)} \right] = -\sqrt{\frac{\pi}{2}}\, \frac{1}{a^2}\, e^{-a\xi} + \sqrt{\frac{\pi}{2}} \cdot \frac{1}{a^2}$$

$$= \sqrt{\frac{\pi}{2}}\, \frac{1}{a^2} \left(1 - e^{-a\xi} \right)$$

i.e
$$\sqrt{\frac{2}{\pi}} \int_0^\infty \frac{\sin \xi\, x}{x(x^2 + a^2)}\, dx = \sqrt{\frac{\pi}{2}} \cdot \frac{1}{a^2} \left(1 - e^{-a\xi} \right)$$

i.e
$$\int_0^\infty \frac{\sin \xi\, x}{x(x^2 + a^2)}\, dx = \frac{\pi}{2a^2} \left(1 - e^{-a\xi} \right).$$

Example 1.16. Find Fourier cosine inverse of $\frac{1}{1+\xi^2}$.

Solution.

Solution. Let
$$f(x) = F_c^{-1} \left[\frac{1}{1 + \xi^2} \; ; \; \xi \to x \right]$$

$$= \sqrt{\frac{2}{\pi}} \int_0^\infty \frac{\cos \xi\, x}{1 + \xi^2}\, d\xi$$

$$\therefore \quad \frac{df(x)}{dx} = -\sqrt{\frac{2}{\pi}} \cdot \frac{\pi}{2} + \sqrt{\frac{2}{\pi}} \int_0^\infty \frac{\sin \xi\, x\, d\xi}{\xi(1 + \xi^2)}$$

$$= -\sqrt{\frac{\pi}{2}} \, e^{-x} \, .$$

Thus, $f(x) = \sqrt{\frac{\pi}{2}} \, e^{-x}$, after evaluating the constant of integration by putting $x = 0$ there.

Example 1.17. Evaluate $F_c[f(x) \; ; \; x \to \xi]$ and $F_c[g(x) \; ; \; x \to \xi]$ where

$$f(x) = x^{-s}, 0 < s < 1, \text{ and } g(x) = \left(1 - x^2\right)^{\nu - \frac{1}{2}} H(1 - x), \left(\nu > -\frac{1}{2}\right)$$

Solution. We find $F_c[f(x) \; ; \; x \to \xi]$ by evaluating the contour integral $\int_\Gamma f(z) \, e^{i\xi z} \, dz$, where Γ is a positively described quarter circle $|z| = R$ with identation at $z = 0$ and as depicted in the adjoining figure.

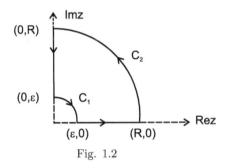

Fig. 1.2

It may be noted here that

$$z \cdot f(z) \, e^{i\xi z} \longrightarrow 0 \text{ as } \epsilon \to 0 \Rightarrow z \to 0$$

and $\qquad z \cdot f(z) \, e^{i\xi z} \longrightarrow 0 \text{ as } R \to \infty, \; 0 < \theta < \frac{\pi}{2}, \text{where } z = Re^{i\theta} \text{ on } C_2$

Therefore, $\displaystyle\int_{C_1} f(z) \, e^{i\xi z} \, dz \longrightarrow 0 \text{ as } |z| \to 0$

and $\qquad \displaystyle\int_{C_2} f(z) \, e^{i\xi z} \, dz \longrightarrow 0 \text{ as } |z| \to \infty$

Then by Cauchy's residue theorem we have

$$\int_0^\infty x^{-s} \, e^{i\xi x} \, dx = e^{\frac{-\pi s i}{2}} \int_0^\infty y^{-s} e^{-\xi y} \, i \, dy$$

since the integrand has no singularity inside the contour depicted above and is regular there.

This gives,

$$\int_0^\infty x^{-s}\, e^{i\xi x}\, dx = i\left(\cos\frac{\pi s}{2} - i\sin\frac{\pi s}{2}\right)\xi^{s-1}\Gamma(1-s), \text{ which implies}$$

$$\left.\begin{aligned}
F_c[f(x)\ ;\ x \to \xi] &= \sqrt{\tfrac{2}{\pi}}\int_0^\infty x^{-s}\cos\xi\, x\, dx\\
&= \sqrt{\tfrac{2}{\pi}}\sin\tfrac{s\pi}{2}\,\Gamma(1-s)\xi^{s-1} \text{ and}\\
F_s[f(x)\ ;\ x \to \xi] &= \sqrt{\tfrac{2}{\pi}}\int_0^\infty x^{-s}\sin\xi\, x\, dx\\
&= \sqrt{\tfrac{2}{\pi}}\cos\tfrac{s\pi}{2}\cdot\Gamma(1-s)\,\xi^{s-1}
\end{aligned}\right\}$$

$$(1.58)$$

Also, from the result in (b) of article 1.14 we get

$$F_c\left[g(x)\ ;\ x \to \xi\right]$$
$$= G_c(\xi)$$
$$= 2^{\nu-\frac{1}{2}}\,\Gamma\left(\nu+\frac{1}{2}\right)\xi^{-\nu}\,J_\nu(\xi)$$

Then by the result (iii) of article 1.12 we have

$$\int_0^\infty F_c(\xi)\, G_c(\xi)\, d\xi = \int_0^\infty f(x)\, g(x)\, dx$$

implying that

$$2^\nu \pi^{-\frac{1}{2}}\sin\frac{s\pi}{2}\,\Gamma(1-s)\,\Gamma\left(\nu+\frac{1}{2}\right)\int_0^\infty \xi^{s-\nu-1}\,J_\nu(\xi)\, d\xi$$

$$= \int_0^1 x^{-s}(1-x^2)^{\nu-\frac{1}{2}}\, dx\ , \qquad \nu > -\frac{1}{2}$$

$$= \frac{1}{2}\int_0^1 y^{-\frac{s}{2}-\frac{1}{2}}(1-y)^{\nu-\frac{1}{2}}\, dy$$

$$= \Gamma\left(\nu+\frac{1}{2}\right)\Gamma\left(\frac{1}{2}-\frac{s}{2}\right)\Big/\left[2\Gamma\left(\nu\ \frac{s}{2}+1\right)\right]$$

Since, $\qquad \pi\operatorname{cosec}\left(\frac{s\pi}{2}\right) = \Gamma\left(\frac{s}{2}\right)\Gamma\left(1-\frac{s}{2}\right)$

and $\qquad \Gamma\left(\frac{1}{2}\right)\Gamma(1-s) = 2^{-s}\,\Gamma\left(\frac{1}{2}-\frac{s}{2}\right)\Gamma\left(1-\frac{s}{2}\right)$

we have

$$\int_0^\infty \xi^{s-\nu-1}\,J_\nu(\xi)\, d\xi = \frac{2^{s-\nu-1}\,\Gamma\left(\frac{s}{2}\right)}{\Gamma\left(\nu-\frac{s}{2}+1\right)}\ ,\quad 0 < s < 1,\ \nu > -\frac{1}{2}\ .$$

Similarly another pair of formulae

$$\int_0^\infty F_s(\xi) G_c(\xi) \sin(\xi x)\, d\xi$$

$$= \frac{1}{2} \int_0^\infty f(u)[g(|u - x|) - g(u + x)]\, du$$

$$= \frac{1}{2} \int_0^\infty g(u)[f(x + u) - f(u - x)]\, du$$

can also be used for the derivation of

$$\int_0^\infty F_s(\xi) G_c(\xi)\, d\xi = \int_0^\infty f(x) g(x)\, dx$$

in this case.

Example 1.18. Find Fourier sine and cosine transforms of the function

$$f(x) = \left(e^{ax} + e^{-ax}\right) / \left(e^{\pi x} - e^{-\pi x}\right) = \frac{ch\ ax}{sh\ \pi x}.$$

Solution. $F_s\,[f(x)\; ;\; x \to \xi] = \sqrt{\dfrac{2}{\pi}} \displaystyle\int_0^\infty \dfrac{ch\ ax}{sh\ \pi x} \sin \xi\, x\, dx$

$$= \sqrt{\frac{2}{\pi}} \int_0^\infty \frac{ch\ ax}{sh\ \pi x} \cdot \frac{e^{i\xi x} - e^{-i\xi x}}{2i}\, dx$$

$$= \sqrt{\frac{2}{\pi}} \cdot \frac{1}{2i} \int_0^\infty \left[\frac{\{e^{(a+i\xi)x} - e^{-(a+i\xi)x}\}}{e^{\pi x} - e^{-\pi x}} - \frac{\{e^{(a-i\xi)x} - e^{-(a-i\xi)x}\}}{e^{\pi x} - e^{-\pi x}} \right] dx$$

$$= \sqrt{\frac{2}{\pi}} \cdot \frac{1}{2i} \left[\frac{1}{2} \tan \frac{a + i\xi}{2} - \frac{1}{2} \tan \frac{a - i\xi}{2} \right]\ ,\ \text{by use of tables of integrals}$$

$$= \sqrt{\frac{1}{2\pi}} \frac{sh\ \xi}{\cos a + ch\ \xi}\ ,\ \text{on simplification.}$$

Similarly, $F_c\,[f(x)\; ;\; x \to \xi] = \sqrt{\dfrac{2}{\pi}} \displaystyle\int_0^\infty \dfrac{ch\ ax}{sh\ \pi x} \cos \xi\, x\, dx$

$$= \sqrt{\frac{2}{\pi}} \frac{1}{2} \int_0^\infty \left[\frac{\{e^{(a+i\xi)x} + e^{-(a+i\xi)x}\}}{e^{\pi x} - e^{-\pi x}} + \frac{\{e^{(a-i\xi)x} + e^{-(a-i\xi)x}\}}{e^{\pi x} - e^{-\pi x}} \right] dx$$

$$= \frac{1}{\sqrt{2\pi}} \left[\sec \frac{a + i\xi}{2} + \sec \frac{a - i\xi}{2} \right]\ ,\ \text{from tables of integrals}$$

$$= \frac{1}{\sqrt{2\pi}} \frac{\cos \frac{a}{2}\ ch\frac{\xi}{2}}{\cos a + ch\ \xi}.$$

Example 1.19. Show that $\displaystyle\int_0^\infty \frac{\cos\, px}{1+p^2}\, dp = \frac{\pi}{2}\, e^{-x}$, $x \geqslant 0$,

Solution. We know that

$$\int_0^\infty e^{-x} \cos\, px\, dx = \frac{1}{1+p^2}$$

This means that

$$F_c[e^{-x}\, ;\, x \to \xi] = \sqrt{\frac{2}{\pi}}\, \frac{1}{1+\xi^2}$$

Therefore, by inverse Fourier cosine transform formula we get

$$F_c^{-1}\left[\sqrt{\frac{2}{\pi}}\, \frac{1}{1+\xi^2}\, ;\, \xi \to x\right] = e^{-x}$$

$$\Rightarrow \quad \sqrt{\frac{2}{\pi}} \int_0^\infty \sqrt{\frac{2}{\pi}}\, \frac{1}{1+\xi^2} \cos\xi\, x\, d\xi = e^{-x}$$

Or, $\displaystyle\int_0^\infty \frac{\cos\xi\, x}{1+\xi^2}\, d\xi = \frac{\pi}{2} \cdot e^{-x}$.

Example 1.20. Find Fourier sine and Fourier cosine transforms of the function

$$f(x) = \begin{cases} \sin\, x & ,\quad 0 < x < a \\ 0 & ,\quad x > a \end{cases}$$

Solution. $F_s\, [\, f(x)\, ;\, x \to \xi\,]$

$$= \sqrt{\frac{2}{\pi}} \int_0^a \sin x\, \sin\xi x\, dx$$

$$= \frac{1}{\sqrt{2\pi}} \int_0^a [\cos(1-\xi)x - \cos(1+\xi)x] dx$$

$$= \frac{1}{\sqrt{2\pi}} \left[\frac{\sin(1-\xi)a}{1-\xi} - \frac{\sin(1+\xi)a}{1+\xi}\right]$$

$F_c\, [f(x)\, ;\, x \to \xi]$

$$= \sqrt{\frac{2}{\pi}} \int_0^a \sin x \cos\xi\, x\, dx$$

$$= \frac{1}{\sqrt{2\pi}} \int_0^a [\sin(1+\xi)x + \sin(1-\xi)x] dx$$

$$= \frac{1}{\sqrt{2\pi}} \left[-\frac{\cos(1+\xi)^x}{1+\xi} - \frac{\cos(1-\xi)x}{1-\xi} \right]_0^a$$

$$= \frac{1}{\sqrt{2\pi}} \left[\frac{-\cos(1+\xi)a+1}{1+\xi} + \frac{-\cos(1-\xi)a+1}{1-\xi} \right]$$

$$= \sqrt{\frac{2}{\pi}} \left[\frac{\sin^2(1+\xi)\frac{a}{2}}{1+\xi} + \frac{\sin^2(1-\xi)\frac{a}{2}}{1-\xi} \right] .$$

Example 1.21. Find $f(x)$, if its Fourier sine transform is $\sqrt{\frac{2}{\pi}}\frac{\xi}{1+\xi^2}$

Solution. By question, $f(x) = F_s^{-1} \left[\sqrt{\frac{2}{\pi}}\frac{\xi}{1+\xi^2} \; ; \; \xi \to x \right]$

Therefore, $\quad f(x) = \frac{2}{\pi} \int_0^\infty \frac{\xi}{1+\xi^2} \sin \xi\, x \, d\xi$

$$= \frac{2}{\pi} \int_0^\infty \left[\frac{\sin \xi x}{\xi} - \frac{\sin \xi x}{\xi(1+\xi^2)} \right] d\xi$$

$$= 1 - \frac{2}{\pi} \int_0^\infty \frac{\sin \xi x}{\xi(1+\xi^2)} \, d\xi \qquad\qquad \text{(i)}$$

$$\therefore \quad \frac{df}{dx} = -\frac{2}{\pi} \int_0^\infty \frac{\cos \xi x}{1+\xi^2} \, d\xi \qquad\qquad \text{(ii)}$$

$$\Rightarrow \quad \frac{d^2 f}{dx^2} = \frac{2}{\pi} \int_0^\infty \frac{\xi \sin \xi x}{1+\xi^2} \, d\xi = f(x)$$

$$\Rightarrow \quad f(x) = A\, e^x + B\, e^{-x} \qquad\qquad \text{(iii)}$$

But $\quad f(0) = 1 = A + B$, by (i) and (ii) above.

Also since $\quad \left[\dfrac{df}{dx} \right]_{x=0} = -1 = A - B$, by (ii) and (iii)

$$\therefore \quad A = 0 \text{ and } B = 1$$

Hence, $\quad f(x) = e^{-x}$.

Example 1.22. Using Fourier transform evaluate the following integrals to prove that

(a) $\displaystyle \int_{-\infty}^{+\infty} \frac{dx}{(x^2+a^2)(x^2+b^2)} = \frac{\pi}{ab(a+b)}$, $a > 0$, $b > 0$

(b) $\displaystyle \int_0^\infty \frac{x^{-p}dx}{a^2+x^2} = \frac{\pi}{2} a^{-(p+1)} \sec\left(\frac{\pi p}{2}\right)$

(c) $\displaystyle \int_0^\infty \frac{x^2\, dx}{(x^2+a^2)(x^2+b^2)} = \frac{\pi}{2(a+b)}$, $a > 0$, $b > 0$

(d) $\displaystyle \int_0^\infty \frac{x^2}{(x^2+a^2)^4} \, dx = \frac{\pi}{(2a)^5}$, $a > 0$

Solution.

(a) Let us take $f(x) = e^{-a|x|}$ and $g(x) = e^{-b|x|}$

Then $F(\xi) = \sqrt{\dfrac{2}{\pi}} \dfrac{a}{\xi^2 + a^2}$, $G(\xi) = \sqrt{\dfrac{2}{\pi}} \dfrac{b}{\xi^2 + b^2}$

Therefore, $\displaystyle\int_{-\infty}^{+\infty} F(\xi)\, G(\xi)\, e^{i\xi x} d\xi = \int_{-\infty}^{+\infty} f(t)\, g(x-t)\, dt$

Hence, by putting $x = 0$ on both sides we get

$$\int_{-\infty}^{+\infty} F(\xi)\, G(\xi)\, d\xi = \int_{-\infty}^{+\infty} f(t)\, g(-t)\, dt$$

$$\Rightarrow \int_{-\infty}^{+\infty} \frac{d\xi}{(\xi^2 + a^2)(\xi^2 + b^2)} = \frac{\pi}{2ab} \int_{-\infty}^{+\infty} e^{-|t|(a+b)}\, dt$$

$$= \frac{\pi}{ab} \int_{0}^{\infty} e^{-(a+b)t}\, dt$$

$$= \frac{\pi}{ab(a+b)}$$

(b) Taking $f(x) = e^{-ax}$, $g(x) = x^{p-1}$ we have

$$F_c(\xi) = F_c\left[e^{-ax} \; ; \; x \to \xi\right] = \sqrt{\frac{2}{\pi}} \frac{a}{a^2 + \xi^2}$$

and $F_c\left[x^{p-1} \; ; \; x \to \xi\right] = \sqrt{\dfrac{2}{\pi}}\, \xi^{-p}\, \Gamma(p)\, \cos \dfrac{\pi p}{2} = G_c(\xi)$

By using the Parseval's relation for Fourier cosine transform we get

$$\int_{0}^{\infty} F_c(\xi)\, G_c(\xi)\, d\xi = \int_{0}^{\infty} f(x)\, g(x)\, dx$$

Hence, $\dfrac{2a}{\pi} \cos \dfrac{\pi p}{2}\, \Gamma(p) \displaystyle\int_{0}^{\infty} \frac{\xi^{-p}\, d\xi}{\xi^2 + a^2}$

$$= \int_{0}^{\infty} x^{p-1} \cdot e^{-ax}\, dx$$

$$= \frac{\Gamma(p)}{a^p} \text{ , after putting } ax = t$$

Therefore, $\displaystyle\int_{0}^{\infty} \frac{\xi^{-p}}{a^2 + \xi^2}\, d\xi = \frac{\pi}{2a^{p+1}} \sec \frac{\pi p}{2}$.

(c) We know that

$$F_s\left[e^{-ax} \; ; \; x \to \xi\right] = \sqrt{\frac{2}{\pi}} \frac{\xi}{\xi^2 + a^2} = F_s(\xi), \text{ say}$$

$$F_s\left[e^{-bx} \; ; \; x \to \xi\right] = \sqrt{\frac{2}{\pi}} \frac{\xi}{\xi^2 + b^2} = G_s(\xi), \text{ say,}$$

if we choose $f(x) = e^{-ax}$ and $g(x) = e^{-bx}$.

Then by convolution theorem we have

$$\int_0^\infty F_s(\xi) \, G_s(\xi) \cos \xi \, x \, d\xi = \frac{1}{2} \int_0^\infty g(\xi) \left[f(\xi + x) + f(\xi - x)\right] d\xi$$

Putting $x = 0$, we get

$$\frac{2}{\pi} \int_0^\infty \frac{\xi^2 \, d\xi}{(\xi^2 + a^2)(\xi^2 + b^2)}$$

$$= \int_0^\infty e^{-(a+b)\xi} \, d\xi$$

$$= \frac{1}{a+b}$$

$$\Rightarrow \int_0^\infty \frac{\xi^2 \, d\xi}{(\xi^2 + a^2)(\xi^2 + b^2)} = \frac{\pi}{2(a+b)} \; .$$

(d) By convolution theorem we have

$$\int_{-\infty}^{+\infty} f(x) \, \overline{f(x)} \, dx = \int_{-\infty}^{+\infty} F(\xi) \, \overline{F(\xi)} \, d\xi \quad \text{as the Parseval's relation,}$$

where $F(\xi)$ is the Fourier transform of $f(x)$.

Let us now write $\;f(x) = \dfrac{1}{2(x^2 + a^2)} \Rightarrow f'(x) = -\dfrac{x}{(x^2 + a^2)^4}$

and $\;F[f(x) \; ; \; x \to \xi] = F(\xi) = \sqrt{\dfrac{\pi}{2}} \dfrac{1}{2a} \exp\left(-a|\xi|\right) \, .$

Hence, $\displaystyle\int_{-\infty}^{+\infty} |f'(x)|^2 \, dx = \int_{-\infty}^{+\infty} |F\{f'(x) \; ; \; x \to \xi\}|^2 \, d\xi$

$$= \int_{-\infty}^{+\infty} |(i\xi) \cdot \frac{1}{2a} \sqrt{\frac{\pi}{2}} \exp\left(-a|\xi|\right)|^2 \, d\xi$$

$$\Rightarrow \int_{-\infty}^{+\infty} \frac{x^2 \, dx}{(x^2 + a^2)^4} = \frac{\pi}{2} \frac{2}{4a^2} \int_0^\infty \xi^2 \exp\left(-2a\xi\right) \, d\xi$$

$$= \frac{2\pi}{(2a)^5} \; .$$

1.17 The solution of Integral Equations of Convolution Type.

An integral equation is an equation in which the unknown function appears under the integral sign. The most general type of integral equation can, in general, be broadly classified into two types - Fredholm integral equation where limits of integral are constants and Volterra integral equation in which one of the limits of integration is variable.

The simplest integral equation that can be tackled by means of Fourier transform technique is the integral equation of convolution type

$$\frac{1}{\sqrt{2\pi}} \int_{-\infty}^{+\infty} f(t) \, k(x-t) \, dt = g(x) \, , \quad -\infty < x < \infty \tag{1.59}$$

Here $k(x)$, $g(x)$ are prescribed functions for all real values of x and are assumed to possess Fourier transforms $K(\xi)$ and $G(\xi)$ respectively.

Taking $F\,T$ of each side of (1.59) and making use of the convolution property we obtain formally the relation

$$F(\xi) \, K(\xi) = G(\xi) \Rightarrow F(\xi) = G(\xi) \, L(\xi), \tag{1.60}$$

where $\quad L(\xi) = \dfrac{1}{K(\xi)}$.

Now, if $\quad F^{-1}[L(\xi) \, ; \, \xi \to x] = l(x)$

exists, then applying the convolution theorem we would obtain the solution of (1.60) and hence from (1.59) that

$$f(x) = \frac{1}{\sqrt{2\pi}} \int_{-\infty}^{+\infty} g(t) \, l(x-t) \, dt \tag{1.61}$$

The main difficulty in the above procedure will arise if $l(x)$, as defined above, does not exist. However if it so happens that for some positive integer n, the inverse transform

$$F^{-1}\left[\frac{1}{(-i\xi)^n \, K(\xi)} \, ; \, \xi \to x\right] = m(x) \tag{1.62}$$

exists, then (1.60) can be expressed as

$$F(\xi) = (-i\xi)^n \, G(\xi) \cdot M(\xi) \quad ,\text{where} \quad M(\xi) = F[m(x) \, ; \, x \to \xi]$$

which finally yeilds

$$f(x) = \frac{1}{\sqrt{2\pi}} \int_{-\infty}^{+\infty} m(x-t)\, g^{(n)}(t)\, dt \; , \tag{1.63}$$

where, $g^{(n)}(t) = \dfrac{d^n}{dt^n}\, g(t)$.

As an example, we consider the integral equation

$$\frac{1}{\sqrt{2\pi}} \int_{-\infty}^{+\infty} |x-t|^{-\frac{1}{2}}\, f(t)\, dt = g(x) \tag{1.64}$$

Here, $k(x) = |x|^{-\frac{1}{2}}$ and hence $K(\xi) = |\xi|^{-\frac{1}{2}}$, from (1.58) of Example 1.17 of article 1.16. Now, since

$$F^{-1}\left[\frac{1}{(-i\xi)\, K(\xi)} \; ; \; \xi \to x\right]$$

$$= \quad F^{-1}\left[i\, \mathrm{sgn}(\xi)\, |\xi|^{-\frac{1}{2}} \; ; \; \xi \to x\right] = |x|^{-\frac{1}{2}}\, \mathrm{sgn}(x)$$

the solution of the integral equation in (1.63) is given by

$$f(x) = \frac{1}{\sqrt{2\pi}} \int_{-\infty}^{x} \frac{g'(t)\, dt}{\sqrt{x-t}} - \frac{1}{\sqrt{2\pi}} \int_{x}^{\infty} \frac{g'(t)\, dt}{\sqrt{t-x}} \; .$$

In cases in which the inverse transform (1.62) does not exist it is sometimes possible to have a solution of eqn. (1.59), provided that the function $g(x)$ has derivatives of all orders. The corresponding method due to Eddington consists essentially in assuming the solution of eqn. (1.59) in the form

$$f(x) = \sum_{n=1}^{\infty} C_n\, g^{(n)}(x) \tag{1.65}$$

where C_n $(n = 0, 1, 2, \cdots)$ are constants to be determined. For such a type of solution we have

$$F[f(x) \; ; \; x \to \xi] = F(\xi) = \sum_{n=0}^{\infty} C_n(-i\xi)^n\, G(\xi),$$

provided $F\,T$ of all the derivatives of $g(x)$ exist. Substituting this form of $F(\xi)$ in eqn. (1.60) we get

$$[K(\xi)]^{-1} = \sum_{n=0}^{\infty} C_n(-i\xi)^n \; ,$$

so that if $1/K(\xi)$ can be expanded as a Maclaurin series in the neighbourhood of $\xi = 0$, then $(-i)^n C_n$ is the coefficient of ξ^n in that expansion.

If there is an Eddington solution of the form (1.65), then C_i's are uniquely determined from $k(x)$; but there exists still some kernels $k(x)$ which exclude, for any infinitely differentiable function $g(x)$, all possibility that an Eddington solution exists.

For the case in which Eddington solution exists, we consider the integral equation

$$\frac{1}{\sigma\sqrt{2\pi}} \int_{-\infty}^{+\infty} f(t) e^{-\frac{1}{2}(x-t)^2/\sigma^2} dt = g(x),$$

where $g^n(x)$ exists for all $n \in I$. Here

$$k(x) = \sigma^{-1} e^{-\frac{1}{2}x^2/\sigma^2} \Rightarrow K(\xi) = e^{-\frac{1}{2}\xi^2\sigma^2}.$$

Assuming the solution due to Eddington we have

$$\sum_{n=0}^{\infty} C_n(-i\xi)^n = e^{\frac{1}{2}\xi^2\sigma^2}.$$

This equation uniquely determines C_n. In fact, here

$$C_{2m-1} = 0$$

and

$$C_{2m} = (-1)^m \frac{\frac{1}{2}\sigma^{2m}}{m!}.$$

Then the solution of the integral equation is

$$f(x) = \sum_{m=0}^{\infty} (-1)^m \left[\frac{\frac{1}{2}\sigma^{2m}}{m!}\right] \cdot g^{2m}(x).$$

Example 1.23. Solve the integral equation

$$\int_0^\infty f(x) \cos \xi\, x\, dx = \begin{cases} (1-\xi) &, \quad 0 \le \xi \le 1 \\ 0 &, \quad \xi > 1 \end{cases}$$

Solution. By definition, we can have

$$
\begin{aligned}
F_c[f(x)\,;\, x \to \xi] &= F_c(\xi) \\
&= \sqrt{\frac{2}{\pi}} \int_0^\infty f(x) \cos \xi\, x\, dx \\
&= \begin{cases} \sqrt{\frac{2}{\pi}} (1-\xi) &, \quad 0 \le \xi \le 1 \\ 0 &, \quad \xi > 1 \end{cases}
\end{aligned}
$$

Therefore, taking inverse Fourier cosine transform we have

$$\sqrt{\frac{2}{\pi}} \int_0^\infty F_c\,[f(x)\,;\; x \to \xi]\cos \xi\, x\, d\xi = f(x)$$

$$\Rightarrow \quad f(x) = \sqrt{\frac{2}{\pi}} \cdot \sqrt{\frac{2}{\pi}} \int_0^1 (1 - \xi)\cos \xi\, x\, d\xi$$

$$= \frac{2}{\pi} \int_0^1 (1 - \xi)\cos \xi\, x\, d\xi$$

$$= \frac{2}{\pi} \left[(1 - \xi)\,\frac{\sin \xi\, x}{x}\Big|_0^1 + \int_0^1 \frac{\sin \xi x}{x}\, d\xi \right]$$

$$= \frac{2}{\pi} \left[-\frac{\cos \xi\, x}{x^2}\Big|_0^1 \right]$$

$$= \frac{2(1 - \cos\ x)}{\pi x^2}\ .$$

Example 1.24. Solve the integral equations

$$\text{(i)} \quad \int_0^\infty f(x)\cos\ \lambda x\ dx = e^{-\lambda}$$

and

$$\text{(ii)} \quad \int_0^\infty f(x)\sin\ \lambda x\ dx = e^{-\lambda}$$

Solution. (i) By definition of F.C. transform, we have

$$F_c[f(x)\,;\; x \to \xi] = F_c(\xi) \quad = \quad \sqrt{\frac{2}{\pi}} \int_0^\infty f(x)\cos \xi x\ dx$$

$$= \quad \sqrt{\frac{2}{\pi}}\, e^{-\xi}\ ,\ \text{by the problem (i)}\ .$$

Thus by Inverse transform, we get

$$f(x) \quad = \quad \frac{2}{\pi} \int_0^\infty e^{-\xi}\cos \xi\, x\ d\xi$$

$$= \quad \frac{2}{\pi} \left[\frac{e^{-\xi}}{1 + x^2}(-\cos \xi\, x + x\sin \xi\, x)\Big|_0^\infty \right]$$

$$= \quad \frac{2}{\pi} \left[\frac{1}{1 + x^2} \right]$$

$$\text{(ii)}\quad F_s[f(x)\,;\; x \to \xi] \quad = \quad F_s(\xi) = \sqrt{\frac{2}{\pi}} \int_0^\infty f(x)\sin \xi\, x\ dx$$

$$= \quad \sqrt{\frac{2}{\pi}}\, e^{-\xi}\ ,\ \text{by the given question}$$

Then on taking the inverse Fourier sine transform, we get

$$f(x) = \frac{2}{\pi} \int_0^\infty e^{-\xi} \sin \xi \, x \, d\xi$$

$$= \frac{2}{\pi} \left[\frac{x}{1+x^2} \right] , \text{ by evaluating the integral.}$$

Example 1.25. Solve the integral equation

$$\int_0^\infty f(x) \sin \xi \, x \, dx = F(\xi) = \begin{cases} 1 & , \quad 0 \leqslant \xi < 1 \\ 2 & , \quad 1 \leqslant \xi < 2 \\ 0 & , \quad \xi \geqslant 2 \, . \end{cases}$$

Solution. As in example (1.23),we can have

$$f(x) = \frac{2}{\pi} \int_0^\infty F(\xi) \sin \xi \, x \, d\xi$$

$$= \frac{2}{\pi} \left[\int_0^1 \sin \xi \, x \, d\xi + \int_1^2 2 \, \sin \xi \, x \, d\xi \right]$$

$$= \frac{2}{\pi} \left[\frac{-\cos \xi \, x}{x} \bigg|_0^1 - \frac{2 \cos \xi \, x}{x} \bigg|_1^2 \right]$$

$$= \frac{2}{\pi} \left[\frac{1 - \cos x}{x} + \frac{2}{x} (\cos x - \cos 2x) \right]$$

$$= \frac{2}{\pi x} [1 + \cos x - 2 \cos 2x]$$

Example 1.26. Solve the following Fredholm integral equation with convolution type kernel

$$\int_{-\infty}^{+\infty} f(t) \, k(x - t) \, dt + \lambda \, f(x) = u(x)$$

for the particular cases

(i) $\quad k(x) = \dfrac{1}{x}$

(ii) $\quad k(x) = \dfrac{1}{2} \dfrac{x}{|x|} , \quad \lambda = 1$

(iii) $\quad k(x) = f(x) , \quad \lambda = 0 \quad \text{and} \quad u(x) = \dfrac{1}{x^2 + a^2}$

(iv) $\quad k(x) = \dfrac{1}{x^2 + a^2} , \quad \lambda = 0 \quad \text{and} \quad u(x) = \dfrac{1}{x^2 + b^2}$

(v) $\quad k(x) = 4 \, e^{-a(x)} , \quad \lambda = 1 \quad \text{and} \quad u(x) = e^{-|x|}$

Solution. Fourier transform of the integral equation gives

$$\sqrt{2\pi}\, F(\xi)\, K(\xi) + \lambda\, F(\xi) = U(\xi)$$

Or, $$F(\xi) = U(\xi) / \left[\sqrt{2\pi}\, K(\xi) + \lambda\right]$$

Inverting the Fourier transformed function we get

$$f(x) = \frac{1}{\sqrt{2\pi}} \int_{-\infty}^{+\infty} \frac{U(\xi)\, e^{i\xi x}}{[\sqrt{2\pi} K(\xi) + \lambda]}\, d\xi$$

(i) Since $k(x) = \frac{1}{x}$, we have $K(\xi) = i \sqrt{\frac{\pi}{2}}\, \mathrm{sgn}\ \xi$.

Therefore, the solution of the particular integral equation is

$$f(x) = \frac{1}{\sqrt{2\pi}} \int_{-\infty}^{+\infty} \frac{U(\xi)\, e^{i\xi x}}{\lambda + i\, \pi\, \mathrm{sgn}\ (\xi)}\, d\xi$$

(ii) If $k(x) = \frac{1}{2} \frac{x}{|x|}$, $K(\xi) = \frac{1}{\sqrt{2\pi}} \frac{1}{i\xi}$, then the solution of the particular integral equation is given by

$$
\begin{aligned}
f(x) &= \frac{1}{\sqrt{2\pi}} \int_{-\infty}^{+\infty} i\xi\, \frac{U(\xi)\, e^{i\xi x}}{(1 + i\xi)}\, d\xi \\
&= \frac{1}{\sqrt{2\pi}} \int_{-\infty}^{+\infty} F\left[u'(x); x \to \xi\right] F\left[\sqrt{2\pi}\, e^{-x}; x \to \xi\right] e^{i\xi x}\, d\xi \\
&= u'(x) * \sqrt{2\pi}\, e^{-x} = \int_{-\infty}^{+\infty} u'(\xi) \exp\ (\xi - x)\, d\xi
\end{aligned}
$$

(iii) In this particular case we have

$$
\begin{aligned}
F(\xi) &= \frac{1}{\sqrt{2a}} \exp\left\{-\frac{1}{2}\, a\, |\xi|\right\} \\
\text{Therefore,} \quad f(x) &= \frac{1}{\sqrt{2\pi}} \frac{1}{\sqrt{2a}} \int_{-\infty}^{+\infty} \exp\left\{i\xi x - \frac{1}{2}\, a\, |\xi|\right\} d\xi \\
&= \frac{1}{2\sqrt{\pi a}} \left[\int_0^\infty \exp\left\{-\xi\left(\frac{a}{2} + ix\right)\right\} d\xi \right.\\
&\qquad\qquad \left. + \int_0^\infty \exp\left\{-\xi\left(\frac{a}{2} - ix\right)\right\} d\xi\right] \\
&= \sqrt{\frac{a}{\pi}}\, \frac{2}{4x^2 + a^2}
\end{aligned}
$$

(iv) For this particular case Fourier transform of the integral equation is

$$\sqrt{2\pi}\, F(\xi)\, F\left[\frac{1}{x^2 + a^2}\;;\; x \to \xi\right] = \sqrt{\frac{\pi}{2}}\,\frac{e^{-b|\xi|}}{b}$$

Or $\quad \sqrt{2\pi}\, F(\xi)\sqrt{\frac{\pi}{2}}\,\frac{e^{-a|\xi|}}{a} = \sqrt{\frac{\pi}{2}}\,\frac{e^{-b|\xi|}}{b}$

Thus, $\quad F(\xi) = \dfrac{1}{\sqrt{2\pi}}\,\dfrac{a}{b}\,\exp\left\{-|\xi|(b-a)\right\}$

Inverting the Fourier transformed function we get

$$
\begin{aligned}
f(x) &= \frac{a}{2\pi b}\left[\int_0^\infty \exp[-\xi\{(b-a)+ix\}]\,d\xi \right.\\
&\qquad\qquad \left. + \int_0^\infty \exp[-\xi\{(b-a)-ix\}]d\xi\right]\\
&= \frac{a}{2\pi b}\left[\frac{1}{(b-a)+ix} + \frac{1}{(b-a)-ix}\right]\\
&= \frac{a}{\pi b}\left[\frac{b-a}{(b-a)^2 + x^2}\right]
\end{aligned}
$$

(v) In this case Fourier transform of the integral equation is

$$F(\xi) = \frac{a^2 + \xi^2}{a^2 + \xi^2 + 8a}\, U(\xi)$$

Inverting, $\quad f(x) = \dfrac{1}{\sqrt{2\pi}}\displaystyle\int_{-\infty}^{+\infty} \dfrac{(a^2 + \xi^2)U(\xi)}{a^2 + \xi^2 + 8a}\, e^{i\xi x}\, d\xi$

If $\quad u(x) = e^{-|x|}, a = 1, U(\xi) = \sqrt{\dfrac{2}{\pi}}\dfrac{1}{1 + \xi^2}$

Therefore, $\quad f(x) = \dfrac{1}{\pi}\displaystyle\int_{-\infty}^{+\infty} \dfrac{e^{i\xi x}}{\xi^2 + 3^2}\, d\xi$

$$= \frac{1}{3}\,\exp\left\{-3|r|\right\},$$

after separately considering the cases $x > 0$ and $x < 0$ and then combining them.

1.18 Fourier Transform of Functions of several variables.

Similar results like those in article 1.13, *FT* of the functions of several variables do hold. For simplicity, we state them for a function $f(x, y)$

of two independent variables. The extension to a higher number of variables is obvious. These extensions are sometimes called **multiple Fourier transforms**.

Let us define

$$F[f(x,y) \; ; \; x \to \xi] = F[\xi,y] = \frac{1}{\sqrt{2\pi}} \int_{-\infty}^{+\infty} f(x,y) \, e^{i\xi x} \, dx$$

$$F_c[f(x,y) \; ; \; x \to \xi] = F_c[\xi,y] = \sqrt{\frac{2}{\pi}} \int_{0}^{\infty} f(x,y) \cos \xi \, x \, dx$$

and $$F_s[f(x,y) \; ; \; x \to \xi] = F_s[\xi,y] = \sqrt{\frac{2}{\pi}} \int_{0}^{\infty} f(x,y) \sin \, \xi \, x \, dx$$

The corresponding inversion formulae are then given by

$$F^{-1}[F(\xi,y) \; ; \; \xi \to x] = F^{-1}(\xi,y) = \frac{1}{\sqrt{2\pi}} \int_{-\infty}^{+\infty} F(\xi,y) \cdot e^{-i\xi x} \, d\xi,$$

$$F^{-1}[F_c(\xi,y) \; ; \; \xi \to x] = F_c^{-1}(\xi,y) = \sqrt{\frac{2}{\pi}} \int_{0}^{\infty} F_c(\xi,y) \cos \xi \, x \, d\xi$$

and $$F^{-1}[F_s(\xi,y) \; ; \; \xi \to x] = F_s^{-1}(\xi,y) = \sqrt{\frac{2}{\pi}} \int_{0}^{\infty} F_s(\xi,y) \sin \xi \, x \, d\xi$$

respectively.

The analogues of the results in 1.13 for continuous functions are

$$F \left[\frac{\partial f}{\partial x} \; ; \; x \to \xi \right] = -i\xi \, F(\xi,y)$$

$$F_c \left[\frac{\partial f}{\partial x} \; ; \; x \to \xi \right] = -f(0,y) + \xi \, F_s(\xi,y)$$

$$F_s \left[\frac{\partial f}{\partial x} \; ; \; x \to \xi \right] = -\xi \, F_c(\xi,y)$$

Applying these results in succession we find that

$$F \left[\frac{\partial^2 f}{\partial x^2} \; ; \; x \to \xi \right] = -\xi^2 \, F(\xi,y)$$

$$F_c \left[\frac{\partial^2 f}{\partial x^2} \; ; \; x \to \xi \right] = -\xi^2 \, F_c(\xi,y) - \frac{\partial f}{\partial x}(0,y)$$

$$F_s \left[\frac{\partial^2 f}{\partial x^2} \; ; \; x \to \xi \right] = -\xi^2 \, F_s(\xi,y) + \xi \, f(0,y)$$

Similarly, *FT* of other higher order partial derivatives with respect to corresponding variables can be deduced. These results will help us in reducing partial differential equation to a lower dimensional equation of variables. Thus, FT can be used in solving the boundary value problems in two or higher dimensions.

The above ideas can be extended to several variables. Let $f(x,y)$ be a function of two independent variables x, y. Temporarily treating $f(x,y)$ as a function of x, we have

$$F[f(x,y) \; ; \; x \to \xi] = \bar{f}(\xi, y) = \frac{1}{\sqrt{2\pi}} \int_{-\infty}^{+\infty} f(x,y) \, e^{i\xi x} \, dx$$

Then treating $\bar{f}(\xi, y)$ as a function of the independent variable y, its FT is given by

$$\bar{\bar{f}}\,(\xi, \eta) = \frac{1}{\sqrt{2\pi}} \int_{-\infty}^{+\infty} \bar{f}(\xi, y) \, e^{i\eta y} \, dy = F[\bar{f}(\xi, y) \; ; \; y \to \eta]$$

Thus, finally we have

$$\bar{\bar{f}}\,(\xi, \eta) = \frac{1}{2\pi} \int_{-\infty}^{+\infty} \int_{-\infty}^{+\infty} f(x,y) \, e^{i(\, \xi x + \eta y \,)} \, dx \, dy$$

and the corresponding inversion formula is

$$f(x,y) = \frac{1}{2\pi} \int_{-\infty}^{+\infty} \int_{-\infty}^{+\infty} \bar{\bar{f}}\,(\xi, \eta) \, e^{-i(\xi x + \eta y)} \, d\xi \, d\eta \; .$$

Such extensions to Fourier sine and Fourier cosine transforms in functions of several variables do exist and the corresponding results can be deduced easily.

Using the above definitions of FT to functions of several variables, we can apply FT to mixed partial derivatives too arising in partial differential equations in solving boundary value problems.

1.19   Application of Fourier Transform to Boundary Value Problems

We first discuss the use of sine and cosine transform and finally discuss the use of complex Fourier transform arising in boundary value problems.

The sine and cosine transforms can be applied when the range of variable selected for exclusion is from 0 to ∞. The choice of sine and cosine transform is decided by the form of the boundary conditions at the lower limit of the variable selected for exclusion.

For example,

$$F_s\left[\frac{\partial^2 u(x,y)}{\partial x^2} ; x \to \xi\right] = \sqrt{\frac{2}{\pi}} \int_0^\infty \frac{\partial^2 u}{\partial x^2} \sin \xi \, x \, dx$$

$$= \sqrt{\frac{2}{\pi}} \xi \int_0^\infty \frac{\partial u}{\partial x} \cos \xi \, x \, dx$$

$$= \sqrt{\frac{2}{\pi}} \xi \, u(0,y) - \sqrt{\frac{2}{\pi}} \xi^2 \, \bar{u}_s(\xi,y), \tag{1.66}$$

provided $u(x,y)$ is known at $x = 0$ and $\dfrac{\partial u}{\partial x} \to 0$ as $x \to \infty$

Similarly,

$$F_c\left[\frac{\partial^2 u(x,y)}{\partial x^2} ; x \to \xi\right] = \sqrt{\frac{2}{\pi}} \int_0^\infty \frac{\partial^2 u}{\partial x^2} \cos \xi \, x \, dx$$

$$= -\sqrt{\frac{2}{\pi}} \left[\frac{\partial u(x,y)}{\partial x}\right]_{x=0} - \sqrt{\frac{2}{\pi}} \xi^2 \int_0^\infty u(x,y) \cos \xi \, x \, dx, \tag{1.67}$$

provided $\dfrac{\partial u(0,y)}{\partial x}$ is known and $u, \dfrac{\partial u}{\partial x} \to 0$ as $x \to \infty$.

Noting carefully the results in (1.66) and (1.67) it may be seen that removing a term $\frac{\partial^2 u(x,y)}{\partial x^2}$ from a partial differential equation requires the knowledge of $u(0,y)$ for use of a sine transform while the use of a cosine transform for the same purpose requires the knowledge of $u_x(0,y)$ to be known.

It may be noted that a term $\frac{\partial u}{\partial x}$ or any odd order partial derivative can not be removed with the help of sine or cosine transforms.

Again, complex Fourier transform will be found useful for the same purpose as above, if the range of variable is from $-\infty$ to $+\infty$ of the partial differential equation.

Example 1.27. The temperature $u(x,t)$ of a semi-infinite rod is determined by the partial differential equation $\frac{\partial u}{\partial t} = \frac{\partial^2 u}{\partial x^2}$, $x > 0$, $t > 0$ subject to the initial condition $u(x,0) = 1$, $0 < x < 1$

$$= 0 , \ x > 1$$

and the boundary condition $u(0,t) = 0$. Find the temperature at any time t at any distance x from $x = 0$.

Solution. Since the variable x varies from 0 to ∞ and since the value of $u(x,t)$ at $x = 0$ is prescribed, Fourier sine transform of both sides of the partial differential equation is to be taken on the variable x which is to be excluded in the transformed equation. Then the given equation becomes

$$\frac{d}{dt}\, \bar{u}_s(\xi,t) = \sqrt{\frac{2}{\pi}} \int_0^\infty \frac{\partial^2 u}{\partial x^2} \, \sin \xi x \, dx$$

$$= \sqrt{\frac{2}{\pi}} [\xi \, u(0,t)] - \xi^2 \, \bar{u}_s(\xi,t)$$

$$= -\xi^2 \, \bar{u}_s(\xi,t)$$

$$\therefore \qquad \bar{u}_s(\xi,t) = c \, e^{-\xi^2 t}, \quad \text{where } c \text{ is an arbitrary constant .} \qquad \text{(i)}$$

It is given that initially

$$u(x,0) \;=\; 1 \,, \; 0 < x < 1$$

$$=\; 0 \,, \; x > 1$$

$$\therefore \; \bar{u}_s(\xi,0) \;=\; \sqrt{\frac{2}{\pi}} \int_0^\infty u(x,0) \cdot \sin \xi \, x \, dx$$

$$=\; \sqrt{\frac{2}{\pi}} \int_0^1 \sin \xi \, x \, dx = \sqrt{\frac{2}{\pi}} \left[\frac{-\cos \xi x}{\xi} \right]_0^1$$

$$=\; \sqrt{\frac{2}{\pi}} \left[\frac{1 - \cos \xi}{\xi} \right] \qquad \text{(ii)}$$

Therefore, from (i) and (ii) we get

$$c = \sqrt{\frac{2}{\pi}} \, \frac{1 - \cos \xi}{\xi}$$

Using this value of c eqn(i) gives

$$u(x,t) = \frac{2}{\pi} \int_0^\infty \frac{1 - \cos \xi}{\xi} \, e^{-\xi^2 t} \sin \xi \, x \, d\xi.$$

This is the required temperature function $u(x,t)$.

Example 1.28. Solve the diffusion equation $\frac{\partial u}{\partial t} = \frac{\partial^2 u}{\partial x^2}, x > 0, t > 0$ subject to the boundary condition $u_x(0,t) = 0$ and $u(x,t)$ is bounded. The initial condition is being

$$u(x,0) = \begin{cases} x & , & 0 \leq x \leq 1 \\ 0 & , & x > 1 \end{cases}$$

Solution. Since the range of x is from 0 to ∞ and the value of $u_x(0, t)$ is prescribed, it is useful to apply Fourier cosine transform to remove the variable x from the PDE.

Thus we get,

$$\frac{d}{dt}\,\bar{u}_c(\xi, t) = \sqrt{\frac{2}{\pi}} \int_0^\infty \frac{\partial^2 u}{\partial x^2} \cos \xi\, x\, dx$$

$$= \sqrt{\frac{2}{\pi}} \left[\cos \xi\, x \frac{\partial u}{\partial x} \Big|_0^\infty + \xi \int_0^\infty \frac{\partial u}{\partial x} \sin \xi\, x\, dx \right]$$

$$= -\sqrt{\frac{2}{\pi}}\xi^2 \int_0^\infty u(x, t) \cos \xi\, x\, dx = -\xi^2\, \bar{u}_c(\xi, t)$$

$$\therefore \quad \bar{u}_c(\xi, t) = A\, e^{-\xi^2 t} \Rightarrow \bar{u}_c(\xi, 0) = A, \text{ arbitrary constant.}$$

Now, $\quad \bar{u}_c(\xi, 0) = \sqrt{\frac{2}{\pi}} \int_0^\infty u(x, 0) \cos \xi\, x\, dx = \sqrt{\frac{2}{\pi}} \int_0^1 x \cos \xi\, x\, dx,$

using given initial condition

i.e $\quad \bar{u}_c(\xi, 0) = \sqrt{\frac{2}{\pi}} \left[\frac{\sin \xi}{\xi} + \frac{\cos \xi - 1}{\xi^2} \right] = A$

This result gives

$$\bar{u}_c(\xi, t) = \sqrt{\frac{2}{\pi}} \left[\frac{\sin \xi}{\xi} - \frac{1 - \cos \xi}{\xi^2} \right] \cdot e^{-\xi^2 t}$$

Taking inverse Fourier cosine transform the required solution is

$$u(x, t) = \frac{2}{\pi} \int_0^\infty \frac{\xi \sin \xi - 1 + \cos \xi}{\xi^2} e^{-\xi^2 t} \cos \xi\, x\, d\xi \ .$$

Example 1.29. Use Fourier transform to determine the displacement $y(x, t)$ of an infinite vibrating string, given that the string is initially at rest and that the initial displacement is $f(x), -\infty < x < \infty$. Show that the required solution can also be put in the form

$$y(x, t) = \frac{1}{2}[f(x + ct) + f(x - ct)]$$

when the one dimensional wave equation of the vibrating string is given by

$$\frac{\partial^2 y}{\partial t^2} = c^2\, \frac{\partial^2 y}{\partial x^2}\ , \quad -\infty < x < \infty\ , \ t > 0\ .$$

Solution. Since the variable x varies from $-\infty$ to ∞, to remove the variable from the PDE, we apply complex FT to the given equation to obtain

$$\frac{d^2\bar{y}(\xi,t)}{dt^2} = \frac{c^2}{\sqrt{2\pi}} \int_{-\infty}^{+\infty} \frac{\partial^2 y(x,t)}{\partial x^2} e^{i\xi x} dx$$

$$= -c^2\xi^2\bar{y}(\xi,t)$$

Therefore, $\qquad \bar{y}(\xi,t) = A\cos(c\xi t) + B\sin(c\xi t) \qquad\qquad$ (i)

and $\qquad \bar{y}_t(\xi,t) = A\ c\ \xi\sin(c\xi t) + B\ c\ \xi\ \cos(c\xi t) \qquad$ (ii)

where A and B are two arbitrary constants. Now, the given initial conditions are

$$y(x,0) = \text{initial displacement } = f(x) \qquad\qquad \text{(iii)}$$

and $\qquad \dfrac{\partial y}{\partial t}(x,0) = 0$, since initial velocity of vibrating string is zero.

(iv)

From (ii) and (iii) we get

$$\bar{y}(\xi,0) = \frac{1}{\sqrt{2\pi}} \int_{-\infty}^{+\infty} f(x)\ e^{i\xi x}\ dx = \bar{f}(\xi)\ , \text{ say} \qquad\qquad \text{(v)}$$

and $\qquad \dfrac{\partial\bar{y}}{\partial t}(\xi,0) = \bar{y}_t(\xi,0) = 0\ .$ (vi)

Then from eqn. (i) we have

$$\bar{y}(\xi,0) = A = \bar{f}(\xi)\ , \text{ by eqn. (v)}$$

Also, from eqn. (ii) we have

$$\bar{y}_t(\xi,0) = B\ c\ \xi = 0, \text{ by eqn. (vi)}$$

Thus, $\qquad B - 0\ .$

Then

$$\bar{y}(\xi,t) = \bar{f}(\xi)\ \cos(c\ \xi\ t)$$

Taking inverse FT to the above equation we get

$$y(x,t) = \frac{1}{\sqrt{2\pi}} \int_{-\infty}^{+\infty} \bar{f}(\xi) \left[\frac{e^{ic\xi t} + e^{-ic\xi t}}{2}\right] e^{-i\xi x}\ d\xi$$

$$= \frac{1}{2} \frac{1}{\sqrt{2\pi}} \int_{-\infty}^{+\infty} \bar{f}(\xi) \cdot \left[e^{-i\xi(x-ct)} + e^{-i\xi(x+ct)} \right] d\xi$$

$$= \frac{1}{2} \left[\frac{1}{\sqrt{2\pi}} \int_{-\infty}^{+\infty} \bar{f}(\xi) \, e^{-\xi(x-ct)} d\xi + \frac{1}{\sqrt{2\pi}} \int_{-\infty}^{+\infty} \bar{f}(\xi) \cdot e^{-i\xi(x+ct)} d\xi \right]$$

$$\Rightarrow \quad y(x,t) = \frac{1}{2} \left[f(x-ct) + f(x+ct) \right] \ , \quad \text{by inverse } F \, T \text{ formula.}$$

Thus, the required result is proved.

Example 1.30. The steady state temperature distribution in a semi-infinite solid $y > 0$ is represented by the two-dimensional Laplace equation

$$\frac{\partial^2 u}{\partial x^2} + \frac{\partial^2 u}{\partial y^2} = 0, \quad 0 < y < \infty \ , \quad -\infty < x < \infty$$

subject to the boundary conditions

$$u(x,0) = 1 \ , \ |x| < a$$

$$\text{and} \quad u(x,0) = 0 \ , \ \text{for } |x| > a$$

Thus the temperature on the insulated surface $y = 0$ is kept at unity over the strip $|x| < a$ and at zero outside the strip. Then using cosine transform show that

$$u(x,y) = \frac{1}{\pi} \left[\tan^{-1} \frac{a+x}{y} + \tan^{-1} \frac{a-x}{y} \right] \ ,$$

after assuming the result that $\int_0^\infty e^{-sx} \dfrac{\sin rx}{x} \, dx = \tan^{-1} \dfrac{r}{s} \ , (r, s > 0).$

Solution. Since the surface $y = 0$ is kept at a fixed temperature, there is no flow of heat normal to it and hence $\frac{\partial u}{\partial x} \to 0$ as $x \to 0$ and as $x \to \infty$. Therefore, taking Fourier cosine transform of the given partial differential equation in variable x we get

$$\sqrt{\frac{2}{\pi}} \int_0^\infty \frac{\partial^2 u}{\partial x^2} \cos \xi \, x \, dx + \sqrt{\frac{2}{\pi}} \int_0^\infty \frac{\partial^2 u}{\partial y^2} \cos \xi x \, dx = 0$$

Or, $\quad -\xi^2 \, \bar{u}_c(\xi, y) + \dfrac{d^2}{dy^2} \, \bar{u}_c(\xi, y) = 0$

Its general solution is

$$\bar{u}_c(\xi, y) = A \, e^{\xi y} + B \, e^{-\xi y}$$

But $u(x,y)$ is finite as $y \to \infty$, $\bar{u}_c(\xi,y)$ is also finite and hence we must have $A = 0$. Thus

$$\bar{u}_c(\xi,y) = B\, e^{-\xi y} \qquad\qquad \text{(i)}$$

Again, $\displaystyle \sqrt{\frac{2}{\pi}} \int_0^\infty u(x,0) \cos \xi\, x\, dx = \sqrt{\frac{2}{\pi}} \int_0^1 \cos \xi\, x\, dx$

$$= \frac{\sin \xi a}{\xi} \cdot \sqrt{\frac{2}{\pi}}$$

Thus $\displaystyle \bar{u}_c(\xi,0) = \sqrt{\frac{2}{\pi}} \frac{\sin \xi\, a}{\xi} = B, \qquad\qquad \text{from eqn. (i)}$

Hence, we have

$$u_c(\xi,y) = \sqrt{\frac{2}{\pi}} \frac{\sin \xi\, a}{\xi}\, e^{-\xi y}$$

Therefore, $\displaystyle u(x,y) = \frac{2}{\pi} \int_0^\infty \frac{\sin \xi\, a}{\xi} \cdot e^{-\xi y} \cos \xi\, x\, d\xi$

$$= \frac{1}{\pi} \int_0^\infty \frac{e^{-\xi y}}{\xi} \left[\sin(a+x)\xi + \sin(a-x)\xi \right] d\xi$$

$$= \frac{1}{\pi} \left[\tan^{-1} \frac{a+x}{y} + \tan^{-1} \frac{a-x}{y} \right].$$

Example 1.31. Solve the one dimensional heat conduction problem where the temperature of a semi-infinite bar satisfies the PDE

$$\frac{\partial u}{\partial t} = k\frac{\partial^2 u}{\partial x^2}, \ t > 0,\ 0 < x < \infty,$$

the boundary and initial prescribed conditions are being

$$u(0,t) = f(t),\, u(x,0) = 0 \ \text{ and } \ u(x,t) \to 0 \text{ as } x \to \infty.$$

Solution. Applying Fourier sine transform over the variable x the PDE transforms to

$$\frac{d}{dt}\, \bar{u}_s(\xi,t) = k\, \xi\sqrt{\frac{2}{\pi}}\, u(0,t) - k\, \xi^2\, \bar{u}_s(\xi,t)$$

Or, $\displaystyle \frac{d\, \bar{u}_s}{dt}(\xi,t) = -k\, \xi^2\, \bar{u}_s(\xi,t) + \sqrt{\frac{2}{\pi}}\, k\, \xi\, f(t)$

This linear ODE in $\bar{u}_s(\xi,t)$ has the solution given by

$$\bar{u}_s(\xi,t) \cdot e^{k\xi^2 t} = \sqrt{\frac{2}{\pi}}\, k\xi \int f(t)\, e^{k\xi^2 t}\, dt + A$$

$$= \sqrt{\frac{2}{\pi}}\, k\xi \int^t f(\varsigma)\, e^{k\xi^2 \varsigma}\, d\varsigma + A \qquad\qquad \text{(i)}$$

Again, $u(x,0) = 0$, by given condition and therefore $\bar{u}_s(\xi,0) = 0$.

Putting $t = 0$ in eqn. (i) we get

$$\bar{u}_s(\xi,0) \cdot 1 = \sqrt{\frac{2}{\pi}} k\xi \int^{\varsigma=0} f(\varsigma) \, e^{k\xi^2\varsigma} d\varsigma + A \qquad (ii)$$

Therefore, $e^{k\xi^2 t}\bar{u}_s(\xi,t) = \sqrt{\frac{2}{\pi}} k\xi \int_0^t f(\varsigma) \, e^{k\xi^2\varsigma} d\varsigma$

Or, $\bar{u}_s(\xi,t) = \sqrt{\frac{2}{\pi}} k\xi \int_0^t f(\varsigma) \, e^{-k\xi^2(t-\varsigma)} d\varsigma$

So, $u(x,t) = F_s^{-1} \left[\sqrt{\frac{2}{\pi}} k\xi \int_0^t f(\varsigma) \, e^{-k\xi^2(t-\varsigma)} d\varsigma \; ; \; \xi \to x \right]$

$$= \sqrt{\frac{2}{\pi}} k \int_0^t f(\varsigma) F_s^{-1}[\xi \, e^{-l\xi^2} \; ; \; \xi \to x] \, , \text{where } l = k(t-\varsigma)$$

From the deduction (iii) of example 1.9 we can have the result

$$F_s^{-1} \left[\xi e^{-l\xi^2} \; ; \; \xi \to x \right] = \frac{1}{\sqrt{8l^3}} \, x \, e^{-\frac{x^2}{4l}} = g(x,\varsigma), \text{ say}$$

Therefore, $u(x,t) = \sqrt{\frac{2}{\pi}} \, k \int_0^t f(\varsigma) \, g(x,\varsigma) \, d\varsigma,$

where $g(x,\varsigma) = \dfrac{x(t-\varsigma)^{-3}}{\sqrt{8k^3}} \cdot e^{\frac{-x^2}{[4k(t-\varsigma)]}}$

gives the required solution of the problem.

Example 1.32. Find the solution of the Laplace's equation

$$\frac{\partial^2 u}{\partial x^2} + \frac{\partial^2 u}{\partial y^2} = 0$$

in the half-plane $y \geqslant 0$, subject to the boundary condition
$u(x,0) = f(x), -\infty < x < \infty$ and the limiting condition $u(x,y) \to 0$ as
$\rho = \sqrt{x^2 + y^2} \to \infty$.

Solution. Introducing Fourier transform to remove the variable x,
we get

$$\frac{d^2 \, \bar{u} \, (\xi,y)}{dy^2} - \xi^2 \, \bar{u}(\xi,y) = 0 \qquad (i)$$

From the given boundary condition we get

$$\bar{u}(\xi,0) = \frac{1}{\sqrt{2\pi}} \int_{-\infty}^{+\infty} f(x) \, e^{i\xi x} dx = F(\xi), \text{ say} \qquad (ii)$$

The limiting condition gives $\bar{u}(\xi, y) \to 0$, $y \to \infty$.   (iii)

Using (ii) and (iii), the solution of (i) is given by

$$\bar{u}(\xi, y) = F(\xi)\, e^{-|\xi| y} \text{ with } F(\xi) = F[f(x)\,;\, x \to \xi]$$

Then using the convolution theoem we can invert the above equation to obtain

$$u(x, y) = \frac{1}{\sqrt{2\pi}} \int_{-\infty}^{+\infty} f(t)\, g(x - t)\, dt \qquad \text{(iv)}$$

where $\quad g(x) = F^{-1}\left[e^{-|\xi| y}\,;\, \xi \to x \right]$

$$= \sqrt{\frac{2}{\pi}}\, \frac{y}{x^2 + y^2}$$

Substituting this result in (iv) we get

$$u(x, y) = \frac{y}{\pi} \int_{-\infty}^{+\infty} \frac{f(t)dt}{(x - t)^2 + y^2}\,,\quad y > 0\,.$$

Example 1.33. Derive the solution of Laplace equation in the infinite strip $-\infty < x < \infty$, $0 \leqslant y \leqslant a$ subject to the boundary conditions $u(x.0) = f(x)$, $u(x, a) = g(x)$ and limiting condition $u(x, y) \to 0$ as $x \to \pm\infty$.

Solution. To remove the variable x, we introduce Fourier transform over the variable x. Then the Laplace equation transforms to

$$\frac{d^2}{dy^2}\, \bar{u}(\xi, y) - \xi^2\, \bar{u}(\xi, y) = 0 \qquad \text{(i)}$$

Also, the boundary conditions under $F\,T$ transform to

$$\bar{u}(\xi, 0) = F(\xi) = F[f(x)\,;\, x \to \xi]\,,\ \bar{u}(\xi, a) = G(\xi) = F[g(x)\,;\, x \to \xi]\,.$$

Therefore, the solution of eqn. (i) under transformed boundary condition is

$$\bar{u}(\xi, y) = F(\xi)\frac{sh\,\xi(a - y)}{sh\,\xi a} + G(\xi)\frac{sh\,\xi y}{sh\,\xi a}\,,\ 0 < y < a\,.$$

By convolution theorem, the required solution of the problem is given by

$$u(x, y) = \frac{1}{2a} \int_{-\infty}^{+\infty} f(t) K_1(x - t\,,\, a - y)\, dt$$

$$+\frac{1}{2a}\int_{-\infty}^{+\infty} g(t)K_1(x-t\ ,\ y)\ dt \qquad \text{(ii)}$$

$$\text{where} \qquad K_1(x,y) = \sqrt{\frac{2}{\pi}}\,a\,F^{-1}\left[\frac{sh\ \xi\ y}{sh\ \xi\ a}\ ;\ \xi \to x\right]$$

$$= \sqrt{\frac{2}{\pi}}\,a\,F_c\left[\frac{sh\ \xi\ y}{sh\ \xi\ a}\ ;\ \xi \to x\right]\ ,$$

By calculus of residues, above Fourier cosine transform is evaluated as

$$K_1\,(x,y) = \frac{\sin\left(\frac{\pi y}{a}\right)}{ch\left(\frac{\pi x}{a}\right) + \cos\left(\frac{\pi y}{a}\right)}$$

Substituting this result in eqn. (ii), we get

$$u(x,y) = \frac{1}{2a}\sin\left(\frac{\pi y}{a}\right)\int_{-\infty}^{+\infty}\frac{f(t)dt}{ch\ \pi\frac{(x-t)}{a} - \cos\frac{\pi y}{a}}$$

$$+\frac{1}{2a}\sin\left(\frac{\pi y}{a}\right)\int_{-\infty}^{+\infty}\frac{g(t)\ dt}{ch\ \pi\frac{(x-t)}{a} + \cos\frac{\pi y}{a}}$$

Example 1.34. Consider the problem as in 1.33 subject to the boundary conditions $u(x,0) = f(x)$ and $u_y(x,a) = 0$. Find the temperature distribution $u(x,y)$ of the problem defined by the Laplace equation

$$\frac{\partial^2 u}{\partial x^2} + \frac{\partial^2 u}{\partial y^2} = 0\ ,\ -\infty < x < \infty\ ,\ 0 \leqslant y \leqslant a$$

Solution. We apply $F\ T$ over variable x to obtain

$$\bar{u}(\xi,y) = F(\xi)\frac{ch\xi(a-y)}{ch\ \xi a}\ ,\ 0 \leqslant y \leqslant a \qquad \text{(i)}$$

$$\text{where} \qquad F(\xi) = \frac{1}{\sqrt{2\pi}}\int_{-\infty}^{+\infty} f(x)\ e^{i\xi x}\ dx \qquad \text{(ii)}$$

after using the transformed value of the boundary conditions $\bar{u}(\xi,0) = F(\xi)$ and $\bar{u}_y(\xi,0) = 0$. Then by convolution theorem we get

$$u(x,y) = \int_{-\infty}^{+\infty} f(t)\ K_2(x-t\ ,\ a-y)\ dt$$

$$\text{where} \qquad K_2(x,y) = \frac{1}{\sqrt{2\pi}}\,F_c\left[\frac{ch\ (\xi y)}{ch\ (\xi a)}\ ;\ \xi \to x\right]$$

This $K_2(x, y)$ may be evaluated by method of contour integral. But this evaluation is being quite involved, we stop here in persuing it. We only mention the result for $K_2(x, y)$ as

$$K_2(x, y) = \frac{ch\left(\frac{\pi x}{2a}\right) \cos\left(\frac{\pi y}{2a}\right)}{ch\left(\frac{\pi x}{a}\right) + \cos\left(\frac{\pi y}{a}\right)}$$

which can also be found from the tables of Integral transform. Then the solution of the problem is given by

$$u(x, y) = \sin\left(\frac{\pi y}{2a}\right) \int_{-\infty}^{+\infty} \frac{f(t)\, ch\frac{\pi(x-t)}{2a}\, dt}{ch\frac{\pi(x-t)}{a} - \cos\frac{\pi y}{a}} .$$

Example 1.35. Illustrate the method of solution of the boundary value problem defined by the Laplace's equation

$$\frac{\partial^2 u}{\partial x^2} + \frac{\partial^2 u}{\partial y^2} = 0 , \ x \geqslant 0 , \ 0 \leqslant y \leqslant a$$

in the semi-infinite strip subjected to the boundary conditions

(i) $u(0, y) = 0, 0 \leqslant y \leqslant a$; $u(x, 0) = f(x)$, $x \geqslant 0$;
     $u(x, a) = g(x)$, $x \geqslant 0$

and (ii) $u(x, 0) = f(x), x \geqslant 0; u_x(0, y) = 0, 0 \leqslant y \leqslant a; u_y(x, a) = 0, x \geqslant 0.$

Solution. **(i)** Physical interpretation of this boundary value problem is the distribution of the temperature $u(x, y)$ in the semi-infinite strip where the faces are maintained at prescribed temperatures e.g. $f(x)$ on $y = 0$ and $g(x)$ on $y = a$ and zero on $x = 0$, $0 \leqslant y \leqslant a$. These conditions tells us that Fourier sine transform over the variable x is found useful in solving the problem as below. Applying the said transform to the PDE we get

$$\left[\frac{d^2}{dy^2} - \xi^2\right] \bar{u}_s(\xi, y) = 0 , \text{ where } \bar{u}_s(\xi, y) = \sqrt{\frac{2}{\pi}} \int_0^\infty u(x, y) \sin \xi\, x\, dx$$

This equation satisfies the boundary conditions on $y = 0$ and $y = a$.

Therefore, in $0 \leq y \leq a$ we have

$$\bar{u}_s(\xi, y) = \Gamma_s(\xi) \frac{sh\, \xi(a - y)}{sh\, \xi a} + G_s(\xi) \frac{sh\, \xi y}{sh\, \xi a}$$

where $F_s(\xi)$ and $G_s(\xi)$ are Fourier sine transforms of $f(x)$ and $g(x)$ respectively. Then proceeding as in example 1.34 we get

$$\bar{u}_s(\xi, y) = \frac{1}{a}\sqrt{\frac{\pi}{2}}[F_s(\xi)F_c\{K_1(x, a-y) \; ; \; x \to \xi\} + G_s(\xi)F_c\{K_1(x, y) \; ; \; x \to \xi\}]$$

Now, using the convolution theorem the solution of the problem is given by

$$u(x, y) = \frac{1}{2a}\int_0^\infty f(t)\{K_1(|t - x| \; , \; a - y) - K_1(t + x \; , \; a - y)\} \; dt$$

$$+ \frac{1}{2a}\int_0^\infty g(t)\{K_1(|t - x| \; , \; y) - K_1(t + x \; , \; y)\} \; dt$$

Solution. (ii) In this case, the given boundary conditions suggest that use of Fourier cosine transform is found useful over variable x. Hence, the PDE transforms to

$$\left[\frac{d^2}{dy^2} - \xi^2\right]\bar{u}_c(\xi, y) = 0 \; ,$$

where $\bar{u}_c(\xi, y) = F_c[u(x, y) \; ; \; x \to \xi]$ (i)

The transformed form of the boundary conditions are then

$$\left.\begin{aligned}\bar{u}_c(\xi, 0) = F_c(\xi) = F_c[f(x) \; ; \; x \to \xi]\\[2mm]\text{and} \qquad \frac{d}{dy}\,\bar{u}_c(\xi, a) = 0\end{aligned}\right\} \qquad \text{(ii)}$$

The appropriate solution of eqn.(i) under eqns. (ii) is given by

$$\bar{u}_c(\xi, y) = F_c(\xi)\frac{ch\;\xi(a - y)}{ch\;\xi a} = \sqrt{2\pi}\,F_c(\xi)\,F_c\,[K_2(x \; , \; a - y) \; ; \; x \to \xi]$$

where $K_2(x, y)$ is same as defined in the solution of the example in 1.34. Now making use of convolution theorem we get

$$u(x, y) = \int_0^\infty f(t)[K_2(x + t \; , \; a - y) + K_2(|x - t| \; , \; a - y)] \; dt$$

A more explicit expression as in examples 1.34 and 1.35 can similarly be obtained. But we do not persue it due to lack of space.

Example 1.36. Solve the PDE

$$\frac{\partial u}{\partial t} = 2\,\frac{\partial^2 u}{\partial x^2}$$

subject to the conditions $u(0,t) = 0, u(x,0) = e^{-x}$, $x > 0$ and $u(x,t)$ is bounded when $x > 0$, $t > 0$.

Solution. Taking Fourier sine transform on x, we get the PDE as

$$\frac{d}{dt}\bar{u}_s(\xi,t) = 2 \cdot \sqrt{\frac{2}{\pi}} \int_0^\infty \frac{\partial^2 u}{\partial x^2} \sin \xi x \, dx$$

$$= -2\xi^2 \bar{u}_s(\xi,t)$$

Integrating, $\qquad \bar{u}_s(\xi,t) = A \, e^{-2\xi^2 t}$ (i)

Now, from the condition $u(x,0) = e^{-x}$, we have

$$\bar{u}_s(\xi,0) = \sqrt{\frac{2}{\pi}} \, \frac{\xi}{(1+\xi^2)}$$ (ii)

Using (ii) in (i) we have $A = \frac{\xi}{1+\xi^2} \sqrt{\frac{2}{\pi}}$ and therefore,

$$\bar{u}_s(\xi,t) = \sqrt{\frac{2}{\pi}} \, \frac{\xi}{1+\xi^2} \, e^{-2\xi^2 t}$$

Now, taking inverse Fourier sine transform of the above result the solution of the PDE under given conditions is

$$u(x,t) = \frac{2}{\pi} \int_0^\infty \frac{\xi}{1+\xi^2} \, e^{-2\xi^2 t} \, \sin \xi x \, d\xi$$

Example 1.37. Prove that the steady state temperature distribution $u(x,y,)$ inside a semi-infinite strip $x > 0$, $0 < y < b$ under the boundary conditions $u(x,0) = f(x)$, $0 < x < \infty$; $u(x,b) = 0$, $0 < x < \infty$ and $u(0,y) = 0$, $0 < y < b$ is given by

$$u(x,y) = \frac{2}{\pi} \int_0^\infty f(u) du \int_0^\infty \frac{sh(b-y)\xi}{sh \, b\xi} \, \sin \xi x \, \sin \xi u \, d\xi$$

Solution. Since $u(x,y)$ is the steady-state temperature, it must satisfy the PDE

$$\frac{\partial^2 u}{\partial x^2} + \frac{\partial^2 u}{\partial y^2} = 0$$ (i)

Taking Fourier sine transform over the variable x, cqn.(i) becomes

$$\frac{d^2 \bar{u}_s}{dy^2} - \xi^2 \, \bar{u}_s(\xi,y) = 0$$, after using the value $u(0,y) = 0$.

Its solution is given by, $\bar{u}_s(\xi, y) = A\ ch\ \xi y + B\ sh(b - y)\xi$,   (ii)
where A, B are arbitrary constants.

Taking Fourier sine transforms of $u(x, 0) = f(x)$ and $u(x, b) = 0$ we get

$$\bar{u}_s(\xi, 0) = \sqrt{\frac{2}{\pi}} \int_0^\infty f(u) \sin \xi\ u\ dx \quad \text{and} \quad \bar{u}_s(\xi, b) = 0\ .$$

Using these results in (ii) we obtain, $A = 0$ and

$$B = \frac{1}{sh\ b\xi}\ [\bar{u}_s(\xi, 0)] = \sqrt{\frac{2}{\pi}} \int_0^\infty \frac{f(u) \sin \xi\ u\ du}{sh\ b\xi}$$

Therefore,

$$\bar{u}_s(\xi, y) = \sqrt{\frac{2}{\pi}}\ \frac{sh(b - y)\xi}{sh\ b\xi} \int_0^\infty f(u) \sin \xi\ u\ du$$

Then taking inverse Fourier sine transform we get

$$
\begin{aligned}
u(x, y) &= \frac{2}{\pi} \int_0^\infty \sin \xi\ x \left[\frac{\sinh(b - y)\xi}{sh\ b\xi} \int_0^\infty f(u) \sin \xi\ u\ du \right] d\xi \\
&= \frac{2}{\pi} \int_0^\infty f(u) du \left[\int_0^\infty \frac{sh(b - y)\xi}{sh\ b\xi} \sin \xi\ x\ \sin \xi\ u\ d\xi \right]
\end{aligned}
$$

Example 1.38. The temperature $u(x, t)$ of a semi-infinite rod $0 \leqslant x < \infty$ satisfies the PDE $\frac{\partial u}{\partial t} = \kappa \frac{\partial^2 u}{\partial x^2}$, subject to conditions $u(x, 0) = 0$, $x \geqslant 0$ and $\frac{\partial u}{\partial x} = -\lambda$, a constant, when $x = 0$, $t > 0$.

Solution. Using Fourier cosine transform over the variable x, the PDE transforms to

$$\frac{d\ \bar{u}_c(\xi, t)}{dt} = \sqrt{\frac{2}{\pi}}\ \kappa\ \lambda - \kappa\ \xi^2\ \bar{u}_c(\xi, t)$$

$$\Rightarrow \qquad \frac{d\ \bar{u}_c}{dt} + \kappa\ \xi^2\ \bar{u}_c(t) = \kappa\ \lambda \sqrt{\frac{2}{\pi}} \qquad\qquad \text{(i)}$$

using given conditions on the boundary.

Now solving the eqn.(i) we have

$$\bar{u}_c(\xi, t) = \frac{\lambda}{\xi^2} \sqrt{\frac{2}{\pi}} + A\ e^{-\kappa \xi^2 t} \qquad\qquad \text{(ii)}$$

Again, $u(x, 0) = 0 \quad \Rightarrow \quad \bar{u}_s(\xi, 0) = 0$ (iii)

Using (iii) in (ii), we get

$$\bar{u}_c(\xi, t) = \sqrt{\frac{2}{\pi}} \frac{\lambda}{\xi^2} \left[1 - e^{-\kappa\xi^2 t} \right]$$

Therefore, by inverse cosine transform we get the required temperature

$$u(x, t) = \frac{2\lambda}{\pi} \int_0^\infty \frac{\cos \xi x}{\xi^2} (1 - e^{-\kappa\xi^2 t}) \, d\xi .$$

Example 1.39. Solve the PDE $\frac{\partial^2 u}{\partial x^2} - \frac{\partial u}{\partial y} = 0$ in $-\infty < x < \infty$ and $y > 0$ under the conditions $u(x, 0) = f(x)$ if $u(x, y)$ is bounded when $|x| \to \infty$.

Solution. Taking $F\,T$ over the variable x, the given PDE gives

$$\frac{1}{\sqrt{2\pi}} \int_{-\infty}^{+\infty} \frac{\partial^2 u}{\partial x^2} e^{i\xi x} \, dx = \frac{1}{\sqrt{2\pi}} \int_{-\infty}^{+\infty} \frac{\partial u(x, y)}{\partial y} \cdot e^{i\xi x} \, dx$$

Thus,
$$- \xi^2 \, \bar{u}(\xi, y) = \frac{d}{dy} \, \bar{u}(\xi, y)$$

Solving this simple ordinary differential equation we get

$$\bar{u}(\xi, y) = A \, e^{-\xi^2 y} \tag{i}$$

The condition $u(x, 0) = f(x)$ gives rise to

$$\bar{u}(\xi, 0) = \frac{1}{\sqrt{2\pi}} \int_{-\infty}^{+\infty} f(x) \, e^{i\xi x} \, dx \tag{ii}$$

Also, from (i), putting $y = 0$ we get

$$\bar{u}(\xi, 0) = A \tag{iii}$$

Comparing (ii) and (iii) and using the result in (i) we finally get

$$\bar{u}(\xi, y) = e^{-\xi^2 y} \frac{1}{\sqrt{2\pi}} \int_{-\infty}^{+\infty} f(u) \, e^{i\xi u} \, du$$

Taking inverse Fourier transform, we get

$$\begin{aligned}
u(x, y) &= \frac{1}{2\pi} \int_{-\infty}^{+\infty} e^{-i\xi x} \left[e^{-\xi^2 y} \int_{-\infty}^{+\infty} f(u) \cdot e^{i\xi u} du \right] d\xi \\
&= \frac{1}{2\pi} \int_{-\infty}^{+\infty} f(u) \left[\int_{-\infty}^{+\infty} e^{-\xi^2 y} \cdot e^{-i\xi(x-u)} d\xi \right] du
\end{aligned}$$

$$= \frac{1}{2\pi} \int_{-\infty}^{+\infty} f(u) \left[\int_{-\infty}^{+\infty} e^{-y\{\xi + \frac{i(x-u)}{2y}\}^2 - \frac{(x-u)^2}{4y}} \, d\xi \right] du$$

$$= \frac{1}{2\pi} \int_{-\infty}^{+\infty} f(u) \cdot e^{\frac{-(x-u)^2}{4y}} \left[\int_{-\infty}^{+\infty} e^{-y\{\xi + \frac{i(x-u)}{2y}\}^2} \, d\xi \right] du$$

Now, putting $\left[\xi + \dfrac{i(x-u)}{2y} \right] = \dfrac{\eta}{\sqrt{y}}$, we get

$$d\xi = \frac{d\eta}{\sqrt{y}}$$

so that the inner integral becomes

$$\int_{-\infty}^{+\infty} e^{-\eta^2} \cdot d\eta = \sqrt{\pi} \ , \text{ a standard result.}$$

Thus we get,

$$u(x, y) = \frac{1}{2\pi} \int_{-\infty}^{+\infty} \frac{f(u)}{\sqrt{y}} \, e^{\frac{-(x-u)^2}{4y}} \sqrt{\pi} \, du$$

$$= \frac{1}{2\sqrt{\pi y}} \int_{-\infty}^{+\infty} f(u) \exp \left[-\frac{(x-u)^2}{4y} \right] du \ .$$

Example 1.40. Solve the boundary value problem for the Laplace equation $u_{xx} + u_{yy} = 0$ in the quarter plane $0 < x < \infty$, $0 < y < \infty$ with the boundary conditions

$$u(0, y) = a \ , \ u(x, 0) = 0$$
$$\nabla u \to 0 \quad \text{as} \quad r = \sqrt{x^2 + y^2} \to \infty$$

where a is a constant. Using Fourier transform solve it.

Solution. Applying Fourier sine transform over x the Laplace equation takes the form

$$\frac{d^2 U_s}{dy^2} - \xi^2 \, U_s + \sqrt{\frac{2}{\pi}} \, \xi a = 0,$$

where $\quad U_s = \sqrt{\dfrac{2}{\pi}} \displaystyle\int_0^\infty u(x, y) \sin \xi x \, dx$

Its solution is

$$U_s(\xi, y) = A \, e^{-\xi y} + \sqrt{\frac{2}{\pi}} \frac{a}{\xi} \ ,$$

where A is a constant. Now, sine transform of $u(x, 0) = 0$ is

$$U_s(\xi, 0) = 0 .$$

Hence, $$U_s(\xi, y) = \frac{a}{\xi} \sqrt{\frac{2}{\xi}} \left[1 - e^{-\xi y}\right] .$$

Inverting, the solution of the boundary value problem is

$$
\begin{aligned}
u(x, y) &= \frac{2a}{\pi} \int_0^\infty \frac{1}{\xi} \left(1 - e^{-\xi y}\right) \sin \xi\, x\, d\xi \\
&= a - \frac{2a}{\pi} \left[\frac{\pi}{2} - \tan^{-1} \frac{y}{x}\right] = \frac{2a}{\pi} \tan^{-1} \left(\frac{x}{y}\right) .
\end{aligned}
$$

Example 1.41. Solve the Laplace equation $u_{xx} + u_{yy} = 0$ in a semi-infinite strip in the xy-plane defined by $0 < x < \infty$, $0 < y < b$ under the boundary conditions

$$u(0, y) = 0 , \quad u(x, y) \to 0 \ \text{ as } x \to \infty \ \text{ for } 0 < y < b$$
$$u(x, b) = 0 , \quad u(x, 0) = f(x) , \quad \text{for } 0 < x < \infty .$$

Solution. Sine transform of the Laplace equation with respect to x under boundary conditions $u(0, y) = 0$ and $u(x, y) \to 0$ as $x \to \infty$ gives rise to

$$\frac{d^2 U_s}{dy^2} - \xi^2 U_s = 0$$

Also sine transforms of $u(x, b) = 0$, $u(x, 0) = f(x)$ become

$$U_s(\xi, b) = 0$$
$$U_s(\xi, 0) = F_s(\xi)$$

Hence, the solution of the transformed differential equation under transformed boundary conditions is

$$U_s(\xi, y) = F_s(\xi) \frac{\sinh \left[\xi(b - y)\right]}{\sinh \xi b}$$

Now, inverting this equation we get

$$
\begin{aligned}
u(x, y) &= \sqrt{\frac{2}{\pi}} \int_0^\infty F_s(\xi) \frac{\sinh[\xi(b - y)]}{\sinh \xi b} \sin \xi\, x\, d\xi \\
&= \frac{2}{\pi} \int_0^\infty \left[\int_0^\infty f(t) \sin \xi t\, dt\right] \frac{\sinh \xi(b - y)}{\sinh \xi b} \sin \xi\, x\, d\xi
\end{aligned}
$$

When $\xi b \to \infty$, $\left[\frac{\text{sh } \xi(b-y)}{\text{sh } \xi b}\right] \sim \exp(-\xi b)$ and so the solution becomes

$$u(x,y) = \frac{1}{\pi} \int_0^\infty f(t) \, dt \int_0^\infty [\cos \xi(x-t) - \cos \xi(x+t)] \, e^{-\xi y} \, d\xi$$

Or, $\quad u(x,y) = \frac{1}{\pi} \int_0^\infty f(t) \left[\frac{y}{(x-t)^2 + y^2} - \frac{y}{(x+t)^2 + y^2}\right] dt$

as the final solution of the corresponding quarter plane problem $0 < x < \infty$, $0 < y < \infty$ of the Laplace equation under the given boundary conditions when $\xi b \to \infty$.

Exercises

(1) Find Fourier Transform of the function $f(x)$, if

$$f(x) = \frac{\sin ax}{x} \ , \quad a > 0 \ . \qquad \begin{bmatrix} \text{Ans. } \pi \ , \text{ when } |\xi| < a \\ 0 \ , \text{ when } |\xi| > a \end{bmatrix}$$

(2) Find Fourier Transform of

$$f(x) = \begin{cases} e^{i\omega x} \ , & 0 < x < a \\ 0 \ , & \text{otherwise} \end{cases}$$

to prove that $\quad F[f(x) \ ; \ x \to \xi] = \frac{i}{\sqrt{2\pi}} \left[\frac{e^{i(\xi+\omega)a} - e^{i(\xi+\omega)b}}{\xi + \omega}\right]$

(3) Prove that $F[H(a - |x|) \ ; \ x \to \xi] = \sqrt{\frac{2}{\pi}} \frac{\sin a\xi}{\xi}$, where $H(x)$ is the Heaviside's unit step function.

(4) Prove that $F_c F_c^{-1} \equiv I$ and $F_c = F_c^{-1}$, where F_c and F_c^{-1} are co-sine transform and inverse cosine transform operators respectively.

(5) Find Fourier transform of each of the following functions :
   (a) $f(x) = \frac{1}{1+x^2}$
   (b) $f(x) = e^x \exp(-e^x)$
   (c) $f(x) = x \exp\left(-\frac{ax^2}{2}\right)$, $a > 0$

[Ans. (a) $\sqrt{\frac{\pi}{2}} \, e^{-|\xi|}$ (b) $\Gamma(1-i\xi)/\sqrt{2\pi}$ (c) Take $f(x) = -\frac{1}{a} \frac{d}{dx}\left(e^{\frac{-ax^2}{2}}\right)$ etc].

(6) (a) Prove that $F[\delta(x - ct) + \delta(x + ct) ; x \to \xi] = \sqrt{\frac{\pi}{2}} \cos \xi \, ct$

(b) Prove that $F[H(ct - |x|) ; x \to \xi] = \sqrt{\frac{\pi}{2}} \frac{\sin \xi \, ct}{\xi}$

(c) Prove that $F[x^n f(x) ; x \to \xi] = i^n \frac{d^n}{d\xi^n} F(\xi)$

(d) If $f(x)$ has a finite discontinuity at $x = a$, show that $F[f'(x) ; x \to \xi] = i\xi \, F(\xi) - \frac{1}{\sqrt{2\pi}} \exp(-i\xi a)[f(a+0) - f(a-0)]$.

(7) (a) Prove that $\sqrt{2\pi}\delta(x) * f(x) = f(x)$

(b) Prove that $\frac{d}{dx}[f(x) * g(x)] = f'(x) * g(x) = f(x) * g'(x)$.

(8) (a) Prove that $F_s\left[\frac{1}{x} ; x \to \xi\right] = \sqrt{\frac{\pi}{2}}$

(b) Prove that $F_s[e^{-ax} ; x \to \xi] = \sqrt{\frac{2}{\pi}} \frac{\xi}{a^2 + \xi^2}$

(c) $F_c[e^{-ax} ; x \to \xi] = \sqrt{\frac{2}{\pi}} \frac{a}{a^2 + \xi^2} , a > 0$

(9) If $F_c[f(x) ; x \to \xi] = F_c(\xi)$, prove that
$$F_c[f(x) \cos \omega x ; x \to \xi] = \tfrac{1}{2}[F_c(\xi + \omega) + F_c(\xi - \omega)]$$

(10) Prove the following :

(a) $F[e^{-x^2} ; x \to \xi] = \frac{1}{\sqrt{2}} e^{\frac{-\xi^2}{2}}$

(b) $F_s\left[\frac{e^{-ax}}{x} ; x \to \xi\right] = \sqrt{\frac{2}{\pi}} \tan^{-1} \frac{\xi}{a}$,

(11) Prove that $F_s^{-1}\left[e^{-\pi\xi} ; \xi \to x\right] = \sqrt{\frac{2}{\pi}} \frac{x}{x^2 + \pi^2}$

(12) Use shift theorem to prove that

$$F\left[e^{-|x-t|} ; t \to \xi\right]$$
$$= e^{i\xi x} F\left[e^{-|t|} ; t \to \xi\right]$$
$$= \sqrt{\frac{2}{\pi}} \frac{e^{i\xi x}}{1 + \xi^2}$$

Hence, show that $e^{i\xi x} = \frac{1}{2}(1 + \xi^2) \int_{-\infty}^{+\infty} e^{-|x-t|+i\xi t} dt$.

(13) Use Parseval's theorem to prove that

$$\int_0^\infty \frac{\sin at}{t(a^2 + t^2)}\, dt = \frac{\pi}{2}\, \frac{1 - e^{-a^2}}{a^2}$$

by choosing $f(x) = e^{-ax}$ and $g(x) = \begin{cases} 1 & , \quad 0 < x < a \\ 0 & , \quad x > a \end{cases}$

(14) Prove that

$$\int_0^\infty \left[\frac{a - b}{\xi} + \frac{\cos a\xi - \cos b\xi}{\xi^2} \right] \sin \xi\, x\, d\xi$$

$$= \begin{cases} \frac{\pi}{2}\, (a - b) & , \quad 0 < x < a \\ \frac{\pi}{2}\, (x - b) & , \quad a < x < b \\ 0 & , \quad x > b \end{cases}$$

(15) From the identity

$$f(x) = -\frac{d}{dx} \int_x^\infty f(u)\, du$$

prove that $F_s[f(x); x \to \xi] = \xi\, F_c\left[\int_x^\infty f(u)\, du; x \to \xi \right]$

by using the result $F_s\left[f'(x)\; ; \; x \to \xi \right] = -\xi\, F_c\left[f(t)\; ; \; t \to \xi \right]$.

(16) If $\delta_n(x) = \frac{n}{\sqrt{\pi}}\, e^{-n^2 x^2}$,   prove that

$$F[\delta_n(x)\; ; \; x \to \xi] = \frac{1}{\sqrt{2\pi}}\, e^{\frac{-\xi^2}{4n^2}}$$

Hence, prove that $\int_{-\infty}^{+\infty} \delta_n(x)\, dx = 1$.

(17) Show that

$$F\left[\frac{\cosh ax}{\cosh \pi x}\; ; \; x \to \xi \right] = \frac{1}{\sqrt{2\pi}}\, \frac{2\cos \frac{a}{2} \cdot \mathrm{ch}\, \frac{\xi}{2}}{\cos a + \mathrm{ch}\, \xi}$$

by using contour integration method round a rectangular contour described by $x = \pm\, R$, $y = 0$, $y = 2\pi$, after suitably indented at points if needed.

(18) Putting $f(x) = x^{-\frac{1}{2}}$ in the sine and cosine forms of Fourier's integral theorem show that

$$\int_0^\infty \frac{\sin x}{\sqrt{x}}\, dx = \int_0^\infty \frac{\cos x}{\sqrt{x}}\, dx = \sqrt{\frac{\pi}{2}}$$

and deduce that $\quad F_s\left[t^{-\frac{1}{2}}\; ;\; t \to \xi\right] = F_c\left[t^{-\frac{1}{2}}\; ;\; t \to \xi\right] = \xi^{-\frac{1}{2}}$

(19) From Fourier cosine transforms of $e^{-ax}\cos ax$ and of $e^{-ax}\sin ax$ deduce F. C. transforms of $(x^4 + k^4)^{-1}$ and $x^2(x^4 + k^4)^{-1}$, $(k > 0)$.

(20) From Fourier sine transforms of $e^{-ax}\cos ax$ and of $e^{-ax}\sin ax$ deduce F.S. transforms of $\quad$ (a) $x(x^4 + k^4)^{-1}$ $\quad$ (b) $x^3(x^4 + k^4)^{-1}$ (c) $x(x^4 + k^4)^{-2}$ $\quad$ (d) $x^3(x^4 + k^4)^{-2}$

(21) Show that if b and ξ are real,

$$F_s\left[\frac{x^2 - b^2}{x(x^2 + b^2)}\; ;\; x \to \xi\right] = \sqrt{2\pi}\left[e^{-|b\xi|} - \frac{1}{2}\right]\text{sgn}\,(\xi)$$

(22) Show that if $\xi > 0$

$$\int_0^\infty \frac{\sin x \cos \xi x}{x}\, dx = \frac{\pi}{2}\, H(1 - \xi)) \;.$$

(23) If $f(x) = (1 - |x|)H(1 - |x|)$, find $F[f(x)\; ;\; x \to \xi]$ and deduce that

(a) $\displaystyle\int_0^\infty \left(\frac{\sin x}{x}\right)^2 dx = \frac{\pi}{2}$

(b) $\displaystyle\int_0^\infty \left(\frac{\sin x}{x}\right)^3 dx = \frac{3\pi}{8}$

(c) $\displaystyle\int_0^\infty \left(\frac{\sin x}{x}\right)^4 dx = \frac{\pi}{3}$

(24) Show that $F\,T$ of $\frac{3}{4a^3}(a^2 - x^2)H\left[1 - \frac{|x|}{a}\right]$, $a > 0$ is $\frac{1}{\sqrt{2\pi}}\varphi(\xi a)$, where $\varphi(t) = 3\frac{\sin t}{t^3} - \frac{\cos t}{t^2}$

(25) If $a > 0, b > 0$ show that

(a) $F_c\left[\frac{e^{-bx} - e^{-ax}}{x}\; ;\; x \to \xi\right] = \frac{1}{\sqrt{2\pi}}\log \frac{a^2 + \xi^2}{b^2 + \xi^2}$, $(a > b)$

(b) $F_s\left[\frac{e^{-bx} - e^{-ax}}{x}\; ;\; x \to \xi\right] = \sqrt{\frac{2}{\pi}}\tan^{-1}\frac{(a - b)\xi}{ab + \xi^2}$, $(a > b)$

(26) If $F(\xi) = F_c\left[e^{-\frac{x^2}{2}} ; x \to \xi\right]$, show that $\frac{dF}{d\xi} = -\xi F$, $F(0) = 1$ and hence duduce that $F(\xi) = e^{-\frac{\xi^2}{2}}$.

(27) Using Parseval's theorem show that if a and b are positive constants

$$\int_0^\infty \sin\, ax \,\sin\, bx \,\frac{dx}{x^2} = \frac{\pi}{2}\, \min\,(a, b).$$

(28) Making use of Fourier cosine transform and the Parseval's relation prove that

$$\int_0^\infty \frac{x^{-s} dx}{1 + x^2} = \frac{\pi}{2} \sec\left(\frac{\pi}{2}\, s\right), \quad 0 < s < 1$$

(29) Use Fourier transform to solve the ODE given below :

(a) $y''(x) - y(x) + 2\, f(x) = 0$, where $f(x) = 0$ when $x < -a$ and when $x > a$ and $y(x)$ and its derivatives vanish at $x \to \pm\infty$

(b) $2y''(x) + xy'(x) + y(x) = 0$

[Ans. (a) $y(x) = e^{-x}\int_{-a}^x e^t f(t)dt + e^x \int_x^a e^{-t} f(t)dt$

(b) $y(x) = A \exp\left(-\frac{x^2}{4}\right)$, A is a constant]

(30) Solve the following integral equation for $f(x)$:

(a) $\displaystyle\int_{-\infty}^{+\infty} e^{-at^2} f(x - t)dt = e^{-bx^2}$, $a, b > 0$

(b) $\displaystyle f(x) + \int_{-\infty}^{+\infty} f(x - t)e^{-at} dt = \frac{1}{x^2 + b^2}$

(c) $\displaystyle\frac{1}{\pi}\int_{-\infty}^{+\infty} \frac{f(t)}{x - t} dt = \varphi(x)$,

the integral is being in Cauchy principle value sense

[Ans. (a) $f(x) = \dfrac{a}{\sqrt{\pi(a - b)}} \exp\left(\dfrac{-abx^2}{a - b}\right)$

(b) $f(x) = \sqrt{\dfrac{2}{\pi}} \left\{\dfrac{b}{a}(a - b)\right\} \dfrac{1}{(a - b)^2 + x^2}$

(c) $f(x) = -\dfrac{1}{\pi}\displaystyle\int_{-\infty}^{+\infty} \dfrac{\varphi(t)}{x - t} dt$ when $F(\xi) = \dfrac{1}{\pi} \cdot \dfrac{i\pi \Phi(\xi)}{\mathrm{sgn}\, \xi}$]

(31) Solve the PDE $u_{tt} + u_{xxx} = 0$, $-\infty < x < \infty$, $t > 0$ under the conditions $u(x,0) = f(x)$, $u_t(x,0) = 0$ for $-\infty < x < \infty$.

[Use $F^{-1}[\cos(\xi^2 t)$; $\xi \to x] = \dfrac{1}{\sqrt{2t}} \cos\left(\dfrac{x^2}{4t} - \dfrac{\pi}{4}\right)$ for solution.]

(32) A function $y(x,t)$ satisfies the diffusion equation $k\dfrac{\partial^2 y}{\partial x^2} = \dfrac{\partial y}{\partial t}$ on the half line $x \geqslant 0$, for $t > 0$, the initial condition $y(x,0) = Cx\, e^{\frac{-x^2}{4a^2}}$, C and a are being constants, and boundary conditions $y(0,t) = 0$ and $y(x,t) \to 0$ as $x \to \infty$. Prove that $y(x,t) = Cx(1 + kt/a^2)^{\frac{-3}{2}} \exp\left[-\dfrac{x^2}{4(a^2+kt)}\right]$.

(33) A harmonic function $u(x,y)$ in $x \geqslant 0$, $y \geqslant 0$ satisfies the boundary conditions $u_x(0,y) = f(y)$, $u(x,0) = 0$. Show that

$$u_y(x,0) = -\frac{2}{\pi} \int_0^\infty \frac{t f(t)\,dt}{x^2 + t^2}$$

Find $u_y(x,0)$ if $f(y) = q\,H(b-y)$, where $q, b > 0$.

(34) Show that $\theta(x,t)$ satisfying $\dfrac{\partial^2 \theta}{\partial x^2} = \dfrac{\partial \theta}{\partial t}$, $x \geqslant 0$, $t \geqslant 0$, $\theta(x,0) = f(x)$ and $\dfrac{\partial \theta}{\partial x} - h\theta = 0$ for $x = 0$ is given by

$$\theta(x,t) = \frac{1}{\sqrt{2\pi}} \int_0^\infty \omega(u,t)[f(|x-u|) + f(x+u) + h\int_{|x-u|}^{x+u} f(v)dv]du,$$

where $\omega(x,t) = \dfrac{1}{\sqrt{2t}} e^{\frac{-x^2}{4t}} - \sqrt{\dfrac{\pi}{2}} h e^{hx+h^2 t}\,\text{Erfc}\left(\dfrac{1}{2}xt^{-\frac{1}{2}} + ht^{-\frac{1}{2}}\right)$.

(35) Prove that Fourier cosine transform of $f(x) = \dfrac{1}{x}$ does not exist but Fourier sine transform of it exists.

(36) Find Fourier sine transform of

(a) $f(x) = x\,e^{-ax}$, $a > 0$

(b) $f(x) = \dfrac{e^{-ax}}{x}$, $a > 0$

[Ans. (a) , (b) : Differentiate with respect to a and integrate w.r.t a both sides the integral $\int_0^\infty e^{-ax} \sin \xi x\, dx = \dfrac{\xi}{\xi^2+a^2}$ to obtain the results respectively.]

(37) If $F_c(\xi) = F_c[e^{-ax^2}$; $x \to \xi]$, prove that $F_c(\xi)$ satisfies the ODE $\dfrac{dF_c}{d\xi} \mid \dfrac{\xi}{2a} F_c - 0$, when $F_c(0) = 1$

(38) Apply Fourier cosine transform to solve $u_{xx} + u_{yy} = 0, 0 < x, y < \infty$ subject to the conditions $u(x, 0) = H(a - x)$, $a > x$; $u_x(0, y) = 0$; $0 < x, y < \infty$.

[Ans. $u(x, y) = -\dfrac{2}{\pi} \displaystyle\int_0^\infty \dfrac{\sin a\xi}{\xi} e^{-\xi y} \cos \xi \, x \, d\xi$]

(39) Use Fourier sine or cosine transform properly to solve the following integral equations :

(a) $\displaystyle\int_0^\infty f(x) \cos \xi \, x \, dx = \sqrt{\dfrac{\pi}{2\xi}}$

(b) $\displaystyle\int_0^\infty f(x) \sin \xi \, x \, dx = \dfrac{a}{a^2 + \xi^2}$

(c) $\dfrac{2}{\pi} \displaystyle\int_0^\infty f(x) \sin \, \xi \, x \, dx = J_0(a\xi)$

(d) $\displaystyle\int_0^\infty f(x) \cos \xi \, x \, dx = \dfrac{\sin a\xi}{\xi}$

$\Big[$ Ans. (a) $f(x) = \dfrac{1}{\sqrt{x}}$ $\qquad$ (b) $f(x) = e^{-ax}$

$\qquad$ (c) $f(x) = \dfrac{H(x - a)}{\sqrt{x^2 - a^2}}$ $\qquad$ (d) $f(x) = H(a - x)\Big]$

(40) Use Parseval formula to evaluate the following integral when $a, b > 0$:

(a) $\displaystyle\int_{-\infty}^{+\infty} \dfrac{dx}{(x^2 + a^2)^2}$ $\quad$ (b) $\displaystyle\int_{-\infty}^{+\infty} \dfrac{\sin ax \, dx}{x(x^2 + b^2)}$ $\quad$ (c) $\displaystyle\int_{-\infty}^{+\infty} \dfrac{\sin^2 ax}{x^2} dx$

$\Big[$ Ans. (a) $\dfrac{\pi a^3}{2}$ (b) $\dfrac{\pi}{b^2}\left(1 - e^{-ab}\right)$ (c) $\pi a\Big]$

(41) Solve the following free vibration problem of a semi-inifinite string defined by $u_{tt} = c^2 \, u_{xx}$, $0 < x < \infty$, $t > 0$ subject to condition $u(0, t) = 0$, $u(x, 0) = f(x)$, $u_t(x, 0) = g(x)$.

[For $x > ct$, $u(x, t) = \dfrac{1}{2}[f(x - ct) + f(x + ct)] + \dfrac{1}{2c} \displaystyle\int_{x-ct}^{x+ct} g(\xi)d\xi$.

Similarly result for $x < ct$ can also be obtained] .

Chapter 2

FINITE FOURIER TRANSFORM

2.1 Introduction.

Integral transform applied to solve the boundary value and initial value problems in mathematical physics in which the range of values of one of the independent variables say x, is finite for reducing the order of the associated partial differential is of the type

$$\int_a^b f(x, \cdots) K(x, \xi) dx,$$

where a, b are finite numbers. Such a transform is termed as a finite Integral transform. Further, if the kernel $K(x, \xi)$ is a Fourier kernel like $\sin x\xi$ or $\cos x\xi$ then the corresponding transforms are called finite Fourier transforms.

2.2 Finite Fourier cosine and sine transforms.

It is well-known result in the theory of Fourier series that, if $f(x)$ satisfies Dirichlet's conditions in $0 \le x \le a$, a being a finite real number and hence $f(x) \in \wp^1 (0, a)$, finite Fourier cosine transform $\bar{f}_c(n)$ of $f(x)$ is defined by

$$f_c[f(x) \; ; \; n] = \bar{f}_c(n) = \int_0^a f(x) \; \cos \frac{n\pi x}{a} \; dx \qquad (2.1)$$

and the corresponding inverse finite cosine transform is given by

$$f_c^{-1} [\bar{f}_c (n); \; x] = f(x) = \frac{1}{a} \bar{f}_c(0) + \frac{2}{a} \sum_{n=1}^{\infty} \bar{f}_c(n) \; \cos \frac{n\pi x}{a} \; dx \qquad (2.2)$$

at a point of continuity of $f(x)$ in $0 \le x \le a$. But at a pointing of discontinuity $x = c$, $0 < c < a$, the right hand side series in (2.2) takes the value $\frac{1}{2}[f(c+0) + f(c-0)]$ and the series takes the value $\frac{1}{2}[f(0) + f(a)]$ both at $x = 0$ or at $x = a$. In the sequel, (2.2) will be referred to as the inversion formula for finite Fourier cosine transform.

Similar results hold for finite sine transform. For a function $f(x) \epsilon \wp^1(0, a)$, and thereby satisfying the Dirichlet's conditions, finite Fourier sine transform $\bar{f}_s(n)$ of $f(x)$ in $(0, a)$ is defined by

$$f_s[f(x); n] = \bar{f}_s(n) = \int_0^a f(x) \sin \frac{n\pi x}{a} \, dx \qquad (2.3)$$

and its corresponding inversion formula is given by

$$f_s^{-1}[\bar{f}_s(n); x] = f(x) = \frac{2}{a} \sum_{n=1}^{\infty} \bar{f}_s(n) \sin \frac{n\pi x}{a} \qquad (2.4)$$

at a point of continuity x of $f(x)$ in $(0, a)$ and at a point of discontinuity $x = c, 0 < c < a$ of $f(x)$, the right hand side series in eqn (2.4) takes the value $\frac{1}{2}[f(c+0) + f(c-0)]$ and the series takes the value $\frac{1}{2}[f(0) + f(a)]$ both at $x = 0$ or $x = a$.

Corollary 2.1. In particular, if $a = \pi$ and that if $f(x) \in \wp^1(0, \pi)$, finite Fourier cosine transform of $f(x)$ and the corresponding inverse transform are given by

$$\bar{f}_c(n) = \int_0^{\pi} f(x) \cos nx \, dx \qquad (2.5)$$

and $f_c^{-1}[\bar{f}_c(n)] = f_c^{-1}[\bar{f}_c(n); x] = f(x) = \frac{1}{\pi}\bar{f}_c(0) + \frac{2}{\pi} \sum_{n=1}^{\infty} \bar{f}_c(n) \cos nx$

$$(2.6)$$

Similarly, finite Fourier sine transform of $f(x)$ and corresponding Inverse transform are given by

$$\bar{f}_s(n) = \int_0^{\pi} f(x) \sin nx \, dx \qquad (2.7)$$

and $\quad f_s^{-1}[\bar{f}_s(n) \ ; \ x] = f(x) = \frac{2}{\pi} \sum_{n=1}^{\infty} \bar{f}_s(n) \sin nx \qquad (2.8)$

The right hand series in eqns (2.6) and (2.8) converge to $\frac{1}{2}[f(x+0) + f(x-0)]$ at a point of discontinuity x and to $f(x)$ at a point of continuity x and to $\frac{1}{2}[f(0) + f(\pi)]$ at points $x = 0$ or $x = \pi$.

2.3 Relation between finite Fourier transforms of the derivatives of a function.

In applications finite Fourier transforms of the derivatives of a function are found useful to the solution of boundary value and initial value problems. Accordingly, the following theorems are important.

Theorem 2.1. Let $f(x)$ be continuous and $f'(x)$ be sectionally continuous on the interval $0 \leq x \leq a$, then

(i) $\bar{f}_c[f'(x) ; x \to n] = (-1)^n f(a) - f(0) + \frac{n\pi}{a} \bar{f}_s(n), n \in Z^*$

(ii) $\bar{f}_s[f'(x) ; x \to n] = -\frac{n\pi}{a} \bar{f}_c(n), n \in N$

Proof.(i) From the definition of finite Fourier cosine transform we have

$$\bar{f}_c[f'(x) ; n] = \int_0^a \frac{df}{dx} \cdot \cos\frac{n\pi x}{a} \, dx$$

$$= \left[f(x) \cdot \cos\frac{n\pi x}{a}\right]_0^a + \frac{n\pi}{a} \int_0^a f(x) \cdot \sin\frac{n\pi x}{a} \, dx$$

$$= (-1)^n f(a) - f(0) + \frac{n\pi}{a} \bar{f}_s(n), \quad n = 0, 1, 2, \ldots\ldots = Z^* \tag{2.9}$$

(ii) Again

$$\bar{f}_s[f'(x) ; n] = \int_0^a \frac{df}{dx} \sin\frac{n\pi x}{a} \, dx$$

$$= -\frac{n\pi}{a} \int_0^a f(x) \cos\frac{n\pi x}{a} \, dx$$

$$= -\frac{n\pi}{a} \bar{f}_c(n) \quad , \quad n = N \tag{2.10}$$

Theorem 2.2. Let $f(x)$ and $f'(x)$ be continuous and $f''(x)$ be sectionally continuous in $0 \leq x \leq a$. Then

(i) $\bar{f}_c[f''(x) ; n] \quad = -f'(0) + (-1)^n f'(a) - \frac{n^2\pi^2}{a^2} \bar{f}_c(n)$

(ii) $\bar{f}_s[f''(x) ; n] \quad = \frac{n\pi}{a} f(0) - (-1)^n \frac{n\pi}{a} f(a) - \frac{n^2\pi^2}{a^2} \bar{f}_s(n)$

Proof.(i) From the definition we have

$$\bar{f}_c[f''(x); n] = \int_0^a \frac{d^2 f}{dx^2} \cos\frac{n\pi x}{a} dx = \left[\frac{df}{dx} \cdot \cos\frac{n\pi x}{a}\right]_0^a + \frac{n\pi}{a} \bar{f}_s[f'(x); n]$$

$$= (-1)^n f'(a) - f'(0) - \frac{n^2\pi^2}{a^2} \bar{f}_c(n), \quad \text{by (2.10) of Th. 2.1.} \tag{2.11}$$

(ii) Again, $\bar{f}_s[f''(x)\; ;\; n] = \int_0^a \dfrac{d^2 f}{dx^2}\, \sin \dfrac{n\pi x}{a}\, dx$

$$= -\frac{n\pi}{a} \int_0^a \frac{df}{dx} \cdot \cos \frac{n\pi x}{a}\, dx$$

$$= -\frac{n\pi}{a} (-1)^n\, f(a) + \frac{n\pi}{a}\, f(0) + \frac{n^2 \pi^2}{a^2} \bar{f}_c(n).$$

$$(2.12)$$

For finite Fourier transform for the derivatives of order greater than 2 of a function may be obtained by induction. In the analysis of boundary value problems we often need to know the results for fourth-order derivatives. It follows that if $\frac{d^2 f}{dx^2}$ is zero at $x = 0$ and at $x = a$, then

$$\int_0^a \frac{d^4 f(x)}{dx^4} \sin \frac{n\pi x}{a}\, dx \;=\; -\frac{n^2 \pi^2}{a^2} \int_0^a \frac{d^2 f}{dx^2} \sin \frac{n\pi x}{a}\, dx$$

$$=\; \frac{n^4 \pi^4}{a^4}\, \bar{f}_s\,(n) \qquad\qquad (2.13)$$

Similarly, if $\frac{d^3 f}{dx^3} = \frac{df}{dx} = 0$ when $x = 0$ and when $x = a$, it is seen that

$$\int_0^a \frac{d^4 f(x)}{dx^4} \cos \frac{n\pi x}{a}\, dx = \frac{n^4 \pi^4}{a^4} \bar{f}_c\,(n) \qquad\qquad (2.14)$$

2.4   Faltung or convolution theorems for finite Fourier transform.

Like ordinary Fourier transform, the convolution theorems for finite Fourier transform can not be put into such simple forms. Accordingly, we introduce two new functions $f_1(x)$ and $f_2(x)$ instead of $f(x)$, which are defined as odd and even periodic extensions respectively of $f(x)$ with period 2π in the sense

$$f_1(x) = \begin{cases} f(x) & , \quad 0 \leqslant x \leqslant \pi \\ -f(-x) & , \quad -\pi < x < 0 \end{cases}$$

$$f_1(x + 2\, r\, \pi) = f_1(x)\, ,\ r = \pm 1\, ,\ \pm 2, \cdots$$

and $$f_2(x) = \begin{cases} f(x) & , \quad 0 \leq x \leq \pi \\ f(-x) & , \quad -\pi < x < 0 \end{cases}$$

$$f_2(x + 2\, r\, \pi) = f_2(x)\, ,\ r = \pm 1\, ,\ \pm 2, \cdots$$

Thus with the above understanding if $f_1(x)$ and $g_1(x)$ are the odd periodic extensions of $f(x)$ and $g(x)$ respectively, then

$$\int_{-\pi}^{\pi} f_1(x-u)g_1(u)du = \int_0^{\pi} f(x-u)g(u)du - \int_0^{\pi} f(u-x)g(u)du$$

(2.15)

Again since $f_1(x)$ is periodic

$$\int_{-\pi}^{\pi} f_1(x-u)g_1(u)du = \int_0^{\pi} f(-x-u)g(u)du - \int_0^{\pi} f(x+u)g(u)du$$

(2.16)

We call

$$\int_{-\pi}^{\pi} f_1(x-u)\ g_1(u)du \equiv f_1(x) * g_1(x),$$

as the convolution of the functions $f_1(x)$ and $g_1(x)$, introducing the notation introduced by Churchill.

If finite Fourier sine transforms of $f(x)$ and $g(x)$ in $(0,\pi)$ be $\bar{f}_s(n)$ and $\bar{g}_s(n)$ respectively, then

$$\frac{2}{\pi}\sum_{n=1}^{m}\bar{f}_s(n)\bar{g}_s(n)\cos nx = \frac{2}{\pi}\int_0^{\pi}d\xi\int_0^{\pi}f(\xi)g(\eta)d\eta\sum_{n=1}^{m}\cos nx\sin n\xi\sin n\eta$$

Making use of

$$4\sin n\xi\ \sin n\eta\ \cos\ nx = \cos(\xi-\eta+x)n + \cos(\xi-\eta-x)n$$
$$-\cos(\xi+\eta+x)n - \cos(\xi+\eta-x)n$$

and $\quad\displaystyle\sum_{r=1}^{n}\cos ry = -\frac{1}{2} + \frac{\sin\left(m-\frac{1}{2}\right)y}{2\sin\left(\frac{1}{2}y\right)}$

we get $\quad\displaystyle\frac{2}{\pi}\sum_{r=1}^{m}\bar{f}_s(n)\bar{g}_s(n)\ \cos\ nx = \frac{1}{4\pi}\int_0^{\pi}f(\xi)d\xi\int_0^{\pi}g(\eta)d\eta$

$$\left[\frac{\sin\ (m-\frac{1}{2})(\xi-\eta+x)}{\sin\frac{1}{2}(\xi-\eta+x)} + \frac{\sin\ (m-\frac{1}{2})(\xi-\eta-x)}{\sin\frac{1}{2}(\xi-\eta-x)}\right.$$

$$\left.-\frac{\sin\ (m-\frac{1}{2})(\xi+\eta+x)}{\sin\frac{1}{2}(\xi+\eta+x)} - \frac{\sin\ (m-\frac{1}{2})(\xi+\eta-x)}{\sin\frac{1}{2}(\xi+\eta-x)}\right]$$

Introducing the transformations

$$\lambda = \xi - \eta + x,\quad \mu = \eta$$

in the above four integrals, we get the first integral

$$\frac{1}{4\pi} \int_0^\pi f(\xi)d\xi \int_0^\pi g(\eta) \frac{\sin\left(m - \frac{1}{2}\right)(\xi - \eta + x)}{\sin\frac{1}{2}(\xi - \eta + x)} \, d\eta$$

$$= \frac{1}{2\pi} \int_0^\pi g(\mu)d\mu \int_{x-\mu}^{x+\pi-\mu} f(\lambda + \mu - x) \frac{\sin\left(m - \frac{1}{2}\right)\lambda}{\lambda} \frac{\frac{1}{2}\lambda}{\sin\frac{1}{2}\lambda} \, d\lambda$$

$$\rightarrow \frac{1}{4} \int_0^\pi g(\mu) f(\mu - x) d\mu \quad \text{as} \quad m \rightarrow \infty .$$

Similarly, for $m \rightarrow \infty$ the other three integrals can be evaluated to give

$$\frac{2}{\pi} \sum_{n=1}^\infty \bar{f}_s(n) \bar{g}_s(n) \cos\, nx = -\frac{1}{4} \int_0^\pi g(\mu)[f(x - \mu) + f(-x - \mu)$$

$$- f(\mu - x) - f(\mu + x)] d\mu$$

Thus, by (2.15) and (2.16) we get from the above equation that

$$\frac{2}{\pi} \sum_{n=1}^\infty \bar{f}_s(n) \, \bar{g}_s(n) \, \cos\, nx \; = \; -\frac{1}{2} \int_{-\pi}^\pi f_1(x - \mu) g_1(\mu) d\mu$$

$$= \; -\frac{1}{2} f_1(x) * g_1(x)$$

Then by inverse finite cosine transform we express the above relation as

$$\bar{f}_c[f_1(x) * g_1(x)] = -2 \, \bar{f}_s(n) \, \bar{g}_s(n) \tag{2.17}$$

Similarly, by considering the limit as $m \rightarrow \infty$ of the sum

$$\frac{2}{\pi} \sum_{n=1}^m \bar{f}_s(n) \, \bar{g}_c(n) \, \sin\, nx$$

we can have the result

$$\bar{f}_s[f_1(x) * g_1(x)] = 2 \, \bar{f}_s(n) \, g_c(n) \tag{2.18}$$

We can also show that

$$\bar{f}_c[f_2(x) * g_2(x)] = 2 \, \bar{f}_c(n) \, \bar{g}_c(n) \tag{2.19}$$

2.5 Multiple Finite Fourier Transform.

Let $f(x, y)$ be a function of two independent variables x and y defined in the rectangle $0 \leqslant x \leqslant a$, $0 \leqslant y \leqslant b$. For the time being, regarding $f(x, y)$ as a function of x satisfying certain necessary conditions, it possess a finite Fourier sine transform. We denote this transform by $\bar{f}_s(m, y)$ and it is defined by

$$\bar{f}_s(m, y) = \bar{f}_s[f(x, y); x \to m] = \int_0^a f(x, y) \sin \frac{m\pi x}{a} \, dx \qquad (2.20)$$

This function will itself have a finite Fourier sine transform over variable y which we may now be denote d by $\bar{f}_{ss}(m, n)$ and defined by

$$\bar{f}_{ss}(m, n) = \bar{\bar{f}}_{ss}[f(x, y) \ ; \ (x, y) \to (m, n)]$$
$$= \int_0^b \int_0^a f(x, y) \sin \frac{m\pi x}{a} \sin \frac{n\pi y}{b} \, dx \, dy \qquad (2.21)$$

We regard this equation as a double finite sine transform of the function $f(x, y)$ defined over the rectangular domain $0 \leqslant x \leqslant a$, $0 \leqslant y \leqslant b$.

The corresponding inversion formula giving $f(x, y)$ in terms of $\bar{f}_{ss}(m, n)$, are easily given by

$$\bar{f}_s(m, y) = \frac{2}{b} \sum_{n=1}^\infty \bar{f}_{ss}(m, n) \sin \frac{n\pi y}{b} \qquad (2.22)$$

$$\text{and} \qquad f(x, y) = \frac{2}{a} \sum_{m=1}^\infty \bar{f}_s(m, y) \sin \frac{m\pi x}{a} \qquad (2.23)$$

and therefore, by (2.22) and (2.23) we finally have

$$f(x, y) = \frac{4}{ab} \sum_{m=1}^\infty \sum_{n=1}^\infty \bar{f}_{ss}(m, n) \sin \frac{m\pi x}{a} \sin \frac{n\pi y}{b} \qquad (2.24)$$

This is the inversion formula for the double finite Fourier transform $\bar{f}_{ss}(m, n)$ of $f(x, y)$.

Proceeding as above, the double finite Fourier cosine transform $\bar{f}_{cc}(m, n)$ of $f(x, y)$ in the region $0 \leqslant x \leqslant a$, $0 \leqslant y \leqslant b$ can be defined as

$$\bar{f}_{cc}(m, n) = \int_0^a \int_0^b f(x, y) \cos \frac{m\pi x}{a} \cos \frac{n\pi y}{b} \, dx \, dy \qquad (2.25)$$

In a similar way we can define double finite Fourier transforms of the types

$$\bar{f}_{cs}(m,n) = \bar{f}_c[\bar{f}_s\{f(x,y); y \to n\}; x \to m]$$

$$= \int_0^a \int_0^b f(x,y) \cos \frac{m\pi x}{a} \sin \frac{n\pi y}{b} \, dx \, dy$$

$$\bar{f}_{sc}(m,n) = \int_0^a \int_0^b f(x,y) \sin \frac{m\pi x}{a} \cos \frac{n\pi y}{b} \, dx \, dy$$

Inversion formulas for double transforms of these types can easily be deduced by successive applications of the inversion formulas for the finite sine and cosine transforms. For example, in the case, in which $f(x,y)$ is defined on the region $0 \leqslant x \leqslant a$, $0 \leqslant y \leqslant b$, we get

$$\bar{f}_{cs}^{-1}[\bar{f}_{cs}(m,n) \; ; \; (m,n) \to (x,y)] \equiv f(x,y)$$

$$= \frac{2}{ab} \sum_{n=1}^{\infty} \bar{f}_{cs}(0,n) \sin \frac{n y \pi}{b} + \frac{4}{ab} \sum_{m=1}^{\infty} \sum_{n=1}^{\infty} \bar{f}_{cs}(m,n) \cos \frac{m\pi x}{a} \sin \frac{n\pi y}{b}$$

2.6   Double Transforms of partial derivatives of functions.

The double finite Fourier sine transforms of the partial derivatives of function $f(x,y) \in C^2(D)$, where

$$D = \{(x,y) : \; 0 < x < a, \; 0 < y < b\}$$

is given by

$$\bar{f}_s\left[\frac{\partial^2 f}{\partial x^2} \; ; \; x \to m\right] = \int_0^a \frac{\partial^2 f}{\partial x^2} \sin \frac{m\pi x}{a} \, dx$$

$$= \frac{m\pi}{a} [f(0,y) + (-1)^{m+1} f(a,y)] - \frac{m^2\pi^2}{a^2} \bar{f}_s [f(x,y) \; ; x \to m]$$

Therefore, $\bar{f}_{ss}\left[\dfrac{\partial^2 f}{\partial x^2} \; ; \; (x,y) \to (m,n)\right] = \Theta_1(m,n) - \dfrac{m^2\pi^2}{a^2} \bar{f}_{ss}(m,n)$

$$(2.26)$$

where $\Theta_1(m,n) = \dfrac{m\pi}{a} [\bar{f}_s\{f(0,y)\} \; ; y \to n\}$

$$+ (-1)^{m+1} \bar{f}_s\{f(a,y); y \to n\}]$$ $$(2.27)$$

Thus, $\bar{f}_{ss}\left[\dfrac{\partial^2 f}{\partial x^2}; (x,y) \to (m,n)\right] = -\dfrac{m^2\pi^2}{a^2} \bar{f}_{ss}(m,n),$

if $f(0,y) = f(a,y) = 0, \; 0 < y < b,$ in particular .

A similar result holds for double finite sine transform of $\frac{\partial^2 f}{\partial y^2}$ as

$$\bar{f}_{ss}\left[\frac{\partial^2 f}{\partial y^2} \; ; \; (x,y) \to (m,n)\right] = \frac{n\pi}{b}\left[\bar{f}_s\{f(x,0) \; ; \; x \to m\}\right]$$

$$+ \frac{n\pi}{b}\left[(-1)^{n+1}\,\bar{f}_s\{f(x,b) \; ; \; x \to m\}\right] - \frac{n^2\pi^2}{b^2}\,\bar{f}_{ss}(m,n)$$

$$= \Theta_2(m,n) - \frac{n^2\pi^2}{b^2}\,\bar{f}_{ss}(m,n), \text{ say.} \tag{2.28}$$

In particular, if $f(x,0) = f(x,b) = 0, \; 0 < x < a$, then

$$\bar{f}_{ss}\left[\frac{\partial^2 f}{\partial y^2} \; ; \; (x,y) \to (m,n)\right] = -\frac{n^2\pi^2}{b^2}\bar{f}_{ss}\,(m,n) \tag{2.29}$$

From (2.26) and (2.28) we immediately get

$$\bar{f}_{ss}\left[\frac{\partial^2 f}{\partial x^2} + \frac{\partial^2 f}{\partial y^2} \; ; \; (x,y) \to (m,n)\right]$$

$$= \Theta_1(m,n) + \Theta_2(m,n) - \pi^2\left[\frac{m^2}{a^2} + \frac{n^2}{b^2}\right]\bar{f}_{ss}(m,n) \tag{2.30}$$

If, in particular, $f(x,y)$ vanishes along the boundary of D, then

$$\bar{f}_{ss}\left[\nabla^2 f(x,y) \; ; \; (x,y) \to (m,n)\right] = -\pi^2\left[\frac{m^2}{a^2} + \frac{n^2}{b^2}\right]\bar{f}_{ss}(m,n) \tag{2.31}$$

The above deductions may be extended for functions of more than two variables similarly in its corresponding region of definitions.

2.7 Application of finite Fourier transforms to boundary value problems.

When the range of one of the variables is finite in the associated PDE of a given boundary value problem, finite Fourier transform over that variable may be applied to remove it. Further, the choice of sine and cosine transform is decided by the given form of the boundary / initial conditions of the problem. We shall now briefly indicate below some applications of finite Fourier transforms to the solution of boundary value and initial value problems.

Example 2.1. The temperature $u(x,t)$ at any point x at any time t in a solid bounded by planes $x = 0$ and $x = 4$ satisfies the heat conduction equation $\frac{\partial^2 u}{\partial x^2} = \frac{\partial u}{\partial t}$, when the end faces $x = 0$ and $x = 4$ are kept at zero temperature. Initially the temperature at x is $2x$. It is required to find $u(x, t)$ for all x and t.

Solution. The initial and the boundary value problem is defined by

$$\frac{\partial^2 u}{\partial x^2} = \frac{\partial u}{\partial t} \ , \ 0 \le x \le 4 \ , \ t \ge 0 \qquad \text{(i)}$$

subject to the initial condition $u(x, 0) = 2x$ and boundary conditions $u(0, t) = u(4, t) = 0$.

Since $u(0, t)$ and $u(4, t)$ are prescribed, we introduce finite Fourier transform over x and then eqn. (i) becomes

$$\int_0^4 \frac{\partial u(x, t)}{\partial t} \ \sin \frac{n\pi x}{4} \ dx = \int_0^4 \frac{\partial^2 u(x, t)}{\partial x^2} \ \sin \frac{n\pi x}{4} \ dx$$

i.e, $\quad \dfrac{d \ \bar{u}_s(n, t)}{dt} = -\dfrac{n\pi}{4} \left[u(4, t) \cos \ n\pi - u(0, t) \right] - \dfrac{n^2 \pi^2}{16} \ \bar{u}_s(n, t)$

$$= -\frac{n^2 \pi^2}{16} \ \bar{u}_s(n, t)$$

Its solution is $\quad \bar{u}_s(n, t) = c \ e^{-\frac{n^2 \pi^2}{16} t} \qquad \text{(ii)}$

Putting $\quad t = 0 , \qquad \bar{u}_s(n, 0) = c$

Now, $\quad \bar{u}_s(n, 0) = \displaystyle\int_0^4 u(x, 0) \sin \frac{n\pi x}{4} \ dx$

$$= \frac{32(-1)^{n+1}}{n\pi} \qquad \text{(iii)}$$

Therefore, from (ii) and (iii), we get

$$\bar{u}_s(n, t) = \frac{32(-1)^{n+1}}{n\pi} \ e^{-\frac{n^2 \pi^2}{16} t} \qquad \text{(iv)}$$

Taking inverse finite Fourier transform of eqn (iv), we get

$$u(x, t) = \frac{2}{4} \sum_{n=1}^{\infty} \bar{u}_s(n, t) \sin \frac{n\pi x}{4}$$

$$= \frac{16}{\pi} \sum_{n=1}^{\infty} \frac{(-1)^{n+1}}{n} \ e^{-\frac{n^2 \pi^2}{16} t} \ \sin \frac{n\pi x}{4} \qquad \text{(v)}$$

as the expression of the temperature of the solid at any point at any time.

Example 2.2. Solve the wave equation

$$\frac{\partial^2 u}{\partial x^2} = \frac{1}{c^2}\frac{\partial^2 u}{\partial t^2} , \quad 0 \le x \le a , \ t > 0 \qquad (i)$$

satisfying the boundary conditions $u(0,t) = u(a,t) = 0$, $t > 0$ and the initial conditions $u(x,0) = \frac{4b}{a^2} x(a-x)$, $\frac{\partial u(x,0)}{\partial t} = 0$, $0 \le x \le a$ to determine the displacement $u(x,t)$.

Solution. Since the boundary condition $u(0,t)$ and $u(a.t)$ are prescribed we may apply finite Fourier sine transform over the variable x. Then the given PDE (i) becomes

$$\left[\frac{d^2}{dt^2} + \frac{c^2 n^2 \pi^2}{a^2}\right] \bar{u}_s(n,t) = 0, \text{ where } \bar{u}_s(n,t) = \bar{f}_s(u(x,t) ; \ x \to n)$$

Its solution is given by

$$\bar{u}_s(n,t) = A \ \cos\frac{n\pi c}{a}t + B \ \sin\frac{n\pi c}{a}t \qquad (ii)$$

Further, taking finite Fourier sine transform over x on initial conditions we get

$$\bar{u}_s(n,0) \ = \ \frac{8ba}{\pi^3}\frac{1}{n^3}\left[1 + (-1)^{n+1}\right], \qquad (iii)$$

$$\frac{d}{dt}\ \bar{u}_s(n,0) \ = \ 0. \qquad (iv)$$

Using (iii), (iv) in eqn. (ii), we get

$$\bar{u}_s(n,t) = \frac{8ab}{\pi^3}\left[\frac{1 + (-1)^{n+1}}{n^3}\right]\cos\frac{n\pi ct}{a} \qquad (v)$$

Making use of inversion formula for finite Fourier transform, we get the required displacement as

$$u(x,t) = \sum_{r=0}^{\infty}\frac{16b(2r+1)^{-3}}{\pi^3}\ \cos\frac{(2r+1)\pi ct}{a}\ \sin\frac{(2r+1)\pi x}{a}\ .$$

Example 2.3. Consider the diffusion equation

$$\frac{\partial^2 u}{\partial x^2} = \frac{1}{k}\frac{\partial u}{\partial t} , \quad 0 < x < a , \ t > 0$$

for its solution satisfying the boundary conditions

$$\frac{\partial u(0,t)}{\partial x} = \frac{\partial u(a,t)}{\partial x} = 0 \ , \ t > 0$$

and the initial condition $u(x,0) = f(x)$, $0 \leqslant x \leqslant a$, using proper finite Fourier transform.

Solution. Since $\frac{\partial u}{\partial x}$ on the boundaries $x = 0$, a are prescribed the proper finite Fourier transform to be applied here is finite cosine transform over x. Hence, the diffusion equation and the initial condition transform to

$$\frac{d\bar{u}_c}{dt} + \frac{kn^2\pi^2}{a^2} \ \bar{u}_c(n,t) = 0 \tag{i}$$

and $\quad \bar{u}_c(n,0) = \bar{f}_c(n) = \bar{f}_c[f(x) \ ; \ x \to n] \tag{ii}$

Solving eqn.(i) under condition (ii), one gets

$$\bar{u}_c(n,t) = \bar{f}_c(n) \ e^{\frac{-k \, n^2 \, \pi^2}{a^2} \, t}$$

Therefore, by inversion theorem for finite Fourier cosine transform, we get

$$u(x,t) = \frac{1}{a}\bar{f}_c(0) + \frac{2}{a} \sum_{n=1}^{\infty} \bar{f}_c(n) \ e^{\frac{-kn^2\pi^2}{a^2}t} \cos \frac{n\pi x}{a} \ .$$

Example 2.4. The displacement $u(x,t)$ describing the vibration of a beam of length a with freely hinged end points satisfies the equation

$$\frac{\partial^4 u}{\partial x^4} + \frac{1}{c^2} \frac{\partial^2 u}{\partial t^2} = \frac{p\,(x,t)}{EI} \ , \ 0 \leq x \leq a \ , \ t > 0$$

The displacement $u(x,t)$ satisfies the initial conditions $u(x,0) = \frac{\partial u}{\partial t}(x,0) = 0$, $0 \leq x \leq a$ and the boundary conditions $u(0,t) = \frac{\partial^2 u}{\partial x^2}(0,t) = 0$ and $u(a,t) = \frac{\partial^2 u}{\partial x^2}(a,t) = 0$, $t > 0$

It is necessary to determine the displacement of the beam anywhere and at any time using finite Fourier transform.

Solution. Taking Fourier sine transform over x, the PDE transforms to

$$\frac{n^4\pi^4}{a^4} \ \bar{u}_s(n,t) + \frac{d^2}{dt^2} \ \bar{u}_s(n,t) = \frac{c^2}{EI} \ \bar{p}_s(n,t), \tag{i}$$

where $\quad \bar{p}_s(n,t) = \bar{f}_s[\,p\,(x,t) \ ; \ x \to n\,]$, after using the four boundary conditions. Again, the transformed form of the two given initial conditions become

$$\bar{u}_s(n,0) = \frac{d\,\bar{u}_s(n,0)}{dt} = 0 \tag{ii}$$

Therefore, the solution of the new initial value problem defined by eqn. (i) and under the initial conditions in eqn. (ii) is given by

$$\bar{u}_s(n,t) = \frac{a^2}{n^2\pi^2 c} \int_0^t \frac{c^2}{EI}\,\bar{p}_s(n,t)\sin\frac{n^2\pi^2 c(t-\varsigma)}{a^2}\,d\varsigma$$

Inverting the above equation we get

$$u(x,t) = \frac{2ac}{\pi^2 EI} \sum_{n=1}^{\infty} \frac{1}{n^2}\sin\frac{n\pi x}{a}\int_0^t \bar{p}_s(n,t)\sin\frac{n^2\pi^2 c}{a^2}(t-\varsigma)\,d\varsigma$$

Example 2.5. Find the solution of the one dimensional heat conduction equation in a bar with the temperature distribution satisfying $\frac{\partial u}{\partial t} = k\frac{\partial^2 u}{\partial x^2}$ under the boundary conditions $\frac{\partial u}{\partial x}(x,t) = 0$ at $x = 0$ and at $x = \pi$ and the given initial condition $u(x,0) = f(x)$, $0 \leqslant x \leqslant \pi$.

Solution. Taking Fourier cosine transform over x, the given PDE under the boundary conditions becomes

$$\frac{d\,\bar{u}_c}{dt} = -kn^2\,\bar{u}_c(n,t)$$

Its solution is, clearly , $\bar{u}_c(n,t) = \bar{u}_c(n,0)\,e^{-kn^2 t}$
Now, since $u(x,0) = f(x)$, we have $\bar{u}_c(n,0) = \int_0^\pi f(x)\cos\,nx\,dx$
Therefore, $\bar{u}_c(n,t) = e^{-kn^2 t}\int_0^\pi f(x)\cos\,nx\,dx$
Now, taking inverse finite Fourier transform, the above eqn. gives

$$u(x,t) = \frac{1}{\pi}\int_0^\pi f(y)dy + \frac{2}{\pi}\sum_{n=1}^{\infty}\left[e^{-kn^2 t}\cos nx\int_0^\pi f(y)\cos\,ny\,dy\right].$$

Example 2.6. The end points of a solid bounded by $x = 0$ and $x = \pi$ are maintained at temperatures $u(0,t) = 1, u(\pi,t) = 3$, where $u(x,t)$ represents its temperature at any point of it at any time t. Initially, the solid was held at 1 unit temperature with its surfaces were insulated. Find the temperature distribution $u(x,t)$ of the solid, given that $u_{xx}(x,t) = u_t(x,t)$.

Solution. The heat conduction equation of the solid is

$$u_{xx} = u_t \qquad\qquad\qquad \text{(i)}$$

with boundary conditions $u(0, t) = 1$, $u(\pi, t) = 3$ and the initial condition $u(x, 0) = 1$.

Taking finite Fourier transform over x, the eqn.(i) becomes

$$\frac{d\bar{u}_s}{dt} = -n\left[u(\pi, t)\cos\ n\pi - u(0, t) + n\bar{u}_s(n, t)\right]$$

i.e,
$$\frac{d\bar{u}_s}{dt} + n^2\bar{u}_s(n, t) = n\left[1 - 3\ \cos\ n\pi\right].$$

Solving this linear ODE we get

$$\bar{u}_s(n, t) = c\ e^{-n^2 t} + (1 - 3\ \cos\ n\pi)/n \qquad\qquad \text{(ii)}$$

Putting $t = 0$, (ii) gives

$$\bar{u}_s(n, 0) = c + (1 - 3\ \cos\ n\pi)/n$$

Also, since
$$\bar{u}_s(n, 0) = \int_0^\pi u(x, 0)\sin\ nx\ dx = \frac{1 - \cos n\pi}{n}$$

we get
$$c = \frac{2\ \cos n\pi}{n} \text{ and hence eqn. (ii) takes the form}$$

$$\bar{u}_s(n, t) = \frac{2\cos n\pi}{n}\ e^{-n^2 t} + \frac{1 - 3\cos n\pi}{n}$$

Therefore, after inversion

$$u(x, t) = \frac{2}{\pi}\sum_{n=1}^\infty \left[\frac{2\cos n\pi}{n}\ e^{-n^2 t} + \frac{1 - 3\cos\ n\pi}{n}\right]\sin nx$$

$$= \frac{4}{\pi}\sum_{n=1}^\infty \frac{\cos n\pi}{n}\ e^{-n^2 t}\sin nx + \frac{2}{\pi}\sum_{n=1}^\infty \frac{1 - 3\cos\ n\pi}{n}\sin\ nx$$

$$\text{(iii)}$$

But
$$\bar{f}_s\left[1\ ;\ x \to n\right] = \int_0^\pi \sin\ nx\ dx = \frac{1 - \cos n\pi}{n} \text{ and therefore}$$

$$1 = \frac{2}{\pi}\sum_{n=1}^\infty \frac{1 - \cos\ n\pi}{n}\sin\ px \qquad\qquad \text{(iv)}$$

Also,
$$\bar{f}_s[x\ ;\ x \to n] = \int_0^\pi x\sin\ nx\ dx = -\frac{\pi\cos\ n\pi}{n} \text{ implying}$$

$$x = \frac{2}{\pi}\sum_{n=1}^\infty \left[\frac{-\pi\cos n\pi}{n}\right]\sin\ nx \qquad\qquad \text{(v)}$$

Thus from (iv) and (v) one gets

$$1 + \frac{2x}{\pi} = \frac{2}{\pi} \sum_{n=1}^{\infty} \frac{1 - \cos n\pi}{n} \sin\ nx \qquad \text{(vi)}$$

Using (vi) in (iii), it gives

$$u(x,t) = 1 + \frac{2x}{\pi} + \frac{4}{\pi} \sum_{n=1}^{\infty} \frac{\cos n\pi}{n} e^{-n^2 t} \sin nx$$

Example 2.7. A uniform string of length π is stretched between its ends and initially one end, say $x = 0$, is given a small oscillation $a \sin wt$ while the other end is kept fixed. Using finite Fourier transform prove that the displacement of the point x of the string at time t is given by

$$a \sin\ wt \sin \frac{w(\pi - x)}{c} \cdot \csc \frac{w\pi}{c} + \frac{2acw}{\pi} \sum_{n=1}^{\infty} \frac{1}{w^2 - n^2 c^2} \sin\ nx\ \sin\ nct,$$

if the displacement $u(x,t)$ of the string satisfies the PDE

$$\frac{\partial^2 u}{\partial t^2} = c^2 \frac{\partial^2 u}{\partial x^2} \qquad \text{(i)}$$

Solution. The boundary and the initial conditions of the problem are given by

$$u(0,t) = a \sin wt \qquad \text{(ii)}$$
$$u(\pi,t) = 0 \qquad \text{(iii)}$$
$$u(x,0) = 0 \qquad \text{(iv)}$$
and
$$\frac{\partial u}{\partial t}(x,0) = 0 \qquad \text{(v)}$$

The form of the boundary conditions in (ii) and (iii) imply that finite Fourier sine transform need to be applied over the variable x to reduce eqn (i) to

$$\frac{d^2\ \bar{u}_s(n,t)}{(dt^2)} = -c^2 n \left[u(\pi,t) \cos n\pi - u(0,t) + n\bar{u}_s(n,t) \right]$$

Using (ii) and (iii), this equation becomes

$$\left[\frac{d^2}{dt^2} + c^2 n^2 \right] \bar{u}_s(n,t) = c^2\ a\ n \sin wt$$

The general solution of this equation is given by

$$\bar{u}_s(n,t) = A \cos\ nct + B \sin\ nct + \frac{c^2 a\ n \sin wt}{c^2 n^2 - w^2} \quad \text{(vi)}$$

The transformed form of (iv) and (v) becomes

$$\bar{u}_s(n,0) = 0 \quad \text{(vii)}$$

and
$$\left[\frac{d\ \bar{u}_s(n,t)}{dt}\right]_{t=0} = 0 \quad \text{(viii)}$$

Using (vii) and (viii), the solution becomes

$$\bar{u}_s(n,t) = \frac{acw}{w^2 - c^2 n^2} \sin cnt - \frac{c^2 an \sin wt}{w^2 - c^2 n^2}$$

Inverting the above transformed solution, we get

$$u(x,t) = \frac{2}{\pi} \sum_{n=1}^{\infty} \left[\frac{acw}{w^2 - c^2 n^2} \sin\ cnt - \frac{c^2 an \sin wt}{w^2 - c^2 n^2}\right] \sin nt \quad \text{(ix)}$$

To write the above solution in the required form we note that

$$\bar{f}_s\left[\sin \frac{w(\pi - x)}{c}\ ;\ x \to n\right] = \int_0^{\pi} \sin \frac{(\pi - x)w}{c} \sin nx\ dx$$

$$= \frac{1}{2}\int_0^{\pi}\left[\cos\left\{\frac{w(\pi - x)}{c} - nx\right\} - \cos\left\{\frac{w(\pi - x)}{c} + nx\right\}\right] dx$$

$$= \sin\left(\frac{w\pi}{c}\right) \cdot \frac{n}{n^2 - \frac{w^2}{c^2}} = \frac{nc^2}{n^2 c^2 - w^2} \sin \frac{w\pi}{c}.$$

This result gives

$$\sin \frac{w(\pi - x)}{c} = \frac{2}{\pi} \sum_{n=1}^{\infty} \frac{nc^2}{n^2 c^2 - w^2} \sin \frac{w\pi}{c} \sin\ nx$$

Thus, $\dfrac{2ac^2 \sin wt}{\pi} \displaystyle\sum_{n=1}^{\infty} \frac{n \sin nx}{n^2 c^2 - w^2} = a\ \sin wt\ \sin \frac{w(\pi - x)}{c} \operatorname{cosec} \frac{w\pi}{c}$

$$\text{(x)}$$

The solution (ix), after use of eqn. (x) becomes

$$u(x,t) = a\ \sin\ wt\ \sin \frac{w(\pi - x)}{c} \operatorname{cosec} \frac{w\pi}{c}$$

$$+ \frac{2acw}{\pi} \sum_{n=1}^{\infty} \frac{1}{w^2 - c^2 n^2} \sin\ nx\ \sin\ nct$$

Example 2.8. Solve the three dimensional Laplace equation

$$\frac{\partial^2 V}{\partial x^2} + \frac{\partial^2 V}{\partial y^2} + \frac{\partial^2 V}{\partial z^2} = 0, \ 0 \leqslant x \leqslant \pi, \ 0 \leqslant y \leqslant \pi, \ 0 \leqslant z \leqslant \pi$$

with the boundary conditions

$$V = V_0 \text{ , when } y = \pi \text{ ; } V = 0 \text{ , when } y = 0 \text{ and}$$
$$V = 0 \text{ , when } x = 0, \pi \text{ and } V = 0 \text{ , when } z = 0, \pi \text{ .}$$

Solution. The boundary conditions can be expressed as

$$V(0, y, z) = V(\pi, y, z) = 0 \qquad \text{(i)}$$
$$V(x, 0, z) = 0 \qquad \text{(ii)}$$
$$V(x, \pi, z) = V_0 \qquad \text{(iii)}$$

and
$$V(x, y, 0) = V(x, y, \pi) = 0 \qquad \text{(iv)}$$

The form of the boundary conditions in (i) and (iv) show that finite Fourier sine transforms over variables x and z successively reduce the given Laplace equation in the following forms :

Firstly, taking finite Fourier sine transform over the variable x we get

$$-m^2 \bar{V}_s(m, y, z) + \left[\frac{\partial^2}{\partial y^2} + \frac{\partial^2}{\partial z^2}\right] \bar{V}_s(m, y, z) = 0, \text{ by (i)}. \quad \text{(v)}$$

Again, taking finite Fourier transform over the variable z, the eqn. (v) gives

$$-m^2 \overline{V}_s(m, y, n) - n^2 \overline{V}_s(m, y, n) + \frac{d^2}{dy^2} \overline{V}_s(m, y, n) = 0, \qquad \cdots \text{(vi)}$$

since $\quad \overline{V}_s(m, y, 0) = \overline{V}_s(m, y, \pi) = 0$

Again, since $\quad \displaystyle\int_0^\pi V(x, \pi, z) \sin mx \ dx = \int_0^\pi V_0 \sin \ mx \ dx$

$$= V_0 (1 - \cos m\pi)/m$$

$$\overline{V}_s(m, \pi, z) = \begin{cases} 0 & , \quad m = \text{even} \\ \frac{2V_0}{m} & , \quad m = \text{odd} \end{cases} \qquad \cdots \text{(vii)}$$

and $\quad \overline{V}_s(m, 0, z) = 0$

The solution of the eqn (vi) is

$$\overline{V}_s(m, y, n) \ = \ c_1 \ ch \ \mu y + c_2 \ sh \ \mu \ y$$

where, $\mu = \sqrt{m^2 + n^2}$

Using the transformed boundary conditions (vii), the above general solution of eqn(vi) becomes,

$$\overline{V}_s(m, y, n) = c_2 \ sh \ \mu y,$$

where c_2 is given by

$$c_2 = \frac{\overline{V}_s(m, \pi, n)}{sh \ \mu \ \pi}$$

But since,

$$\int_0^\pi \overline{V}_s(m, \pi, z) \ \sin \ nz \ dz = 0, \text{when } m = \text{even}$$

and $\displaystyle\int_0^\pi \overline{V}_s(m, \pi, z) \ \sin \ nz \ dz$

$$= \int_0^\pi \frac{2 \ V_0}{m} \ \sin \ nz \ dz = \frac{2 \ V_0}{mn} [1 - \cos \ n\pi]$$

$$= \begin{cases} 0, & \text{if} \quad m = \text{odd}, \quad n = \text{even} \\ \frac{4V_0}{mn}, & \text{if} \quad m = \text{odd}, \quad n = \text{odd}. \end{cases}$$

$$\overline{V}_s(m, \pi, n) = \frac{4V_0}{mn} \ , \text{ when } \ m, n \ \text{ are both odd}.$$

Hence, $\overline{V}_s(m, y, n) = \dfrac{4V_0}{mn} \dfrac{\sin \mu y}{\sin \mu \pi}$,after substituting the value of c_2 .

Thus, $\overline{V}_s(m, y, n) = \dfrac{4V_0}{(2m + 1)(2n + 1)} \dfrac{sh \ \mu \ y}{sh \ \mu \ \pi}$,

where, $\mu = \left[\sqrt{m^2 + n^2}\right]_{m,n=\text{odd}} = \sqrt{(2m + 1)^2 + (2n + 1)^2}; m, n = 0, 1, 2, \cdots$

Inverting firstly over z, we get

$$\overline{V}_s(m, y, z) = \frac{2}{\pi} \sum_{n=0}^\infty \overline{V}_s(m, y, n) \ \sin(2n + 1)z$$

Again, inverting over the variable x.

$$V(x, y, z) = \frac{16}{\pi^2} \sum_{m=0}^\infty \sum_{n=0}^\infty \frac{sh \ \mu y}{sh \ \mu \pi} \cdot \frac{\sin(2m + 1)x}{2m + 1} \cdot \frac{\sin \ (2n + 1)z}{(2n + 1)} \ ,$$

where $\mu = \sqrt{\left[(2m + 1)^2 + (2n + 1)^2\right]}$.

Example 2.9. The deflection $y(x)$ of a uniform beam of finite length l satisfies the well known ODE $\frac{d^4y}{dx^4} = \frac{w(x)}{EI} = W(x)$, $0 \leqslant x \leqslant l$. Find

$y(x)$ if the beam is freely hinged at $x = 0$ and at $x = l$, implying

$$y(x) = y''(x) = 0 \quad \text{at} \quad x = 0, l.$$

Solution. Application of finite Fourier transform the given equation under the given boundary condition gives $\bar{y}(n) = \left(\frac{l}{n\pi}\right)^4 \bar{W}_s(n)$. Inverting the result we get

$$y(x) = \frac{2l^3}{\pi^4} \sum_{n=1}^{\infty} \frac{1}{n^4} \sin \frac{n\pi x}{l} \, \bar{W}_s(n)$$

Or $\quad y(x) = \frac{2l^3}{\pi^4} \sum_{n=1}^{\infty} \frac{1}{n^4} \sin \frac{n\pi x}{l} \int_0^l W(\xi) \sin \frac{n\pi\xi}{l} \, d\xi.$

In particular, if $W(x) = W_0 \, \delta(x - \alpha)$, where W_0 is a constant, then

$$y(x) = \frac{2l^3 W_0}{\pi^4} \sum_{n=1}^{\infty} \frac{1}{n^4} \sin \frac{n\pi x}{l} \sin \frac{n\pi\alpha}{l}, \quad \text{where} \ \ 0 < \alpha < l.$$

Example 2.10. The transverse displacement of an elastic membrane $u(x, y, t)$ satisfies the PDE $\frac{\partial^2 u}{\partial x^2} + \frac{\partial^2 u}{\partial y^2} = \frac{1}{c^2}\frac{\partial^2 u}{\partial t^2}$, under the boundary conditions $u = 0$ on the boundary, $u = f(x, y), u_t = g(x, y)$ at $t = 0$. Find the displacement $u(x, y, t)$ after utilising finite Fourier transform.

Solution. Applying double finite Fourier sine transform we get

$$\bar{u}_s(m, n) = \int_0^a \int_0^b u(x, y) \sin \frac{m\pi x}{a} \sin \frac{n\pi y}{b} \, dx \, dy \ .$$

Then the PDE gives $\quad \dfrac{d^2 \bar{u}_s}{dt^2} + c^2 \pi^2 \left[\dfrac{m^2}{a^2} + \dfrac{n^2}{b^2} \right] \bar{u}_s = 0$

with transformed boundary conditions

$$\bar{u}_s(m, n, 0) = \bar{f}_s(m, n), \quad \left[\frac{du_s}{dt} \right]_{t=0} = \bar{g}_s(m, n).$$

The solution of this transformed problem is

$$\bar{u}_s(m, n, t) = \bar{f}_s(m, n) \cos (c\pi w_{mn} t) + \bar{g}_s(m, n) \sin (c\pi w_{mn} t)$$

where $\quad w_{mn} = \left[\dfrac{m^2}{a^2} + \dfrac{n^2}{b^2} \right]^{1/2}$

Inverting this equation we get the solution of the problem as

$$u(x, y, t) = \frac{4}{ab} \sum_{m=1}^{\infty} \sum_{n=1}^{\infty} \sin \frac{m\pi x}{a} \cdot \sin \frac{n\pi x}{b} \left[\bar{f}_s(m, n) \right.$$
$$\left. \cos (c\pi w_{mn} t) + \bar{g}_s(m, n) \sin (c\pi w_{mn} t) \right]$$

where

$$\left[\bar{f}_s(m, n), \bar{g}_s(m, n) \right] = \int_o^a \int_0^b \sin \frac{m\pi \xi}{a} \cdot \sin \frac{n\pi \eta}{b} \left[f(\xi, \eta), g(\xi, \eta) \right] d\xi \, d\eta$$

Exercises.

(1) Find finite sine and cosine transforms of $f(x) = x^2, 0 < x < \pi$.

$$\text{[Ans. } \bar{f}_s(n) \quad = \quad \frac{2}{n^3}(\cos n\pi - 1) - \frac{\pi^2}{n} \cos n\pi, \; n = 1, 2, \cdots$$
$$= \quad \frac{\pi^2}{3}, \; n = 0 \; ;$$
$$\bar{f}_c(n) \quad = \quad 2\pi(\cos n\pi - 1)/n^2]$$

(2) Find $f(x)$ if (i) $\bar{f}_s(n) = (1 - \cos n\pi)/n^2\pi^2$ and $0 < x < 1$

$$\left[\text{Ans.} \quad \frac{2}{\pi^3} \sum_{n=1}^{\infty} \frac{1 - \cos n\pi}{n^2} \sin nx \right]$$

(3) Calculate $\bar{f}_s(n)$, if $f(x) = \begin{cases} x & , \quad 0 \leqslant x \leqslant \frac{\pi}{2} \\ \pi - x & , \quad \frac{\pi}{2} \leqslant x \leqslant \pi \end{cases}$

$$\left[\text{Ans.} \quad \frac{2}{n^2} \sin \frac{n\pi}{2} \right]$$

(4) Find $\bar{f}_c(n)$, if $f(x) = \frac{ch \, c(\pi - x)}{sh \, \pi c}$.   [Ans. $c/(c^2 + n^2)$]

(5) Find $\bar{f}_s(n)$, if $f(x) = \frac{2}{\pi} \tan^{-1} \frac{2b \sin x}{1 - b^2}$

$$\text{[Ans. } b^n \left[1 - (-1)^n \right]/n, |b| \leq 1]$$

(6) Prove that, if $f(x)$ is defined in $(0, a)$

(a) $\bar{f}_s [1 ; n] = \frac{a}{n\pi} \left[1 + (-1)^{n+1} \right]$

(b) $\bar{f}_s [x(a - x) ; n] = \frac{2a^3}{n^3\pi^3} \left[1 + (-1)^{n+1} \right]$

(c) $\bar{f}_s^{-1} \left[\frac{1}{n} ; x \right] = \frac{\pi}{a} \left[1 - \frac{x}{a} \right]$

(7) By finding finite Fourier sine and cosine transform of

$f(x) = e^{kx}, 0 \leqslant x \leqslant a$, deduce that

$$\bar{f}_s^{-1}\left[\frac{n}{k^2 + n^2 \frac{\pi^2}{a^2}} \; ; \; n \to x\right] = \frac{a}{\pi} \frac{sh\{k(a-x)\}}{sh\{ka\}} \; , \; 0 \leqslant x \leqslant a$$

(8) Show that the solution $u(x,t)$ of the diffusion equation

$\frac{\partial^2 u}{\partial x^2} = \frac{1}{k} \frac{\partial u}{\partial t}, 0 \leqslant x \leqslant a$, $t > 0$ satisfying boundary conditions

$$u(0,t) = u(a,t) = 0, \; t > 0$$

and the initial condition

$$u(x,0) = f(x), \; 0 \leqslant x \leqslant a$$

can be expressed as

$$u(x,t) = \frac{2}{\pi} \sum_{n=1}^{\infty} \bar{f}_s(n) \; e^{\frac{-k \; n^2 \pi^2 t}{a^2}} \; \sin\left(\frac{n\pi x}{a}\right).$$

(9) The function $u(x,y)$ satisfying $\frac{\partial^2 u}{\partial x^2} + \frac{\partial^2 u}{\partial y^2} = 0$ in the semi-infinite

strip $\{ (x,y) : 0 \leq x < \infty , 0 \leq y \leq b\}$, and the boundary

conditions

$$u(x,0) = 0 \; , \; u(x,b) = 0 \; , \; x \geqslant 0$$
$$u(0,y) = f(y) \; , \qquad\qquad 0 \leq y \leq b$$
$$u(x,y) \to 0 \text{ as } x \to \infty \; , \; 0 \leq y \leq b$$

can be expressed as

$$u(x,y) = \frac{2}{b} \sum_{n=1}^{\infty} \bar{f}_s(n) \cdot e^{-n\pi x/b} \; \cos \frac{n\pi y}{b} \; .$$

(10) $u(x,y)$ satisfies the PDE

$$\frac{\partial^2 u}{\partial x^2} + \frac{\partial^2 u}{\partial y^2} = 0$$

in the rectangle $R = \{(x,y) : 0 \leq x \leq a \; , \; 0 \leq y \leq b\}$
and the boundary conditions

$$\frac{\partial u(0,y)}{\partial x} = \frac{\partial u(a,y)}{\partial x} = 0, \; 0 < y \leq b$$
$$u(x,0) = f(x) \; , \; u(x,b) = 0 \; , \; 0 \leq x \leq b$$

Then prove that taking Fourier cosine transform over the variable x given by $\bar{f}_c[u(x,y) \; ; \; x \to m] = \bar{f}_c(m)$, the solution of the boundary value problem is given by

$$u(x,y) = \bar{f}_c^{-1}\left[\bar{f}_c(m) \; \frac{\sinh\{m\pi(b-y)/a\}}{\sinh(m\pi b/a)} \; ; \; m \to x\right]$$

(11) Find the steady state temperature in a long square bar of side π when one face of it is kept at constant temperature u_0 and the other face at zero temperature. Assume that the temperature function $u(x,y)$ is bounded and satisfies the PDE $u_{xx} + u_{yy} = 0$.

[Hint : Take the boundary conditions as $u(0,y) = u(\pi,y) = 0 \, , u(x,0) = 0$ and $u(x,\pi) = u_0$. Also take finite Fourier transform over x] .

$$\left[\text{Ans. } u(x,y) = \frac{4u_0}{\pi} \sum_{n=0}^{\infty} \frac{sh(2n+1)y \; \sin(2n+1)x}{(2n+1)sh(2n+1)\pi}\right]$$

(12) Prove that

(a) $F_c\left[x^2 \; ; \; n\right] = \begin{cases} \frac{a^3}{3} \, , & \text{if } \; n = 0 \\ 2\left(\frac{a}{n\pi}\right)^2 a(-1)^n, & \text{if } \; n = 1,2,3,\cdots \end{cases}$

(b) $F_s\left[x^2; n\right] = \frac{a^3}{n\pi}(-1)^{n+1} - 2\left(\frac{a}{n\pi}\right)^3\left[1 + (-1)^{n+1}\right]$
    when $0 < x < a$

(13) Prove that the initial-boundary value problem defined by

$u_t = k\,u_{xx}, \;\; 0 \le x \le a, \; t > 0$ under the conditions $u(x,0) = 0,$

$u(0,t) = f(t), \; u(a,t) = 0$ is given by

$$u(x,t) = \frac{2\pi k}{a} \sum_{n=1}^{\infty} n \, \sin \frac{n\pi x}{a} \int_0^t f(\varsigma)\exp\left[-k(t - \varsigma)\left(\frac{n\pi}{a}\right)^2\right] d\varsigma$$

(14) Prove the following results defined over $(0, \pi)$

(i) $F_s\left[\frac{x}{2}(\pi - x) \; ; \; x \to n\right] = \frac{1}{n^3}\left[1 + (-1)^{n+1}\right]$

(ii) $F_s\left[\frac{sh\,a(\pi-x)}{sh\,\pi a} \; ; \; x \to n\right] = \frac{n}{n^2+a^2} \, , \; a \neq 0$

(iii) $F_c\left[(\pi - x)^2 \; ; \; x \to n\right] = \frac{2\pi}{n^2}$ for $n \neq 0$ and $\frac{\pi^3}{3}$ for $n = 0$

(iv) $F_c\left[ch\ a(\pi - x)\ ;\ x \rightarrow n\right] = \frac{a\ sh\ (a\pi)}{n^2 + a^2}$, for $a \neq 0$.

(v) $F_s\left[\frac{d}{dx}\{f(x) * g(x)\}\ ;\ x \rightarrow n\right] = 2n\bar{f}_s(n)\ \bar{g}_s(n)$

(vi) $F_s\left[\int_0^x f(t) * g(t)dt\ ;\ x \rightarrow n\right] = \frac{2}{n}\bar{f}_s(n)\ \bar{g}_s(n)$

(15) If $\bar{f}_c(p) = \int_0^\pi f(x)\cos\ px\ dx$, $\bar{f}_s(p) = \int_0^\pi f(x)\sin\ px\ dx$, where p is not necessarily an integer, show for any constant α that

(i) $F_c\left[\{2f(x)\ \cos\ \alpha x\}\ ;\ x \rightarrow n\right] = \bar{f}_c(n - \alpha) + \bar{f}_c(n + \alpha)$

(ii) $F_c\left[\{2f(x)\ \sin\ \alpha x\}\ ;\ x \rightarrow n\right] = \bar{f}_s(n + \alpha) - \bar{f}_s(n - \alpha)$

(iii) $F_s\left[\{2f(x)\ \cos\ \alpha x\}\ ;\ x \rightarrow n\right] = \bar{f}_s(n + \alpha) + \bar{f}_s(n - \alpha)$

(iv) $F_s\left[\{2f(x)\ \sin\ \alpha x\}\ ;\ x \rightarrow n\right] = \bar{f}_c(n - \alpha) - \bar{f}_c(n + \alpha)$.

Chapter 3

THE LAPLACE TRANSFORM

3.1 Introduction.

In article 1.3 of chapter 1 it is noted that if $\int_{-\infty}^{+\infty} |f(t)| dt$ is not convergent, Fourier transform $F(\xi)$ of the function $f(t)$ need not exist for all real ξ. For example, when $f(t) = \sin \omega t$, $\omega = $ real, $F(\xi)$ does not exist. But such situations do arise occasionally in practice. To handle this situation, we consider a new function $f_1(t)$ connected to $f(t)$ defined by

$$f_1(t) = e^{-\gamma t} f(t) H(t),$$

where γ is an arbitrary real positive constant and $H(t)$ is the Heaviside unit step function. Clearly, here $f_1(t) \in P^1(R)$ and $f_1(t) \in A_1(R)$ for all $t \in R$ and therefore, Fourier transform of $f_1(t)$ exists, since

$$\int_{-\infty}^{+\infty} f_1(t) dt = \int_0^\infty e^{-\gamma t} f(t) \ dt$$

is convergent. In fact, in this case, by Fourier Integral theorem

$$f_1(t) = \frac{1}{2\pi} \int_{-\infty}^{+\infty} e^{-i\xi t} \ d\xi \int_{-\infty}^{+\infty} f_1(u) \ e^{i\xi u} \ du$$

implying

$$f(t) = \frac{1}{2\pi} \int_{-\infty}^{+\infty} e^{-i\xi t} \ d\xi \cdot e^{\gamma t} \int_0^\infty f(u) \cdot e^{-(\gamma - i\xi)u} \ du$$

If we write $p = \gamma - i\xi$, the above relation can be expressed as

$$
\begin{aligned}
f(t) &= \frac{1}{2\pi i} \int_{\gamma - i\infty}^{\gamma + i\infty} e^{pt} \left[\int_0^\infty f(u) \ e^{-pu} \ du \right] dp \\
&= \frac{1}{2\pi i} \int_{\gamma - i\infty}^{\gamma + i\infty} e^{pt} \ \bar{f}(p) \ dp
\end{aligned}
\tag{3.1}
$$

where $\quad \bar{f}(p) = \int_0^\infty f(u) e^{-pu} \ du, \quad Re \ p = \gamma > 0$ $\tag{3.2}$

Eqns. (3.1) and (3.2) constitute a transform with $K(p, u) = e^{-pu}$ as the kernel of it.

3.2 Definitions

We now define Laplace transform of a piecewise continuous function $f(t)$ of the real variable t defined on the semi-axis $t \geqslant 0$ by

$$L\left[f(t) \; ; \; t \to p\right] = L\left[f(t)\right] = F(p) = \bar{f}(p) = \int_0^\infty f(t)e^{-pt}dt \qquad (3.3)$$

and inverse Laplace transform by

$$f(t) = L^{-1}\left[\bar{f}(p), \; p \to t\right] = L^{-1}\left[F(p)\right] = \frac{1}{2\pi i}\int_{\gamma-i\infty}^{\gamma+i\infty} e^{pt}\bar{f}(p)\, dp \quad (3.4)$$

where $\gamma = \operatorname{Re} p > 0$.

Note 1. Some authors use variable s in place of p.

Note 2. The function $f(t)$ must be of exponential order for the existence of its Laplace transform.

Note 3. In operator notation eqn.(3.3) is expressed as

$$L\left[f(t) \; ; \; t \to p\right] = \bar{f}(p) \qquad (3.5)$$

and relation in eqn.(3.4) is expressed as

$$L^{-1}\left[\bar{f}(p) \; ; \; p \to t\right] = f(t) \qquad (3.6)$$

Thus,
$$L^{-1}[L(f(t))] = f(t) = I\left[f(t)\right]$$
$$\Rightarrow \quad L^{-1}\, L \equiv I$$

Also,
$$L[L^{-1}(\bar{f}(p))] = \bar{f}(p) = I[\bar{f}(p)]$$
$$\Rightarrow \quad LL^{-1} \equiv I$$

This means that $L^{-1}L \equiv LL^{-1} \equiv I$, showing that these operators L and L^{-1} are commutative.

3.3 Sufficient conditions for existence of Laplace Transform.

Theorem 3.1. If $f(t)$ is of some exponential order for large t and is piecewise continuous over $o \leqslant t \leqslant \infty$, then Laplace transform of $f(t)$ exists.

Proof. Let $f(t)$ be of exponential order σ such that

$$|f(t)| < Me^{\sigma t} , \quad \text{for} \quad t \geqslant t_0$$

Then we have

$$L\left[f(t); t \to p\right] = \int_0^\infty e^{-pt} f(t) \, dt$$

$$= \int_0^{t_0} e^{-pt} f(t) \, dt + \int_{t_0}^\infty e^{-pt} f(t) \, dt$$

$$= I_1 + I_2 , \quad \text{say}$$

Since $f(t)$ is a piecewise continuous function on every finite interval $0 \leqslant t \leqslant t_0$, I_1 exists and it is convergent Also, we have

$$|I_2| = \left| \int_{t_0}^\infty e^{-pt} f(t) \, dt \right|$$

$$\leqslant \int_{t_0}^\infty e^{-pt} |f(t)| \, dt$$

$$< M \int_{t_0}^\infty e^{-(p-\sigma)t} dt = \frac{Me^{-(p-\sigma)t_0}}{(p-\sigma)} , \quad \text{if } p > \sigma$$

Therefore, $|I_2|$ is finite for all $t_0 > 0$ and $p > \sigma$ and hence, I_2 is convergent. Thus, $L[f(t)]$ exists for all $p > \sigma$.

Note 3.1 Though the conditions stated in the theorem above are sufficient for the existence of Laplace transform, but these are not . This means that even if the above conditions are not satisfied by a function, Laplace transform of that function may or may not exist. This can be shown by considering the example $f(t) = \frac{1}{\sqrt{t}}$.

Here, $f(t) \to \infty$ as $t \to 0$. Hence, $f(t)$ is not a piecewise continuous function on every finite interval for $t \geqslant 0$.

Now, $L[\frac{1}{\sqrt{t}}] = \int_0^\infty e^{-pt} \frac{1}{\sqrt{t}} dt = \frac{2}{\sqrt{p}} \int_0^\infty e^{-x^2} dx = \sqrt{\frac{\pi}{p}}$, $p > 0$. This proves the existence of Laplace transform of $f(t)$.

3.4 Linearity property of Laplace Transform.

Theorem 3.2. If $L[f_1(t) ; t \to p]$ and $L[f_2(t) ; t \to p]$ both exist and c_1, c_2 are constants, then

$$L[c_1 f_1(t) + c_2 f_2(t) ; t \to p] = c_1 L[f_1(t) ; t \to p] + c_2[f_2(t) ; t \to p]$$

Proof. By definition,

$$L[c_1 f_1(t) + c_2 f_2(t) \; ; t \to p\,] = \int_0^\infty e^{-pt}[c_1 f_1(t) + c_2 f_2(t)]dt$$

$$= c_1 \int_0^\infty e^{-pt} f_1(t)dt + c_2 \int_0^\infty e^{-pt} f_2(t)dt$$

$$= c_1 \; L[f_1(t); \; t \to p] + c_2 \; L \; [f_2(t); \; t \to p\,]$$

3.5 Laplace Transforms of some elementary functions

We calculate below Laplace transforms of some elementary functions from definition.

Example 3.1. $L[H(t) \; ; \; t \to p]$

$$= \int_0^\infty e^{-pt} H(t)dt = \int_0^\infty e^{-pt} dt = \frac{1}{p}.$$

Therefore, $L[H(t-a) \; ; t \to p] = \int_0^\infty H(t-a)e^{-pt}dt$

$$= \int_a^\infty e^{-pt} dt = \frac{e^{-ap}}{p}$$

Example 3.2. $L[t^\nu \; ; \; t \to p\,]$

$$= \int_0^\infty t^\nu e^{-pt} dt = p^{-\nu-1} \int_0^\infty u^\nu e^{-u} du, \text{when } u = p\,t$$

$$= p^{-\nu-1} \; \Gamma(\nu + 1), \text{ which exists when Re } \nu > -1.$$

Thus in particular,

$$L[t^n \; ; \; t \to p\,] = \frac{n!}{p^{n+1}} \; , \quad n = 1, 2, 3, \cdots \; , \; p > 0$$

so that $L[t \; ; \; t \to p] = \frac{1}{p^2} \; , \quad L[t^2 \; ; \; t \to p] - \frac{2}{p^3} \; , \cdots \text{ ctc., } p > 0$

Example 3.3. $L[e^{at} \; ; \; t \to p\,]$

$$= \int_0^\infty e^{-pt} e^{at} \; dt = \frac{1}{p-a} \; , \; p > a.$$

Example 3.4.

$$L \; [\sin at \; ; \; t \to p\,] = \int_0^\infty e^{-pt} \sin \; at \; dt$$

$$= \int_0^\infty e^{-pt} \left[\frac{e^{ait} - e^{-ait}}{2i} \right] dt$$

$$= \left[\int_0^\infty \frac{e^{-(p-ai)t} - e^{-(p+ai)t}}{2i} \right] dt$$

$$= \frac{1}{2i} \left[\frac{1}{p - ai} - \frac{1}{p + ai} \right], \quad \text{by linearity property}$$

$$= \frac{a}{p^2 + a^2}, \quad p > 0,$$

Example 3.5. $L [\cos at ; t \to p]$

$$= \int_0^\infty e^{-pt} \frac{e^{ait} + e^{-ait}}{2} dt$$

$$= \frac{1}{2} \left[\int_0^\infty e^{-(p-ai)t} dt \right] + \frac{1}{2} \left[\int_0^\infty e^{-(p+ai)t} dt \right], \quad \text{by linearity property}$$

$$= \frac{1}{2} \left[\frac{1}{p - ai} + \frac{1}{p + ai} \right]$$

$$= \frac{p}{p^2 + a^2}, \quad p > 0$$

Example 3.6. $L [\cosh at ; t \to p]$

$$= \int_0^\infty e^{-pt} \left[\frac{e^{at} + e^{-at}}{2} \right] dt$$

$$= \frac{1}{2} \int_0^\infty e^{-(p-a)t} dt + \frac{1}{2} \int_0^\infty e^{-(p+a)t} dt, \quad \text{by linearity property}$$

$$= \frac{1}{2} \left[\frac{1}{p - a} + \frac{1}{p + a} \right]$$

$$= \frac{p}{p^2 - a^2}, \quad p > a > 0$$

Example 3.7. $L [\sinh at ; t \to p]$

$$= \int_0^\infty e^{-pt} \left[\frac{e^{at} - e^{-at}}{2} \right] dt$$

$$= \frac{1}{2} \int_0^\infty e^{-(p-a)t} dt - \frac{1}{2} \int_0^\infty e^{-(p+a)t} dt, \quad \text{by linearity property}$$

$$= \frac{1}{2} \left[\frac{1}{p - a} - \frac{1}{p + a} \right]$$

$$= \frac{a}{p^2 - a^2}, \quad p > a > 0$$

3.6 First shift theorem.

Theorem 3.3. If Laplace transform of $f(t)$ is $\bar{f}(p)$, then Laplace transform of $e^{at} f(t)$ is $\bar{f}(p-a)$.

Proof. Given that

$$L[\, f(t)\, ;\, t \to p\,] = \int_0^\infty f(t)\, e^{-pt}\, dt$$

$$= \bar{f}(p)$$

Therefore, $\quad L[\, e^{at} f(t)\,] = \int_0^\infty e^{at} f(t)\, e^{-pt}\, dt$

$$= \int_0^\infty f(t) \cdot e^{-(p-a)t}\, dt$$

$$= \bar{f}(p-a). \tag{3.7}$$

3.7 Second shift theorem.

Theorem 3.4. If Laplace transform of $f(t)$ is $\bar{f}(p)$, then Laplace transform of $f(t-a)H(t-a)$ is $e^{-ap}\bar{f}(p)$.

Proof. Given that

$$L[f(t)\, ;\, t \to p] = \int_0^\infty f(t)\, e^{-pt}\, dt$$

$$= \bar{f}(p)$$

Therefore , $\quad L[\, f(t-a)H(t-a)\, ;\, t \to p\,]$

$$= \int_0^\infty f(t-a)H(t-a)e^{-pt}dt$$

$$= \int_a^\infty f(t-a)\, e^{-pt}\, dt$$

$$= \int_0^\infty f(x)\, e^{\, px} \cdot e^{\, pa}\, dx$$

$$= e^{-pa} \cdot \bar{f}(p). \tag{3.8}$$

3.8 The change of scale property.

Theorem 3.5. If $L[f(t); t \to p] = \bar{f}(p)$, then $L[f(at); t \to p] = \frac{1}{a}\bar{f}\left(\frac{p}{a}\right)$

Proof. Given that, $\displaystyle\int_0^\infty f(t)\, e^{-pt}\, dt = \bar{f}(p)$

Therefore, $\displaystyle L\left[f(at)\,;\ t \to p\,\right] = \int_0^\infty f(at)\, e^{-pt}\, dt$

$$= \frac{1}{a}\int_0^\infty f(x)\cdot e^{\frac{-P}{a}x}\, dx$$

$$= \frac{1}{a}\int_0^\infty f(x)\cdot e^{-\left(\frac{p}{a}\right)x}\, dx$$

$$= \frac{1}{a}\,\bar{f}\left(\frac{p}{a}\right). \qquad\qquad (3.9)$$

3.9   Examples

Example 3.8. Evaluate

$$L[f(t)]\,,\ \text{where}\ \ f(t) = \begin{cases} t/a & ,\quad 0 < t < a \\ 1 & ,\quad t > a \end{cases}$$

Solution.

$$\begin{aligned} L\left[f(t)\right] &= \int_0^\infty e^{-pt} f(t)\, dt \\[2mm] &= \int_0^a e^{-pt}\cdot \frac{t}{a}\, dt + \int_a^\infty e^{-pt}\, dt \\[2mm] &= \frac{1 - e^{-ap}}{ap^2} \end{aligned}$$

Example 3.9. Evaluate $L\left[\dfrac{1}{\sqrt{\pi t}}\right]$

Solution. $\displaystyle L\left[\frac{1}{\sqrt{\pi t}}\right] = \int_0^\infty e^{-pt}\frac{1}{\sqrt{\pi t}}\, dt$

$$= \frac{\sqrt{p}}{\sqrt{\pi}}\int_0^\infty e^{-x}\, x^{\frac{1}{2}-1}\,\frac{dx}{p}$$

$$= \frac{1}{\sqrt{\pi p}}\,\Gamma\left(\frac{1}{2}\right)$$

$$= \frac{\sqrt{\pi}}{\sqrt{\pi p}} = \frac{1}{\sqrt{p}}\,.$$

Example 3.10. Evaluate $L\left[\,t^{\nu}e^{-at}\ ;\ t \to p\,\right]$, if Re $\nu + 1 > 0$

Solution. It is known that $L\left[\,t^{\nu};\ t \to\ p\,\right] = \frac{\Gamma(\nu+1)}{p^{\nu+1}}$.
Therefore, by shift theorem, we get

$$L\left[\,t^{\nu}e^{-at};\ t \to p\,\right] = \frac{\Gamma(\nu+1)}{(p+a)^{\nu+1}}$$

Example 3.11. Evaluate $L\left[\sin(t-a)H(t-a)\ ;\ t \to p\right]$.
Hence, evaluate $L\left[e^{(t-a)k}\sin(t-a)\ H(t-a)\ ;\ t \to p\right]$.

Solution. We know that $L\left[\sin t\ ;\ t \to p\right] = \frac{1}{p^2+1}$

Hence, by the second shift theorem, we get

$$L\left[\sin(t-a)\ H(t-a)\right] = e^{-pa}/(1+p^2)$$

Thus, $\quad L\left[e^{k(t-a)}\ \sin(t-a)H(t-a)\ ;\ t \to p\right]$

$$= e^{-(p-k)}/\left[1+(p-k)^2\right],\ \text{after using the first shift theorem.}$$

Example 3.12. Evaluate $L\left[\,\sinh bt\ ;\ t \to p\,\right]$. Hence evaluate
$L\left[\,e^{at}\ \sinh\ bt\ ;\ t \to p\,\right]$

Solution. Since, $L\left[\,\sinh t\ ;\ t \to p\,\right] = \frac{1}{p^2-1}$, we have by the change of
scale property that

$$L\left[\,\sinh\ bt\ ;\ t\ \to p\,\right] = \frac{b}{p^2-b^2}$$

Hence, $\quad L\left[e^{at}\ \text{sh}\ bt\ ;\ t \to p\right] = \frac{b}{(p-a)^2+b^2}$

Example 3.13. Evaluate $L\left[\,e^{at}\cos\ bt\ ;\ t \to p\,\right]$. Hence, evaluate
$L\left[e^{a(t-c)}\cos b(t-c)H(t-c)\ ;\ t \to p\right]$

Solution. We know that $L\left[\cos bt\ ;\ t \to p\right] - \frac{p}{p^2+b^2}$

So, $\quad L\left[\,e^{at}\ \cos\ bt\ ;\ t \to p\,\right] = \frac{p-a}{(p-a)^2+b^2}$, by shift theorem.

Hence , $L\left[\,e^{a(t-c)}\ \cos\ b\ (t-c)H\ (t-c)\ ;\ t \to p\,\right]$

$$= e^{-pc}\frac{(p-a)}{(p-a)^2+b^2}$$

Example 3.14. Evaluate $L\left[(\pi t)^{-\frac{1}{2}}\ e^{\frac{t}{2}}\right]$.

Solution. We know that $L\left[\frac{1}{\sqrt{\pi t}}\right] = \frac{1}{\sqrt{p}}$

Hence, $L\left[\frac{1}{\sqrt{\pi t}}\, e^{\frac{t}{2}} \; ; \; t \to p\right] = \frac{1}{\sqrt{p - \frac{1}{2}}}$.

Example 3.15. Evaluate

(i) $L\left[t\, H\,(t-1) \; ; \; t \to p\,\right]$

and (ii) $L\left[\,H(t-1)\, \cos\, at \; ; \; t \to p\,\right]$

Solution. From definition $L\left[\,f(t)H(t-a) \; ; \; t \to p\,\right]$

$$= \int_a^\infty e^{-pt}\, f(t)\, dt$$

$$= \int_0^\infty e^{-p(u+a)}\, f(u+a)\, du$$

$$= e^{-pa} \int_0^\infty e^{-pu}\, f(u+a)\, du$$

$$= e^{-pa} L\left[f(t+a) \; ; \; t \to p\right]$$

(i) $L\left[t\, H\,(t-1) \; ; \; t \to p\right]$

$$= \int_1^\infty e^{-pt}\, t\, dt \qquad\qquad , \; a = 1$$

$$= e^{-p} \int_0^\infty e^{-pu}\, (t+1)\, dt$$

$$= e^{-p}\left[\frac{1}{p^2} + \frac{1}{p}\right]$$

$$= \frac{p+1}{p^2}\, e^{-p}$$

(ii) $L\left[\,H(t-1)\, \cos\, at\,\right]$

$$= e^{-p}\, L\,\left[\,\cos a(t+1)\right]$$

$$= e^{-p}\, L\,\left[\,\cos at.\, \cos\, a -\, \sin\, at.\, \sin\, a\right]$$

$$= \frac{e^{-p}}{p^2 + 1}\,\left[\,p\cos\, a - a\, \sin\, a\,\right].$$

3.10   Laplace Transform of derivatives of a function.

Theorem 3.9.   Let $f(t)$ be a continuous function of $t \geqslant 0$ and it is of exponential order for large t and if $f'(t)$ is a piecewise continuous function for $t \geqslant 0$, then Laplace transform of the derivative $f'(t)$ exists when $p > \sigma$ and is given by

$$L\left[\, f'(t)\right] \; = \; p\, L\left[\, f(t)\,\right] - f(0)$$

Proof. By definition $L[\,f'(t)\,]$

$$= \int_0^\infty e^{-pt}\, f'(t)\, dt$$

$$= [\,e^{-pt}\cdot f(t)\,]_0^\infty - \int_0^\infty (-p)\, e^{-pt}\, f(t)\, dt$$

$$= \lim_{t\to\infty} e^{-pt} f(t) - f(o) + p\, L[\,f(t)\,] \tag{3.10}$$

Since, $f(t)$ is of exponential order σ for $t\to\infty$, we have

$$|f(t)| \leqslant M\, e^{\sigma t}\,,\ t \geqslant 0$$

Hence, $|f(t)\, e^{-pt}| = e^{-pt}\, |f(t)| \leqslant M\, e^{-pt}\cdot e^{\sigma t} = M\, e^{-(p-\sigma)t}$

So, $\lim_{t\to\infty} e^{-pt}\, f(t) = 0$, as $p - \sigma > 0$

Thus from (3.10), we get

$$L[\,f'(t)\,] = p\, L[\,f(t)\,] - f(0). \tag{3.11}$$

Corollary 3.1. If $f''(t)$ exists for $t \geqslant 0$ and is a piecewise continuous function, then proceeding as above we can extend the result of the above theorem as

$$\begin{aligned}
L[\,f''(t)\,] &= p\, L[\,f'(t)\,] - f'(0). \\
&= p[\,p\{L(f(t))\} - f(0)] - f'(0) \\
&= p^2\, L[(f(t)]\,] - pf(0) - f'(0) \tag{3.12}
\end{aligned}$$

Corollary 3.2. In general, if $f^n(t)$ exists for $t \geqslant 0$ and is a piecewise continuous function of t, then

$$L[f^n(t)] = p^n\, L[f(t)] - p^{n-1}\, f\,(0) - p^{n-2}f'(0) - \cdots - f^{(n-1)}(0)$$

$$\tag{3.13}$$

Theorem 3.10. Suppose $f(t)$ is not continuous at $t = a, 0 < a < \infty$ and $f(a - 0)$, $f(a + 0)$ exist and $f(t)$ is of requisite order for existence of $L[f(t)]$. Further if $f'(t)$ exists, for $t \neq a > 0$ then

$$L[f'(t)\,] = p[\,f(p)\,] - f(0) - e^{-ap}\,[f(a + 0) - f(a - 0)]$$

Proof. By definition

$$L[\,f'(t)\,] = \int_0^\infty e^{-pt}\cdot f'(t)\, dt$$

$$= \int_0^{a-0} e^{-pt} \cdot f'(t) \, dt + \int_{a+0}^{\infty} e^{-pt} \, f'(t) \, dt$$

$$= \left[e^{-pt} \, f(t) \right]_0^{a-0} + p \int_0^{a-0} e^{-pt} f(t) \, dt$$

$$+ \left[e^{-pt} \, f(t) \right]_{a+0}^{\infty} + p \int_{a+0}^{\infty} e^{-pt} f(t) \, dt$$

$$= \left[e^{-p(a-0)} \, f(a-0) - f(0) \right] + \left[0 - e^{-p(a+0)} \, f(a+0) \right]$$

$$+ p \left[\int_0^{a-0} e^{-pt} f(t) \, dt + \int_{a+0}^{\infty} e^{-pt} f(t) \, dt \right]$$

$$= p \, L \left[f(t) \right] - f(0) - e^{-ap} \left[f(a+0) - f(a-0) \right] \qquad (3.14)$$

Corollary 3.3. If $f(t)$ is discontinuous at $t = a_1, a_2, \cdots a_k$, the above theorem 3.10 can easily be extended as

$$L \left[f'(t) \right] = p \, L \left[f(t) \right] - f(0) - \sum_{i=1}^{k} e^{-a_i \, p} \{ f(a_i + 0) - f(a_i - 0) \}$$

$$(3.15)$$

Remark. 3.1 In the light of the above Theorem 3.9 and Theorem 3.10, one can easily get the extension of the theorems for the Laplace transforms of $f''(t), f'''(t). \cdots$ etc, provided that the functions satisfy the requisite properties for the existence of their Laplace transforms.

3.11 Laplace Transform of Integral of a function

. **Theorem 3.11.** If Laplace transform of a function $f(t)$ is $\bar{f}(p)$. then Laplace transform of $\int_0^t f(\tau) d\tau$ is $\bar{f}(p)/p$.

Proof. Given that

$$L \left[f(t) \; ; \; t \to p \right] = \int_0^{\infty} e^{-pt} \, f(t) \, dt = \bar{f}(p)$$

Now, $L \left[\int_0^t f(\tau) d\tau \right]$

$$= \int_0^{\infty} e^{-p \, t} \left[\int_0^t f(\tau) \, d \, \tau \right] dt$$

$$= \int_0^{\infty} f(\tau) \left[\int_{\tau}^{\infty} e^{-pt} \, dt \right] d\tau,$$

by changing order of integration

$$= \int_0^\infty f(\tau) . \frac{1}{p} e^{-p\tau} d\tau$$

$$= \frac{1}{p} \bar{f}(p). \tag{3.16}$$

An alternative proof.

Let $\int_0^t f(\tau) d\tau = g(t)$. Then, $g'(t) = f(t)$ and $g(0) = 0$.

Now, $L\left[\int_0^t f(\tau) d\tau\right] . = L[g(t)]$.

But, by the theorem 3.9 of article 3.10, we know that

$$L[g'(t)] = p L[g(t)] - g(0) = p L[g(t)].$$

Therefore, $L[g(t)] = \frac{1}{p} L[g'(t)]$.

$$= \frac{1}{p} L[f(t)] = \frac{1}{p} \bar{f}(p)$$

Hence, $L\left[\int_0^t f(\tau) d\tau\right] = \frac{1}{p} \bar{f}(p).$

3.12   Laplace Transform of $t^n f(t)$

Theorem 3.12. If Laplace transform of a piecewise continuous function $f(t)$ is $\bar{f}(p)$, then Laplace transform of $t^n f(t)$ is $(-1)^n \frac{d^n \bar{f}(p)}{dp^n}$, when n is a positive integer.

Proof.   Before proving the above result, we may restate it as the derivative of the Laplace transform of a given function satisfying requisite conditions for existence of the necessary results.

Given that, $\bar{f}(p) = \int_0^\infty e^{-pt} f(t) dt$

Differentiating both sides of the above equation with respect to p, we get

$$\frac{d\,\bar{f}(p)}{dp} = \frac{d}{dp} \int_0^\infty e^{-pt} f(t)dt$$

$$= -\int_0^\infty t\, e^{-pt} f(t)dt\;,$$

valid for assumed nature of the function $f(t)$

$$= -L\,[\,t\,f(t)\,]$$

Thus, $L\,[tf(t)] = (-1)^1\,\dfrac{d}{dp}\,\bar{f}(p)\;,$ \hfill (3.17)

which is the case of the theorem, when $n = 1$.

To prove the general case for any positive integral value of n, let us assume that for $n = k$

$$L\,[\,t^n\,f(t)] = (-1)^n\,\frac{d^n}{dp^n}\,\bar{f}(p) \hspace{2cm} (3.18)$$

Hence, we have

$$L\left[\,t^k\,f(t)\right] = (-1)^k\,\frac{d^k}{dp^k}\,\bar{f}(p)$$

implying, $\displaystyle \int_0^\infty e^{-pt}\{t^k f(t)\}\,dt = (-1)^k\,\frac{d^k}{dp^k}\bar{f}(p)$ \hfill (3.19)

Differentiating both sides of (3.19) with respect to p once, we get

$$\int_0^\infty e^{-pt}\,\{t^{k+1} f(t)\}\,dt = (-1)^{k+1}\,\frac{d^{k+1}}{dp^{k+1}}\,\bar{f}(p)$$

implying, $\displaystyle L\left[t^{k+1} f(t)\right] = (-1)^{k+1}\,\frac{d^{k+1}}{dp^{k+1}}\,\bar{f}(p)$ \hfill (3.20)

Equation (3.20) shows that the result of the theorem is true for $n = k+1$, if it is true for $n = k$. But it is seen in equation (3.17) that it is true for $n = 1$. Hence, it is true for $n = 1 + 1 = 2, n = 2 + 1 = 3, \cdots$. Therefore, by the method of mathematical induction the required general result of the theorem is true for all positive integral value of n.

This result is sometimes known as the derivative of the transformed function.

3.13   Laplace Transform of $\frac{f(t)}{t}$

Theorem 3.13. If Laplace transform of a piecewise continuous function $f(t)$ be $\bar{f}(p)$, then Laplace transform of $f(t)/t$ is $\int_p^\infty \bar{f}(p)dp$.

Proof. Given that

$$\int_0^\infty f(t)\, e^{-pt}\, dt = \bar{f}(p)$$

Integrating both sides of the above equation with respect to p from p to ∞, we get

$$\int_p^\infty \bar{f}(p_1)dp_1 = \int_p^\infty \left[\int_0^\infty e^{-p_1 t} f(t)\, dt \right] dp_1,$$

$$= \int_0^\infty f(t) \left\{ \int_p^\infty e^{-p_1 t}\, dp_1 \right\} dt, \text{ by changing order of integration.}$$

$$= \int_0^\infty \frac{f(t)}{t}\, e^{-pt}\, dt$$

$$= L\,[\,f(t)/t\,]. \tag{3.21}$$

Thus, the result of the theorem is proved to be true. This result is sometimes known as the integral of Laplace transformed function.

3.14   Laplace Transform of a periodic function.

Theorem 3.14. Let $f(t)$ be a periodic function of period τ, so that $f(t + n\tau) = f(t)$, for $n = 1, 2, 3 \cdots$. If $f(t)$ is a piecewise continuous function for $t > 0$, then

$$L\,[\,f(t)\,] = \frac{1}{1 - e^{-p\tau}} \int_0^\tau e^{-pt}\, f(t)\, dt$$

Proof.

$$L\,[\,f(t)\,] = \int_0^\infty e^{-pt}\, f(t)\, dt$$

$$= \int_0^\tau e^{-pt}\, f(t)dt + \int_\tau^{2\tau} e^{-pt}\, f(t) + \int_{2\tau}^{3\tau} e^{-pt}\, f(t)\, dt \mid$$

$$= \sum_{n=0}^\infty \int_{n\tau}^{(n+1)\tau} e^{-pt}\, f(t)\, dt$$

$$= \sum_{n=0}^\infty \int_0^\tau e^{-p(x+n\tau)}\, f(x + n\tau)dx, \text{ putting } t = x + n\tau$$

$$= \sum_{n=0}^\infty e^{-pn\tau} \int_0^\tau e^{-px}\, f(x)\, dx$$

$$= \left[\int_0^\tau e^{-px} f(x)dx \right] \sum_{n=0}^\infty e^{-pn\tau}$$

$$= \int_0^\tau e^{-px} f(x)dx \cdot \frac{1}{1 - e^{-p\tau}} \quad , \text{ since } p > 0, e^{-p\tau} < 1.$$

$$= \frac{1}{1 - e^{-p\tau}} \int_0^\tau e^{-pt} f(t)dt, \text{ changing } x \text{ to } t. \tag{3.22}$$

3.15　The initial-value theorem and the final-value theorem of Laplace Transform.

Theorem 3.15.

Let $f(t)$ be a continuous function for all $t \geqslant 0$ and is of exponential order for large t and $f'(t)$ be a piecewise continuous function of t. Then,

$$\lim_{t \to 0} f(t) = \lim_{p \to \infty} p\, L\, [f(t); t \to p]$$

This result is called the **initial-value theorem.**

Proof. We know by theorem 3.9 of article 3.10 that

$$L\,[f'(t)\,] = \int_0^\infty e^{-pt} f'(t)\, dt = p\, L\,[f(t)] - f(0)$$

Taking $p \to \infty$, the above equation gives

$$\lim_{p \to \infty} \int_0^\infty e^{-pt} f'(t)dt = \int_0^\infty \lim_{p \to \infty} [e^{-pt} f'(t)]\, dt = \lim_{p \to \infty} [p\, L\, (f(t) - f(0)]$$

Since $f'(t)$ is sectionally continuous and of exponential order, we have from above that

$$\lim_{p \to \infty} [pL\{f(t)\} - f(0)] = 0$$

which implies　　　　$$f(0) = \lim_{p \to \infty} p\,[L\{f(t)\}]$$

Thus,　　　　$$\lim_{t \to 0} f(t) = \lim_{p \to \infty} p\,[L\{\, f(t)\}] \tag{3.23}$$

Theorem 3.16. Let $f(t)$ be a continuous function for all $t \geqslant 0$ and is of exponential order and $f'(t)$ be a piecewise continuous function of t. Then the result of the **final-value theorem** states that

$$\lim_{t \to \infty} f(t) = \lim_{p \to 0} p\, L\, [\, f(t)\,]$$

Proof. As in theorem 3.15 above we have

$$\lim_{p \to 0} \int_0^\infty e^{-pt} f'(t) \, dt = \lim_{p \to 0} \left[pL\{f(t)\} - f(0) \right]$$

This result gives

$$\int_0^\infty \left[\lim_{p \to 0} e^{-pt} \right] f'(t) \, dt = \lim_{p \to 0} \; p \, L \left[f(t) \right] - f(0)$$

Or $\quad \displaystyle\int_0^\infty f'(t) \, dt = \left[\, f(t) \, \right]_0^\infty = \lim_{p \to 0} \; p \, L \left[f(t) \right] - f(0)$

Or $\quad \displaystyle\lim_{t \to \infty} f(t) - f(0) = \lim_{p \to 0} \; p \, L \; \left[\, f(t) \, \right] - f(0)$

Or $\quad \displaystyle\lim_{t \to \infty} f(t) = \lim_{p \to 0} \; p \, L \; \left[\, f(t) \, \right] \qquad\qquad (3.24)$

Thus the result of the final value theorem of Laplace transform is proved.

3.16 Examples

Example 3.16 Find the Laplace transform of $f(t) = \sin at - at \cos at$

Solution. By linearity property of Laplace transform, we have

$$L \left[\sin at - at \cos at \right] = L \left[\sin at \right] - L \left[at \cos at \right]$$

$$= \frac{a}{p^2 + a^2} + a \frac{d}{dp} \, L \left[\cos at \right], \qquad \text{by article 2.12}$$

$$= \frac{a}{p^2 + a^2} + a \frac{d}{dp} \left[\frac{p}{p^2 + a^2} \right]$$

$$= \frac{a}{p^2 + a^2} + a \left\{ \frac{1}{p^2 + a^2} - \frac{2p^2}{(p^2 + a^2)^2} \right\}$$

$$= \frac{2a}{(p^2 + a^2)^2} \left[\, a^2 \, \right] = \frac{2a^3}{(p^2 + a^2)^2}$$

Example 3.17. Evaluate $L \left[\frac{sh \; t}{t} \right]$.

Solution. From article 3.13 that $L \left[\frac{f(t)}{t} \right] = \int_p^\infty \bar{f}(p) \, dp$, where $\bar{f}(p) = L \left[\, f(t) \right]$. Now choosing $f(t) = sh \; t$, we have

$$L \left[\frac{sh \; t}{t} \right] = \int_p^\infty L \left[sh t \; ; \; t \to p \right] dp$$

$$= \int_p^\infty \frac{1}{p^2 - 1} \, dp = \frac{1}{2} \, \log \frac{p - 1}{p + 1}$$

Example 3.18. Evaluate $L[\sin at/t]$. Does $L[\cos at/t]$ exist ?

Solution. As in the above example 2, we have

$$L[\sin at/t] = \int_p^\infty L[\sin at; t \to p] dp$$

$$= \int_p^\infty \frac{a}{p^2 + a^2} dp = \left[\tan^{-1} \frac{p}{a}\right]_p^\infty = \cot^{-1} \frac{p}{a}$$

Thus, $\int_0^\infty e^{-pt} \frac{\sin at}{t} = \cot^{-1}(p/a)$.

Putting $p = 0$ and $a = 1$ we have,

$$\int_0^\infty \frac{\sin t}{t} dt = \lim_{p \to 0} \left(\frac{\pi}{2} - \tan^{-1} p\right) = \frac{\pi}{2}, \text{ an important result}$$

Again, $L[\cos at/t] = \int_p^\infty \frac{p}{p^2 + a^2} dp$

$$= \frac{1}{2} [\log(p^2 + a^2)]_p^\infty$$

But, $\lim_{p \to \infty} \log(p^2 + a^2)$ is unbounded and therefore $L\left[\frac{\cos at}{t}\right]$ does not exist.

Example 3.19. Evaluate $L\left[t^{\frac{1}{n}}\right]$ and then evaluate $L[\sin \sqrt{t}]$ and finally evaluate $L\left[\cos \sqrt{t} / \text{ i.e } \frac{\cos \sqrt{t}}{\sqrt{t}} \sqrt{t}\right]$.

Solution. From definition, we have

$$L\left[t^{\frac{1}{n}}\right] = \int_0^\infty e^{-pt}. t^{\frac{1}{n}} dt = \frac{\Gamma\left(\frac{1}{n} + 1\right)}{p^{\frac{1}{n}+1}}$$

Again, since $\sin \sqrt{t} = t^{\frac{1}{2}} - \frac{t^{\frac{3}{2}}}{3!} + \frac{t^{\frac{5}{2}}}{5!} - \frac{t^{\frac{7}{2}}}{7!} \cdots\cdots,$

we have

$$L\left[\sin \sqrt{t}\right] = \frac{\Gamma(\frac{3}{2})}{p^{\frac{3}{2}}} - \frac{1}{3!} \frac{\Gamma(\frac{5}{2})}{p^{5/2}} + \frac{1}{5!} \frac{\Gamma(\frac{7}{2})}{p^{7/2}} \cdots\cdots$$

$$= \frac{\sqrt{\pi}}{2p^{\frac{3}{2}}} \left[1 - \frac{1}{4p} + \frac{1}{2!}\left(\frac{1}{4p}\right)^2 - \frac{1}{3!}\left(\frac{1}{4p}\right)^3 + \cdots\cdots\right]$$

$$= \frac{\sqrt{\pi}}{2p^{\frac{3}{2}}} e^{-\frac{1}{4p}}.$$

Now considering $f(t) = \sin \sqrt{t}$, we get

$$f'(t) = \frac{\cos \sqrt{t}}{2\sqrt{t}}$$

and also $L\left[f'(t)\right] = p\,L\left[f(t)\right] - f(0)$ implying

$$L\left[\frac{\cos\sqrt{t}}{2\sqrt{t}}\right] = p\,\frac{\sqrt{\pi}}{2p^{\frac{3}{2}}}\,e^{-\frac{1}{4p}}$$

Thus, $L\left[\dfrac{\cos\sqrt{t}}{\sqrt{t}}\right] = \sqrt{\dfrac{\pi}{p}}\,e^{-\frac{1}{4p}}$

Example 3.20. Let $f(t)$ be a periodic function of period 4, where

$$f(t) = \begin{cases} 3t & , \quad 0 < t < 2 \\ 6 & , \quad 2 < t < 4 \end{cases}$$

Evaluate. $L[f(t)]$

Solution. By the result of article 3.14 we have here

$$L[f(t)\;;\;t \to p] = \frac{1}{1 - e^{-4p}}\int_0^4 f(t)\cdot e^{-pt}dt$$

$$= \frac{1}{1 - e^{-4p}}\left[\int_0^2 3t\,e^{-pt}dt + \int_2^4 6\,e^{-pt}dt\right]$$

$$= \frac{1}{1 - e^{-4p}}\left[\frac{3(1 - e^{-2p} - 2p\,e^{-4p^2})}{p^2}\right]\;,\;\text{after simplification.}$$

Example 3.21. Evaluate $L[\cos\,at\;;\;t \to p]$ and hence deduce $L[\sin at]$.

Solution. From the definition

$$L[\cos at] = \int_0^\infty e^{-pt}\cos at\,dt = \frac{p}{p^2 + a^2}\;,\;\text{by actual integration.}$$

Again, $\displaystyle\int_0^t \cos at\,dt = \frac{\sin at}{a}\,.$

Hence, $L\left[\dfrac{\sin at}{a}\right] = L\left[\displaystyle\int_0^t \cos at\,dt\right]$

$$= \frac{1}{p}\,L[\cos at] = \frac{1}{p}\cdot\frac{p}{p^2 + a^2}$$

This result implies,

$$L[\sin at] = \frac{a}{p^2 + a^2}\,.$$

Example 3.22. Verify the result of (a) initial-value theorem and
(b) final-value theorem for the function $f(t) = e^{-at}\cos bt$, where $a > 0$.

Solution. (a) It is seen that $f(t)$ satisfy the necessary condition for application of these theorems. We have now

$$\lim_{t \to 0} f(t) = \lim_{t \to 0} e^{-at} \cos bt = 1$$

Also, $L[f(t)] = \dfrac{(p+a)}{(p+a)^2 + b^2}$. So, $pL[f(t)] = \dfrac{p(p+a)}{(p+a)^2 + b^2}$

Therefore, $\lim_{p \to \infty} pL\,[\,f(t)\,;\,t \to p\,] = \lim_{p \to \infty} \dfrac{p(p+a)}{(p+a)^2 + b^2} = 1$

Thus, the initial value theorem is verified.

(b) Again, $\lim_{t \to \infty} f(t) = \lim_{t \to \infty} e^{-at} \cos bt = 0$

  Also, $\lim_{p \to 0} pL\,[\,f(t)\,;\,t \to p\,]$

  $= \lim_{p \to 0} \dfrac{p(p+a)}{(p+a)^2 + b^2} = 0$

Therefore, the final-value theorem of the above function is verified.

Example 3.23. Is it possible to use the initial-value or the final-value theorem to evaluate $\lim_{t \to 0} f(t)$ or to evaluate $\lim_{t \to \infty} f(t)$, when $f(t) = e^t \sin t$? Give reason for your answer. Also comment on your answer.

Solution.

$$L[f(t)] = L[e^t \sin t] = \frac{1}{(p-1)^2 + 1}$$

Now, $\lim_{t \to 0} f(t) = \lim_{t \to 0} e^t \sin t = 0$

  $\lim_{p \to \infty} pL[f(t)] = \lim_{p \to \infty} \dfrac{p}{(p-1)^2 + 1} = 0$

For this function, IVT is applicable.

Again, $\lim_{t \to \infty} f(t) = \lim_{t \to \infty} e^t \sin t =$ does not exist

Also, $\lim_{p \to 0} pL\,[f(t)] = \lim_{p \to 0} \dfrac{p}{(p-1)^2 + 1} = 0$

This shows that FVT is not applicable here.

   The reason for it is that $f(t)$ does not satisfy the condition of exponential order of $f(t)$ for large t.

   The above two results show that the conditions in the theorems are though sufficient but not necessary.

Example 3.24. Verify the result $L[f''(t)] = p^2 \bar{f}(p) - pf(0) - f'(0)$ for the case $f(t) = e^t \sin t$.

Solution. We have $f'(t) = e^t(\sin t + \cos t)$, $f''(t) = 2e^t \cos t$

Now, $\qquad L[f''(t)] = L[2e^t \cos t] = \dfrac{2(p-1)}{(p-1)^2 + 1}$

Also, $\qquad \bar{f}(p) = L[f(t)] = \dfrac{1}{(p-1)^2 + 1}$, $\quad f(0) = 0$, $\quad f'(0) = 1$

Therefore, $p^2 \bar{f}(p) - p\, f(0) - f'(0)$

$$= \frac{p^2}{(p-1)^2 + 1} - p \times 0 - 1 = \frac{p^2 - (p-1)^2 - 1}{(p-1)^2 + 1}$$

$$= \frac{2(p-1)}{(p-1)^2 + 1} \ . \ \text{Thus, the result is verified.}$$

Example 3.25. If $y(t) = \frac{1}{2k}[e^{kt} \int_0^t e^{-ku} du - e^{-kt} \int_0^t e^{ku} du]$, and noting $y''(t) - k^2 y(t) = 1$, verify that $\bar{y}(p) = \frac{1}{p(p^2-k^2)}$ when $k \neq 0$, after using Laplace transform in article 2.10.

Solution. We have,

$$L[y''(t) - k^2 y(t)] = L[y''(t)] - k^2 L[y(t)]$$
$$= p^2 \bar{y}(p) - py(0) - y'(0) - k^2 \bar{y}(p)$$
$$= (p^2 - k^2)\, \bar{y}(p) \equiv L[1] = \frac{1}{p}$$

$$\Rightarrow \bar{y}(p) = \frac{1}{p(p^2 - k^2)}$$

Also, $\bar{y}(p) = L\,[y(t)]$

$$= \frac{1}{2k^2} \left[-\frac{2}{p} + \frac{2p}{p^2 - k^2} \right]$$

$$= \frac{1}{p\,(p^2 - k^2)}$$

Thus, value of $y(p)$ is verified.

3.17 Laplace Transform of some special functions.

I. The Heaviside function
We have already introduced the unit step function $H(t)$ defined by

$$H(t) = \begin{cases} 0 \ , \ t < 0 \\ 1 \ , \ t > 0, \end{cases}$$

This is usually known as "Heaviside unit function". It may further be noted that any step function $f(t)$ can always be expressed as a combination of different Heaviside functions. For example, let a step function be defined by the relation

$$f(t) = \begin{cases} g(t) , & 0 < t < a \\ k & , & a < t < b \\ 0 & , & t > b \end{cases}$$

Then we can express it as

$$f(t) = g(t)H(a-t) + k[H(t-a) - H(t-b)]$$

Further, $$L[H(t)] = \int_0^\infty H(t)e^{-pt}dt$$

$$= \int_0^\infty 1 \cdot e^{-pt} = \frac{1}{p}$$

implying, $$L^{-1}\left[\frac{1}{p}\right] = H(t)$$

II. The Dirac delta function.

A practical problem that arises in application in finding the Laplace inversion of unity or $H(t)$. Also, similar situation arises in handling unit impulse function. Dirac introduced it as, $\delta(t)$ which acts for a very short time duration τ with a finite impulse 1 satisfying the relations

$$\delta(t) = 0 , \quad \text{for } t \neq 0 \tag{3.25}$$

and $$\int_{-\infty}^\infty \delta(t)dt = \int_0^\tau \delta(t)dt = 1 \tag{3.26}$$

In mathematical analysis, such function is unlike ordinary one and introduction of such a function sometimes leads to inconsistency in analysis. For this reason, Dirac called it as an "improper function" . It may be used only when no inconsistency do follow afterwards in analysis. Thus, entirely formally such function may be used in classical analysis without searching for rigour in treatment. The equations (3.25) and (3.26) above do not depict any clear picture for $\delta(t)$. Since, the precise variation of $\delta(t)$ with t in the interval is not an important issue but its effect is the point of study, we can use it properly with assumption that it does not have any unnecessary violent oscillations in the neighbourhood of $t = 0$.

For example, if we define a step function $\delta_\in(t)$ by

$$\delta_\in(t) \quad = \quad \begin{cases} 0 & , \quad t < 0 \\ \frac{1}{\in} & , \quad 0 < t < \in \\ 0 & , \quad t > \in \end{cases} \tag{3.27}$$

then since $\displaystyle\int_{-\infty}^{+\infty} \delta_\in(t)\, dt = \int_0^\in \frac{1}{\in}\, dt = 1 ,$ (3.28)

it follows that $\displaystyle\lim_{\in \to 0} \delta_\in(t) = \delta(t)$ (3.29)

as defined in (3.25) and (3.26)

If $f(t)$ be a continuous function in the neighbourhood of $t = 0$, then

$$\int_{-\infty}^{+\infty} \delta_\in(t) f(t) dt = \int_0^\in \delta_\in(t) f(t) dt = \frac{1}{\in} \int_0^\in f(t) dt$$

$$= \frac{f(\theta \in)}{\in} \int_0^\in dt = f(\theta \in) , \quad 0 < \theta < 1 , \quad \text{by mean value theorem.}$$

Thus, $\displaystyle\int_{-\infty}^{+\infty} f(t)\delta_\in(t)\, dt = f(\theta \in) , \quad 0 < \theta < 1$

Now, letting $\in \to 0$ we get

$$\int_{-\infty}^{+\infty} f(t)\, \delta(t)\, dt = f(0) \tag{3.30}$$

In general, putting

$$\int_{-\infty}^{+\infty} f(\xi)\, \delta(\xi - a)\, d\xi$$

$$= \int_{-\infty}^{+\infty} f(t + a)\, \delta(t)\, dt , \quad \text{putting } \xi - a = t$$

$$= f(0 + a) = f(a) \tag{3.31}$$

This is an important result which can be used formally in analysis whenever no confusion do arise.

The Laplace transform of $\delta(t)$ is then given by definition that

$$L[\delta(t) \ ; \ t \to p] \quad = \quad \int_0^\infty e^{-pt} \delta(t) dt$$

$$= \quad e^{-p.0} = 1 = H(t) , \quad t > 0 \tag{3.32}$$

Thus, $\qquad\qquad L[\delta(t)] = H(t)$

and in turn $\qquad\qquad L^{-1}[H(t)] = \delta(t)$ (3.33)

Hence, (3.32) and (3.33) are the required result obtained as proposed in the beginning of this analysis.

III. Laplace Transform of the sine and the cosine integrals.

The **sine integral** **Si(t)** is denoted and defined by

$$Si(t) = \int_0^t \frac{\sin u}{u}\, du$$

Now, $L\,[\,Si(t)\,;\,t \to p\,] = \int_0^\infty e^{-pt} \left[\int_0^t \frac{\sin u}{u}\, du \right] dt$

$$= \frac{1}{p} \int_0^\infty \frac{\sin t}{t}\, e^{-pt} dt$$

$$= \frac{1}{p} \left[\int_p^\infty L\,[\,\sin t\,]\, dp \right] ,\ \text{by 3.13}$$

$$= \frac{1}{p} \left[\int_p^\infty \frac{1}{p^2+1}\, dp \right] = \frac{1}{p} \left[\frac{\pi}{2} - \tan^{-1} p \right]$$

$$= \frac{1}{p} \cot^{-1} p. \tag{3.34}$$

Again, the **cosine integral** $Ci(t)$ is denoted by

$$Ci(t) = -\int_t^\infty \frac{\cos u}{u}\, du$$

Therefore, $L\,[\,Ci(t)\,;\,t \to p\,]$

$$= -\int_0^\infty e^{-pt} \left[\int_t^\infty \frac{\cos u}{u}\, du \right] dt$$

$$= -\int_0^\infty \frac{\cos u}{u} \left[\int_0^u e^{-pt} dt \right] du ,$$

$$\text{by changing the order of integration}$$

$$= -\frac{1}{p} \int_0^\infty \frac{\cos u}{u} (1 - e^{-pu})\, du \tag{3.35}$$

Again, $\bar{f}(p_1) = \int_0^\infty f(u)\, e^{-p_1 u} dp_1$

Integrating this result with respect to p_1 between the limits 0 to p, we get

$$\int_0^p \bar{f}(p_1)\, dp_1 = \int_0^p \left[\int_0^\infty f(u)\, e^{-p_1 u}\, du \right] dp_1$$

Therefore,

$$\int_0^p \bar{f}(p_1) dp_1 = \int_0^\infty f(u) \left[\int_0^p e^{-p_1 u} dp_1 \right] du \ ,$$

by changing the order of integration.

$$= \int_0^\infty \frac{f(u)}{u} (1 - e^{-pu}) du$$

Assuming, $f(u) = \cos u$, we have $\bar{f}(p) = L[f(u) \ ; \ u \to p] = \dfrac{p}{p^2 + 1}$

and hence,

$$\int_0^p \frac{p_1}{p_1^2 + 1} \ dp_1 = \int_0^\infty \frac{\cos u}{u} \ (1 - e^{-pu}) \ du \qquad (3.36)$$

Using (3.36) in (3.35) above, we get

$$L[Ci \ (t) \ ; \ t \to p] = -\frac{1}{p} \int_0^p \frac{p_1 \ dp_1}{p_1^2 + 1}$$

$$= -\frac{1}{2p} \ \log \ (1 + p^2)$$

An alternative method.

Since, $Ci \ (t) = - \displaystyle\int_t^\infty \frac{\cos u}{u} \ du$

$t \ Ci' \ (t) = \cos t$

So, $L[t \ Ci' \ (t)] = \dfrac{p}{p^2 + 1} \Rightarrow -\dfrac{d}{dp}[L\{C_i' \ (t)\}] = \dfrac{p}{p^2 + 1}$

or, $\dfrac{d}{dp} \ [p \ L\{Ci \ (t)\} - Ci \ (0)] = -\dfrac{p}{p^2 + 1}$

or, $\dfrac{d}{dp}[p \ L\{Ci \ (t)\}] = -\dfrac{p}{p^2 + 1}$

Integrating both sides with respect to p, we get

$$p \ L[Ci \ (t)] = -\frac{1}{2} \ \log \ (p^2 + 1) + c \text{ where } c \text{ is a constant}$$

Now, by the final-value theorem

$$\lim_{p \to 0} p \ L[Ci \ (t)] = \lim_{t \to \infty} Ci(t)$$

we get, $\displaystyle\lim_{p \to 0} \left[-\frac{1}{2} \ \log \ (p^2 + 1) + c \right] = - \lim_{t \to \infty} \int_t^\infty \frac{\cos u}{u} \ du = 0$

Thus $c = 0 \Rightarrow p \ L[Ci \ (t)] = -\dfrac{1}{2} \ \log \ (p^2 + 1)$

$$\Rightarrow L[Ci \ (t)] = -\frac{1}{2p} (p^2 + 1) \ .$$

IV. Laplace Transform of the Exponential integral.

The exponential integral is defined by

$$Ei\ (t) = \int_t^\infty \frac{e^{-u}}{u}\ du$$

Then, $Ei'\ (t) = -\dfrac{e^{-t}}{t} \Rightarrow t\ E_i'\ (t) = -e^{-t}$

Thus as before,

$$L[t\ Ei'\ (t)] = -\frac{1}{p+1} \Rightarrow -\frac{d}{dp}\ [\ L\{Ei'\ (t)\}\ ] = -\frac{1}{p+1}$$

Or, $\dfrac{d}{dp}\ [\ p\ L\{Ei\ (t)\} - Ei\ (0)\ ] = \dfrac{1}{p+1}$

Or, $\dfrac{d}{dp}\ [\ p\ L\{Ei\ (t)\}\ ] = \dfrac{1}{p+1}$

Integrating both sides with respect to p , we get

$$p\ L[Ei\ (t)] = \log\ (p+1) + c\ ,\ \text{a constant}$$

Now, by final-value theorem, we get

$$\lim_{p \to 0}\ [\ \log\ (p+1) + c\ ] = \lim_{t \to \infty} Ei(t) = 0 \Rightarrow c = 0$$

Thus, $L[Ei\ (t)] = \dfrac{1}{p}\ \log\ (p+1)$ (3.37)

V. Laplace Transform of the Error function.

The **error function** $Erf(t)$ is defined by

$$Erf(t) = \frac{2}{\sqrt{\pi}} \int_0^t e^{-x^2} dx$$

and the **error complementary function** $Erfc(t)$ is also defined by

$$Erfc(t) = \frac{2}{\sqrt{\pi}} \int_t^\infty e^{-x^2} dx = 1 - \frac{2}{\sqrt{\pi}} \int_0^t e^{-x^2} dx \qquad (3.38)$$
$$= 1 - Erf(t),$$

since, $\dfrac{2}{\sqrt{\pi}} \displaystyle\int_0^\infty e^{-x^2} dx = 1$, by use of Gamma function.

Thus, by evaluating Laplace transform of $Erf(t)$, the same for $Erfc(t)$ can easily be evaluated.

To make the evaluation steps relatively simpler, we evaluate Laplace transform of $Erf(\sqrt{t})$ instead of $Erf(t)$ in the following steps. We know that

$$Erf(\sqrt{t}) = \frac{2}{\sqrt{\pi}} \int_0^{\sqrt{t}} e^{-x^2} dx$$

$$= \frac{2}{\sqrt{\pi}} \int_0^{\sqrt{t}} \left[1 - x^2 + \frac{x^4}{2!} - \frac{x^6}{3!} + \cdots \right] dx,$$

$$= \frac{2}{\sqrt{\pi}} \left[t^{1/2} - \frac{t^{3/2}}{3} + \frac{t^{5/2}}{5.2!} - \frac{t^{7/2}}{7.3!} + \cdots \right]$$

Thus,

$$L[Erf(\sqrt{t})] = \frac{2}{\sqrt{\pi}} \left[\frac{\Gamma(3/2)}{p^{3/2}} - \frac{\Gamma(5/2)}{3 \cdot p^{5/2}} + \frac{\Gamma(7/2)}{5.2! \cdot p^{7/2}} - \frac{\Gamma(9/2)}{7.3! p^{9/2}} + \cdots \right]$$

$$= \frac{1}{p^{3/2}} \left[1 - \frac{1}{2} \cdot \frac{1}{p} + \frac{1.3}{2.4} \frac{1}{p^2} - \frac{1.3.5}{2.4.6} \frac{1}{p^3} + \cdots \right]$$

$$= \frac{1}{p^{3/2}} \left(1 + \frac{1}{p} \right)^{-\frac{1}{2}} = \frac{1}{p(p+1)^{1/2}} \qquad (3.39)$$

Therefore,

$$L[Erfc(\sqrt{t})] = L[1 - Erf(\sqrt{t})]$$

$$= \frac{1}{p} - \frac{1}{p(p+1)^{1/2}} = \frac{(p+1)^{1/2} - 1}{p(p+1)^{1/2}} \qquad (3.40)$$

VI. Laplace Transforms of Bessel functions.

The **Bessel function** of first kind of order ν, $J_\nu(t)$ is defined by

$$J_\nu(t) = \sum_{r=0}^{\infty} \frac{(-1)^r \left(\frac{1}{2} t \right)^{\nu+2r}}{r! \, \Gamma(\nu+r+1)} \, .$$

Its Laplace transform is given by the equation

$$L[t^\nu J_\nu(t) \, ; \, t \to p] = \sum_{r=0}^{\infty} \frac{(-1)^\nu 2^{-\nu-2r}}{r! \, \Gamma(r+\nu+1)} L[t^{2\nu+2r} \, ; \, t \to p] \qquad (3.41)$$

The duplication formula for the Gamma function is

$$2^{2z-1} \, \Gamma(z) \, \Gamma\left(z + \frac{1}{2} \right) = \sqrt{\pi} \, \Gamma(2z) \qquad (3.42)$$

Using (3.42) we have the relation

$$\frac{1}{\Gamma(r+\nu+1)} L[t^{2\nu+2r}; t \to p] = \frac{\Gamma(2\nu+2r+1)}{\Gamma(\nu+r+1)} p^{-2\nu-2r-1}$$

$$= \pi^{-\frac{1}{2}} \, 2^{2\nu+2r} \, \Gamma\left(\nu + r + \frac{1}{2}\right) p^{-2\nu-2r-1},$$

$$\left(\operatorname{Re} p > 0 \;,\; \nu > -\frac{1}{2}\right) \qquad (3.43)$$

Therefore, results in eqns.(3.41) and (3.43) give

$$L[t^\nu \, J_\nu(t) \;;\; t \to p] \; = 2^\nu \, \pi^{-\frac{1}{2}} \, p^{-2\nu-1} \sum_{r=0}^{\infty} \frac{\Gamma\left(\nu + \frac{1}{2} + r\right)}{r!} \left(-p^{-2}\right)^r$$

If $|p| > 1$, the above series is convergent and it gives

$$L[t^\nu \, J_\nu(t) \;;\; t \to p] = \frac{2^\nu \, \Gamma\left(\nu + \frac{1}{2}\right)}{\Gamma\left(\frac{1}{2}\right)} (p^2 + 1)^{-\nu-\frac{1}{2}} \;,\; \operatorname{Re} p > 1 \qquad (3.44)$$

after using the Binomial theorem

$$(1 - x)^{-\alpha} = \sum_{r=0}^{\infty} \frac{\Gamma(\alpha + r)x^r}{r! \, \Gamma(\alpha)} \;,\; |x| < 1.$$

In particular, putting $\nu = 0$ in (3.44), we get

$$L[J_0(t) \;;\; t \to p] = (p^2 + 1)^{-\frac{1}{2}} \;,\; \operatorname{Re} p > 1$$

Letting $p \to 0$, it follows that $\int_0^\infty J_0(t)dt = 1$.

Again by the result in eqn. (3.44), we have

$$L\left[\, t^\nu \, J_\nu(at) \;;\; t \to p \, \right]$$
$$= \frac{(2a)^\nu \, \Gamma\left(\nu + \frac{1}{2}\right)}{\sqrt{\pi}(p^2 + a^2)^{\nu+\frac{1}{2}}} \;,\; \nu > -\frac{1}{2} \;,\; \operatorname{Re} p > a > 0$$

and $\qquad L[J_0(at) \;;\; t \to p] = (p^2 + a^2)^{-\frac{1}{2}} \;,\; \operatorname{Re} p > a > 0$

Also, $\qquad L[e^{-at} J_0(at) \;;\; t \to p] = \dfrac{1}{\sqrt{p^2 + 2ap + 2a^2}}$

Deduction 3.1.

If $\operatorname{Re} p > 1$, like eqn. (3.41) one can write

$$L\left[\, t^{\nu+1} \, J_\nu(at) \;;\; t \to p \, \right] = \sum_{r=0}^{\infty} \frac{(-1)^r \, 2^{-\nu-2r} \, \Gamma(2\nu + 2r + 2)}{r! \, \Gamma(\nu + r + 1) \, p^{2\nu+2r+2}}$$

At this stage, using the duplication formula of the Gamma function , it is found that

$$\frac{2^{-\nu-2r}\ \Gamma(2\nu+2r+2)}{\Gamma(\nu+r+1)} = 2^{\nu+1}\frac{1}{\sqrt{\pi}}\ \Gamma\left(\nu+r+\frac{3}{2}\right),$$

and so,

$$L\left[t^{\nu+1}J_\nu(t)\ ;\ t\to p\right] = 2^{\nu+1}\cdot\frac{1}{\sqrt{\pi}}\cdot\frac{1}{p^{2\nu+2}}\sum_{r=0}^{\infty}\left(-\frac{1}{p^2}\right)^\nu\frac{\Gamma\left(\nu+r+\frac{3}{2}\right)}{r!}$$

$$= 2^{\nu+1}\frac{1}{\sqrt{\pi}}\ p^{-2\nu-2}\left(1+p^{-2}\right)^{-\nu-\frac{3}{2}}\ \Gamma\left(\nu+\frac{3}{2}\right),$$

after using Binomial theorem. Thus,

$$L\left[t^{\nu+1}J_\nu(t)\ ;\ t\to p\right] = \frac{2^{\nu+1}p\ \Gamma\left(\nu+\frac{3}{2}\right)}{\sqrt{\pi}(p^2+1)^{\nu+\frac{3}{2}}}\ ,\quad \text{Re } p>1$$

Therefore, $L\left[t^{\nu+1}J_\nu(at)\ ;\ t\to p\right] = \dfrac{2^{\nu+1}\ a^\nu p\ \Gamma\left(\nu+\frac{3}{2}\right)}{\sqrt{\pi}(p^2+a^2)^{\nu+\frac{3}{2}}}\ ,\quad \text{Re } p>a>0$

$$(3.45)$$

In particular, if $\nu=0$,

$$L\left[tJ_0(at)\right] = \frac{p}{(p^2+a^2)^{\frac{3}{2}}}\ .$$

Deductions 3.2. By principle of analytic continuation we can extend the result of the definition of Bessel function of first kind for the permissible range of argument of the variable as

$$t^{\frac{\nu}{2}}\ J_\nu(2\sqrt{t}) = \sum_{r=0}^{\infty}\frac{(-1)^r t^{\nu+r}}{r!\ \Gamma(\nu+r+1)}$$

and hence

$$L\left[t^{\frac{\nu}{2}}\ J_\nu(2\sqrt{t})\ ;\ t\to p\right] = \sum_{r=0}^{\infty}\frac{(-1)^r}{r!\ \Gamma(r+\nu+1)}\ L\left[t^{\nu+r}\ ;\ t\to p\right]$$

$$= p^{-\nu-1}\sum_{r=0}^{\infty}\frac{\left(-\frac{1}{p}\right)^r}{r!}\ ,\quad \text{if Re } p>0$$

$$= p^{-\nu-1}\ e^{-\frac{1}{p}}\ ,\quad \text{Re } p>0$$

Also, if $a>0$, Re $p>0$. $\hspace{5cm}(3.46)$

$$L\left[t^{\frac{\nu}{2}}J_\nu(2\sqrt{at})\ ;\ t\to p\right] = a^{\frac{\nu}{2}}p^{-\nu-1}e^{-\frac{a}{p}} \hspace{2cm}(3.47)$$

and $L\left[J_0(2\sqrt{at})\ ;\ t\to p\right] = \dfrac{1}{p}e^{-\frac{a}{p}}\ .$

Deduction 3.3. Modified Bessel function of the first kind $I_\nu(t)$ of the variable t is given by

$$I_\nu(t) = \sum_{r=0}^{\infty} \frac{\left(\frac{1}{2} t\right)^{\nu+2r}}{r!\, \Gamma(\nu+r+1)}$$

Therefore, by the same procedure as above, we get

$$L\left[t^\nu\, I_\nu(t)\,;\, t \to p\right] = \pi^{-\frac{1}{2}}\, 2^\nu p^{-2\nu-1} \sum_{r=0}^{\infty} \frac{\Gamma\left(\nu+r+\frac{1}{2}\right)}{r!}\, p^{-2r}$$

This result implies that if $\operatorname{Re} p > 1$, $\nu > -\frac{1}{2}$,

$$L\left[t^\nu\, I_\nu(t)\,;\, t \to p\right] = \frac{1}{\sqrt{\pi}}\, 2^\nu\, \Gamma\left(\nu+\frac{1}{2}\right)(p^2-1)^{-\nu-\frac{1}{2}}$$

Also, $\quad L\left[t^\nu\, I_\nu(at)\,;\, t \to p\right] = \dfrac{(2a)^\nu\, \Gamma\left(\nu+\frac{1}{2}\right)}{\sqrt{\pi}(p^2-a^2)^{\nu+\frac{1}{2}}}$, for $\operatorname{Re} p > a > 0$

$$(3.48)$$

In a similar way one can prove that if $\nu > -\frac{1}{2}$

$$L\left[t^{\nu+1} I_\nu(at)\,;\, t \to p\right] = \frac{2^{\nu+1} a^\nu p\, \Gamma\left(\nu+\frac{3}{2}\right)}{\sqrt{\pi}(p^2-a^2)^{\nu+\frac{3}{2}}}\ ,\quad \text{for } \operatorname{Re} p > a > 0 \quad (3.49)$$

Again,

$$L\left[t^{\frac{\nu}{2}} I_\nu(2\sqrt{at})\,;\, t \to p\right] = a^{\frac{\nu}{2}} p^{-\nu-1} e^{\frac{a}{p}}\ ,\quad \text{for } \operatorname{Re} p > 0\ ,\ a > 0\ ,\ \nu > -\frac{1}{2}$$

$$(3.50)$$

by similar arguments made above.

The special cases for $\nu = 0$ are given by

$$L\left[I_0(at)\,;\, t \to p\right] = (p^2-a^2)^{-\frac{1}{2}}\ ,\ \operatorname{Re} p > a > 0 \qquad (3.51)$$

$$L\left[t\, I_0(at)\,;\, t \to p\right] = p(p^2-a^2)^{-\frac{1}{2}}\ ,\ \operatorname{Re} p > a > 0 \quad (3.52)$$

$$\text{and } \ L\left[I_0(2\sqrt{at})\,;\, t \to p\right] = p^{-1}\, e^{\frac{a}{p}}\ ,\ \operatorname{Re} p > 0\ ,\ a > 0 \qquad (3.53)$$

An important result due to Tricomi [cf Tricomi (1935)] can be stated as

$$L\left[t^{\frac{\nu}{2}} \int_0^{\infty} x^{-\frac{\nu}{2}}\, J_\nu\left(2\sqrt{xt}\right) f(x)\, dx\,;\, t \to p\right] = p^{-\nu-1}\, \bar{f}\left(\frac{1}{p}\right) \qquad (3.54)$$

We do not persue the derivation of this last result due to lack of scope of this book.

3.18　The Convolution of two functions.

Let $f(t)$ and $g(t)$ be two piecewise continuous functions and are of some exponential order for large t and for every $t \geqslant 0$. Then the convolution of these functions is denoted by $f * g(t)$ and is defined by

$$f * g(t) = \int_0^t f(u)g(t-u)du \tag{3.55}$$

The above relation in eqn. (3.55) is also known as faltung or resultant of $f(t)$ and $g(t)$. By definition,

$$
\begin{aligned}
f * g(t) &= \int_0^t f(u)g(t-u)\ du \\
&= \int_0^t f(t-\tau)\ g(\tau)\ d\tau \\
&= g * f(t)
\end{aligned}
\tag{3.56}
$$

Therefore, convolution of two functions satisfies the commutative law . It also satisfies the distributive law

$$f * [g + h](t) = f * g(t) + f * h(t) \tag{3.57}$$

and also the associative law

$$[f * (g * h)](t) = [(f * g) * h](t)\ , \tag{3.58}$$

The two results in eqns. (3.57) and (3.58) can be directly verified from the definition (3.55). For example,

$$
\begin{aligned}
[(f * g) * h](t) &= \int_0^t h(t-\tau)\ d\tau \left[\int_0^\tau f(\xi)\ g(\tau-\xi)\ d\xi \right] \\
&= \int_0^t f(\xi)d\xi \int_\zeta^t g(\tau-\xi)h(t-\tau)d\tau, \ \text{Changing the order of integration} \\
&= \int_0^t f(\xi)\ d\xi \int_0^{t-\xi} g(\eta)\ h(t-\xi-\eta)\ d\eta \\
&= f * (g * h)\ (t)
\end{aligned}
$$

We now turn our attention to evaluate Laplace transform of the convolution of two functions $f(t)$ and $g(t)$.

$$L[(f * g)(t)\ ;\ t \to p] = L \left[\int_0^t f(\tau)\ g(t-\tau)\ d\tau\ ;\ t \to p \right]$$

$$= \int_0^\infty e^{-pt} \left[\int_0^t f(\tau)\, g(t-\tau)\, d\tau \right] dt$$

$$= \int_0^\infty f(\tau) \left[\int_\tau^\infty e^{-pt}\, g(t-\tau)\, dt \right] d\tau$$

$$= \int_0^\infty f(\tau)\, e^{-p\tau} \left[\int_0^\infty e^{-p\eta}\, g(\eta)\, d\eta \right] d\tau$$

$$= \bar{f}(p)\, \bar{g}(p) , \tag{3.59}$$

where $\bar{f}(p)$ and $\bar{g}(p)$ are Laplace transforms of $f(t)$ and $g(t)$ respectively.

In particular, $\qquad L[f * f(t) \; ; \; t \to p] = \left[\bar{f}(p) \right]^2$

As an illustration, if $\quad f(t) = t^{a-1}, a > 0, \qquad g(t) = t^{b-1}, b > 0$

$$L\left[t^{a-1}; t \to p \right] = \frac{\Gamma(a)}{p^a} , \quad L\left[t^{b-1} \; ; \; t \to p \right] = \frac{\Gamma(b)}{p^b}$$

then $\qquad L\left[(t^{a-1} * t^{b-1}) \; ; \; t \to p \right] = \frac{\Gamma(a)\,\Gamma(b)}{p^{a+b}} \tag{3.60}$

Also, it is known that

$$L\left[t^{a+b-1} \; ; \; t \to p \right] = \frac{\Gamma(a+b)}{p^{a+b}} \tag{3.61}$$

Dividing the results of (3.60) by that of (3.61) one gets

$$\beta(a, b) = \frac{\Gamma(a)\,\Gamma(b)}{\Gamma(a+b)} = \frac{L\left[(t^{a-1} * t^{b-1}) \; ; \; t \to p \right]}{L\left[t^{a+b-1} \; ; \; t \to p \right]}$$

3.19   Applications

Example 3.26. By considering $Si(t)$, find its Laplace transform and hence prove that

$$\int_0^\infty \frac{\sin u}{u}\, du = \frac{\pi}{2}$$

Solution. We have, from III of article 3.17 that

$$L[Si(t) \; ; \; t \to p] = \int_0^\infty e^{-pt} \left[\int_0^t \frac{\sin u}{u}\, du \right] dt$$

$$= \frac{1}{p} \int_0^\infty \frac{\sin t}{t}\, e^{-pt}\, dt$$

$$= \frac{1}{p} \left[\int_p^\infty L\{\sin t \; ; \; t \to p\} dp \right]$$

$$= \frac{1}{p} \int_p^\infty \frac{1}{p^2 + 1} \, dp = \frac{1}{p} \left[\frac{\pi}{2} - \tan^{-1} p \right]$$

Thus,
$$\int_0^\infty \frac{\sin t}{t} \, e^{-pt} \, dt = \frac{\pi}{2} - \tan^{-1} p \qquad (3.62)$$

In eqn.(3.62) making $p \to 0$, we get

$$\int_0^\infty \frac{\sin t}{t} \, dt = \frac{\pi}{2} \, .$$

Example 3.27. Prove that $L \left[e^{-\frac{t^2}{4}} \, ; \, t \to p \right] = \sqrt{\pi} \, e^{p^2} \, Erfc \, (p)$.

Solution. We know that

$$L \left[\, Erfc\{g(t)\} \, ; \, t \to p \, \right] = L \left[\, 1 - Erf\{g(t)\} \, ; \, t \to p \, \right]$$
$$= \frac{1}{p} - L \left[Erf\{g(t)\} \, ; \, t \to p \right]$$

Now,
$$L \left[e^{-\frac{t^2}{4}} \, ; \, t \to p \right] = \int_0^\infty e^{-pt} \, e^{-\frac{t^2}{4}} \, dt$$
$$= \int_0^\infty e^{-\left\{ \left(\frac{t}{2} + p \right)^2 - p^2 \right\}} dt = e^{p^2} \int_0^\infty e^{-\left(\frac{t}{2} + p \right)^2} dt$$
$$= e^{p^2} \int_p^\infty e^{-u^2} \cdot 2 \, du \, , \quad \text{putting} \quad u = p + \frac{t}{2}$$
$$= 2 \, e^{p^2} \times Erfc(p) \times \frac{\sqrt{\pi}}{2} \, , \quad \text{by eqn. (3.38) of article 3.17 (iv)}$$
$$= \sqrt{\pi} \, e^{p^2} \, Erfc(p) \, .$$

Example 3.28. Find $L \left[\, Erf \, (t) \, ; \, t \to p \, \right]$ to show that its value is expressed as $\frac{1}{p} \, e^{\frac{p^2}{4}} \, Erfc \, \left(\frac{1}{2} \, p \right)$.

Solution. We have from definition

$$L \left[\, Erf(t) \, ; \, t \to p \, \right] = \int_0^\infty e^{-pt} \left[\frac{2}{\sqrt{\pi}} \int_0^t e^{-u^2} \, du \right] dt$$
$$= \frac{2}{\sqrt{\pi}} \int_0^\infty e^{-u^2} \left[\int_u^\infty e^{-pt} \, dt \right] du$$
$$= \frac{2}{p \sqrt{\pi}} \int_0^\infty e^{-u^2 - pu} \, du$$
$$= \frac{2}{p\sqrt{\pi}} \int_0^\infty e^{-\left(u + \frac{p}{2} \right)^2} \cdot e^{\frac{p^2}{4}} \, du$$
$$= \frac{2e^{p^2/4}}{p\sqrt{\pi}} \int_{p/2}^\infty e^{-v^2} \, dv$$

$$= \frac{2e^{p^2/4}}{p\sqrt{\pi}} \cdot Erfc\left(\frac{p}{2}\right) \cdot \frac{\sqrt{\pi}}{2}$$

$$= \frac{1}{p}\, e^{p^2/4}\, Erfc\left(\frac{1}{2}p\right)$$

Deduction.

$$L\,[\,Erf\,(bt)\,;\,t \to p\,] = \frac{1}{b}\, L\,\left[\,Erf\,(t);\,t \to \frac{p}{b}\,\right]$$

$$= \frac{b}{p} \cdot \frac{1}{b}\, e^{\frac{p^2}{4b^2}} \cdot Erfc\left(\frac{p}{2b}\right) \quad , \text{ by the above result.}$$

Example 3.29. If $f(t) = |\sin t|$, prove that $L\,[\,|\sin t|\,;\,t \to p\,]$ is equal to $\coth\,(p\frac{\pi}{2})/(1 + p^2)$.

Solution. Clearly, $f(t) = |\sin t|$ is a periodic function of principal period $\tau = \pi$. Hence

$$L\,[\,f(t)\,;\,t \to p\,] = L\,[\,|\sin t|\,;\,t \to p\,]$$

$$= \frac{1}{1 - e^{-p\pi}} \int_0^{\pi} e^{-pt}\,|\sin t|\,dt$$

$$= \frac{1}{1 - e^{-p\pi}} \int_0^{\pi} e^{-pt}\,\sin t\,dt = \frac{1}{1 - e^{-pn}} \cdot \frac{1 + e^{-p\pi}}{1 + p^2}$$

$$= \frac{1}{1 + p^2}\, \coth\left(\frac{p\pi}{2}\right)$$

Example 3.30. Find the Laplace transform of $f(t) = \int_0^{\infty} \cos\,tx^2\,dx$

Solution. We have, by definition

$$L\,\left[\int_0^{\infty} \cos\,tx^2 dx\right] = \int_0^{\infty} e^{-pt}\left[\int_0^{\infty} \cos\,tx^2\,dx\right] dt$$

$$= \int_0^{\infty}\left[\int_0^{\infty} \cos\,t\,x^2\,.e^{-pt}\,dt\right] dx\,,$$

after changing order of integration.

$$= \int_0^{\infty} \frac{p}{p^2 + x^4}\,dx$$

$$= \frac{1}{2\sqrt{p}} \int_0^{\frac{\pi}{2}} \sqrt{\cot\theta}\,d\theta, \text{ where } x^2 = p\,\tan\,\theta$$

$$= \frac{\pi}{2\sqrt{2p}}$$

Example 3.31. Find (i) $L[\,f(t)\,]$ and (ii) $L[\,f'(t)\,]$ where $f(t)$ is defined by

$$f(t) = \begin{cases} 2t &, \quad 0 \leqslant t < 1 \\ t &, \quad t > 1 \end{cases}$$

Solution.

(i) $L[f(t)\,;\,t \to p]$

$$= \int_0^1 2t \cdot e^{-pt}\,dt + \int_1^\infty t\,e^{-pt}\,dt$$

$$= \frac{2}{p^2} - \left(\frac{1}{p} + \frac{1}{p^2}\right)e^{-p}\,, \quad \text{after evaluation of integrals}$$

(ii)   Since $f(t) = \begin{cases} 2t, & 0 \leqslant t < 1 \\ t, & t > 1 \end{cases}$, we have $f'(t) = \begin{cases} 2, & 0 \leqslant t < 1 \\ 1, & t > 1 \end{cases}$

Therefore, $L[f'(t)\,;\,t \to p] = \int_0^1 2e^{-pt}\,dt + \int_1^\infty e^{-pt}\,dt$

$$= \frac{2 - e^{-p}}{p}\,.$$

Example 3.32. The **Laguerre polynomial** $L_n(t)$ is defined by

$$L_n(t) = \frac{e^t}{n!}\frac{d^n}{dt^n}\left[\,e^{-t}\,t^n\,\right]\,, \quad n = 0, 1, 2, \ldots$$

Prove that

$$L[L_n(t)\,;\,t \to p\,] = \frac{(p-1)^n}{p^{n+1}}\,, \quad n = 0, 1, 2, \ldots$$

Solution. By definition,

$$L\,[\,L_n(t)\,;\to p\,] = \int_0^\infty \frac{e^{-t(p-1)}}{n!}\frac{d^n}{dt^n}\left(e^{-t}\,t^n\right)dt$$

$$= \frac{1}{n!}(p-1)\int_0^\infty e^{-t(p-1)}\frac{d^{n-1}}{dt^{n-1}}\left(e^{-t}\cdot t^n\right)dt$$

$$= \cdots\cdots$$

$$= \frac{(p-1)^n}{n!}\int_0^\infty e^{-pt}\cdot t^n\,dt$$

$$= \frac{(p-1)^n}{n!}\frac{\Gamma(n+1)}{p^{n+1}}$$

$$= \frac{(p-1)^n}{p^{n+1}}\,.$$

Example 3.33. If $f(t) = |\sin t|$, prove that $f(t)$ is a periodic function of t of principal period $\tau = \pi$. Find the Laplace transform of $f(t)$.

Solution. Here

$$f(t) = |\sin(t)|, f(t+\tau) = f(t+\pi) = |\sin(\pi+t)|$$
$$= |-\sin t| = |\sin t| = f(t)$$
$$f(t+n\tau) = |\sin t| = f(t), \ n = 0, 1, 2, \cdots$$

Hence, $f(t)$ is a periodic function of period $\tau = \pi$. Therefore,

$$L\,[\,f(t)\,;\,t \to p\,]$$
$$= L\,[\,|\sin t|\,;\,t \to p\,]$$
$$= \frac{1}{1-e^{-p\pi}} \int_0^\pi f(t)e^{-pt}\,dt$$
$$= \frac{1}{1-e^{-p\pi}} \int_0^\pi |\sin t|\,e^{-pt}\,dt$$
$$= \frac{1}{1-e^{-p\pi}} \int_0^\pi \sin t \cdot e^{-pt}\,dt$$
$$= \frac{e^{-p\pi}+1}{1-e^{-p\pi}} \cdot \frac{1}{p^2+1}\,, \qquad\qquad Re\ p > 0$$
$$= \coth \frac{p\,\pi}{2} / (p^2+1)\,, \qquad\qquad Re\ p > 0$$

Example 3.34.

$$\text{Let } f(t) = \begin{cases} 1\,, & 0 \leqslant t \quad < \tau/2 \\ -1\,, & \tau/2 < t \quad < \tau \end{cases}$$

and f(t) is a periodic function of principal period τ. Prove that

$$L\,[\,f(t)\,;\,t \to\ p] = \frac{1}{p}\tanh\frac{p\tau}{2}\,\cdot$$

Solution. We can express $f(t) = H(t) - 2\,H\left(t-\frac{\tau}{2}\right)$, $0 < t < \tau$,

and $\qquad f(t+r\tau) = f(t)$, for all $t > 0$.

$$L\,[\,f(t)\,;\,t \to p\,] = \frac{1}{1-e^{-p\tau}} \int_0^\tau f(t)e^{-pt}\,dt$$
$$= \frac{1}{1-e^{-p\tau}} \left[\int_0^{\frac{\tau}{2}} e^{-pt}\,dt - \int_{\tau/2}^\tau e^{-pt}\,dt \right]$$
$$= \frac{1}{p\,(1-e^{-p\tau})} \left[1-e^{-p\tau/2} \right]^2 = \frac{1}{p}\tanh\frac{p\tau}{2}\,, \quad Re\ p > 0$$

In particular, if $\tau = \pi$ we get

$$L\left[\, f(t)\,;\ t \to p\,\right]_{\tau=\pi} = \frac{1}{p}\ \tanh\ \frac{p\pi}{2}, \qquad Re\ p > 0$$

Example 3.35. By considering the Fourier series of a periodic function $f(t)$ of principal period 2π, find Laplace transform of $f(t)$ to deduce

$$\frac{a_0}{2p} + \sum_{n=1}^{\infty} \frac{pa_n + n\,b_n}{p^2 + n^2} = \frac{1}{1 - e^{-2p\pi}} \int_0^{2\pi} f(u)\,e^{-pu}\,du$$

Hence, deduce the Mittag-Leffler identity

$$\frac{1}{2} + \frac{2p}{\pi} \sum_{r=1}^{\infty} \left[p^2 + (2r-1)^2\right]^{-1} = \left(1 + e^{-p\pi}\right)^{-1}$$

$$\text{for } f(t) = \begin{cases} 1, & 0 < t < \pi \\ 0, & \pi < t < 2\pi \end{cases}$$

Solution. The Fourier series corresponding to $f(t)$ is given by

$$f(t) \simeq \frac{a_0}{2} + \sum_{n=1}^{\infty} \left[\, a_n \cos nt + b_n \sin\ nt\,\right] \tag{1}$$

where
$$\left.\begin{array}{l} a_n = \frac{1}{\pi} \int_0^{2\pi} f(u) \cos\ nu\ du \\[2mm] b_n = \frac{1}{\pi} \int_0^{2\pi} f(u) \sin\ nu\ du \end{array}\right\} \tag{2}$$

Taking Laplace transform of both sides of eqn.(1) one gets

$$\frac{1}{1 - e^{-2p\pi}} \int_0^{2\pi} f(u)e^{-pu}du = \frac{a_0}{2p} + \sum_{n=1}^{\infty} \frac{pa_n + nb_n}{p^2 + n^2} \tag{3}$$

where a_n and b_n are given in eqns.(2).

$$\text{As an example let } f(t) = \begin{cases} 1, & 0 < t < \pi \\ 0, & \pi < t < 2\pi \end{cases}$$

be a periodic function of period 2π. Then,

$$a_0 = 1, \qquad a_n = 0 \quad \text{for} \quad n = 1, 2, \cdots$$

$$\text{and} \quad b_{2n} = 0, \quad b_{2n-1} = \left(n - \frac{1}{2}\right)^{-1} \pi, \quad n = 1, 2, 3, \cdots$$

$$\text{Also,} \quad \int_0^{2\pi} f(u)e^{-pu}du = \int_0^{\pi} e^{-pu}du = \frac{1}{p}\left(1 - e^{-p\pi}\right),$$

Substituting these results in eqn. (3), we get

$$\frac{1}{2} + \frac{2p}{\pi} \sum_{n=1}^{\infty} \frac{1}{p^2 + (2n-1)^2} = \frac{1}{1 + e^{-p\pi}} \ ,$$

which is the required result.

Example 3.36. Using the definition of convolution of two functions, find

$$L\left[\int_0^t J_0(\tau) \, J_0(t - \tau) \, d\tau \; ; \; t \to p\right]$$

in its simplest form and hence evaluate $\int_0^t J_0(\tau) J_0(t - \tau) d\tau$.

Solution. Let $f(t) = J_0(t)$ and $g(t) = J_0(t)$. Then

$$f(t) * g(t) = \int_0^t f(\tau) \, g(t - \tau) \, d\tau$$

$$= \int_0^t J_0(\tau) \, J_0(t - \tau) \, d\tau$$

$$\therefore \quad L\left[\int_0^t J_0(\tau) \, J_0(t - \tau) d\tau\right] = L\left[f * g(t)\right]$$

$$= \bar{f}(p) \, \bar{g}(p)$$

$$= \left[\bar{f}(p)\right]^2, \text{ since } f(t) = g(t) \Rightarrow \bar{f}(p) = \bar{g}(p)$$

$$\therefore \quad L\left[\int_0^t J_0(\tau) \, J_0(t - \tau) d\tau \; ; \; t \to p\right] = \frac{1}{p^2 + 1}$$

This result leads to

$$L^{-1}\left[L\left\{\int_0^t J_0(\tau) \, J_0(t - \tau) \, d\tau\right\}\right] = L^{-1}\left[\frac{1}{p^2 + 1}\right] = \sin t$$

Or $\quad \int_0^t J_0(\tau) \, J_0(t - \tau) \, d\tau = \sin t$

Excercises

(1) Find the Laplace transform of

$$f(t) = \begin{cases} 0 & , \quad 0 < t < 1 \\ (t-1)^2 & , \quad t > 1 \end{cases}$$

$\left[\text{Ans. } \frac{2e^{-p}}{p^3}\right]$

(2) Evaluate

$L[f(t) ; t \rightarrow p]$ where

(i) $f(t) = \begin{cases} \sin t & , \quad 0 < t < \pi \\ 0 & , \quad t > \pi \end{cases}$

(ii) $f(t) = \sin \sqrt{t}$

$\left[\text{Ans.} \quad \text{(i)}(1 + e^{-p\pi}) / (p^2 + 1) \quad \text{(ii)} \quad \sqrt{\pi} \, e^{-1/4p} / (2p^{\frac{3}{2}}) \right]$

(3) Find Laplace transform of $f(t)$, where

(i) $f(t) = sh \, at \cos at$

(ii) $f(t) = t \, e^{-t}(3 \, sh \, 2t - 5 \, ch \, 2t)$

(iii) $f(t) = [\, g(t) \cos \, wt \,]$ if $L[g(t)] = \bar{g}(p)$

$\left[\text{Ans.} \quad \text{(i)} \; a(p^2 - 2a^2)/(p^4 + 4a^4) \quad \text{(ii)} \quad -\frac{d}{dp}\left[\frac{1-5p}{p^2+2p-3}\right] \right.$

$\left. \text{(iii)} \quad \frac{1}{2}[\, \bar{g}(p - iw) + \bar{g}(p + iw) \,] \,] \right]$

(4) Evaluate $L[(\sin \, at - at \cos \, at) ; t \rightarrow p]$ and hence find $\lim\limits_{t \to 0} L[\sin \, at - at \cos \, at ; t \rightarrow p]$.

[Ans. $2a^3/(p^2 + a^2)^2 , \; p > 0 ; \; 0$]

(5) Using Laplace transform evaluate $I = \int_0^\infty t \, e^{-2t} dt$ to prove that $I = \frac{3}{25}$.

(6) Prove that $\int_0^\infty (e^{-at} - e^{-bt})/t \; dt = \log(b/a)$ .Hence, find the value of $\int_0^\infty (e^{-t} - e^{-3t}) \frac{dt}{t}$

[Ans. $\log 3$]

(7) Prove that $L[J_1(t) ; t \rightarrow p] = 1 - \frac{p}{\sqrt{p^2+1}}$

and $L[t \, J_1(t) ; t \rightarrow p] = \frac{1}{(p^2+1)^{3/2}}$

(8) Prove that

(i) $L[t \; \text{erf}(2\sqrt{t}) ; t \rightarrow p] = \frac{(3p \mid 8)}{p^2(p^2+4)^{3/2}}$

(ii) $L[e^{3t}\mathrm{erf}\sqrt{t} \; ; \; t \to p] = \frac{1}{(p-3)\sqrt{p-2}}$

(9) Prove that $\int_0^\infty \frac{f(t)}{t} \, dt = \int_0^\infty \bar{f}(p) dp$,   provided that the integrals converge.

(10) If $L[f(t) \; ; \; t \to p] = \bar{f}(p)$, show that $L\left[\int_0^\infty \frac{f(u)}{u} \, du \; ; \; t \to p\right]$
$= \frac{1}{p} \int_p^\infty \bar{f}(x) dx$. Hence, show that $L\left[\int_0^t \frac{\sin u}{u} \, du \; ; \; t \to p\right] = \frac{1}{p}\cot^{-1} p$

(11) Prove that $L[Erf \; (t) \; ; \; t \to p] = \frac{2}{p\sqrt{\pi}} \int_0^\infty e^{-u^2 - pu} \, du$

(12) Prove that $L[t^n \, e^{iat} \; ; \; t \to p] = n!(p+ia)^{n+1}(p^2+a^2)^{-n-1}$, where n is a positive integer.

Chapter 4

THE INVERSE LAPLACE TRANSFORM
AND
APPLICATION OF LAPLACE TRANSFORM

4.1   Introduction.

If $f(t)$ belongs to a class A (meaning that it is a piecewise continuous function over $0 \leq t < \infty$ and is of some exponential order), its Laplace transform $\bar{f}(p)$ exists. This is denoted symbolically by

$$L\ [f(t);\ t \to p]\ =\ \int_0^\infty\ e^{-pt}\ f(t)\ dt \equiv \bar{f}(p) \qquad (4.1)$$

Then, in turn $f(t)$, the object function is the inverse Laplace transform of the image function $\bar{f}(p)$ and is given symbolically by

$$L^{-1}\ \left[\bar{f}(p);\ p \to t\right]\ \equiv\ f(t) \qquad (4.2)$$

Thus by eqn. (4.1), one can evaluate $\bar{f}(p)$ for a given $f(t)$, since $\bar{f}(p)$ exists. We shall now consider the inverse problem - deriving information from the prescribed $\bar{f}(p)$ that enable us to derive the original function $f(t)$ through some formula, called the Laplace inversion formula.

Before answering this basic problem raised above, we shall at first present heuristic approach of determining the inverse Laplace transform of some given $\bar{f}(p)$. Also as per need we define a new function, called the **null function** in this connection together with a connected theorem, known as Lerch's theorem.

Definition.
Null function. A function $N(t)$ satisfying the condition $\int_0^t N\ (\tau)\ d\tau = 0$, for all $t > 0$ is called a null function.

 Theorem 4.1.
Lerch's Theorem. If $L[\ f_1(t)\ ;\ t \to p]$, $Re\ p > c_1$ and

$L[\,f_2(t)\;;\;t \to p\,]\;,\;Re\,p > c_2$ both exists and if

$$L[\,f_1(t)\;;\;t \to p\,]\;=L[\,f_2(t)\;;\;t \to p\,]$$

for $Re\,p > c = max(c_1, c_2)$, then $f_2(t) - f_1(t) = N(t)$

In addition, if $f_1(t)$ and $f_2(t)$ are continuous on the whole real line, then $f_1(t) = f_2(t)\,,\;\;t > 0$.

Proof. We shall not persue here the proof of this theorem. But the comment that the inverse Laplace transform of a given $\bar{f}(p)$ is unique except for a null function, should be kept in mind.

In the proposed heuristic approach, we reduce the given function $\bar{f}(p)$ as a combination of some functions which are the Laplace transforms of known functions. Then applying the idea of the Lerch's theorem, the inverse Laplace transform of the given $\bar{f}(p)$ can be obtained by using the inverse results already discussed in the previous article of this chapter. This approach will be just like evaluation of the integration as an inverse process of differentiation of a given function, called the integrand, after reducing it to sum of functions which are the derivatives of known standard functions. Sometimes for reduction of $\bar{f}(p)$ into sum of standard expressions for manipulation of Laplace inversion, partial fraction rules are found useful. Here, again the linearity property of Laplace inversion holds.

Theorem 4.2. If $\bar{f}_1(p) = L\,[f_1(t);\;t \to p]$ and $\bar{f}_2(p) = L\,[f_2(t);\;t \to p]$ and c_1, c_2 be two arbitrary constants, then

$$L^{-1}\left[\,c_1\;\bar{f}_1(p) + c_2\;\bar{f}_2(p)\;;p \to t\right]$$
$$= c_1\;L^{-1}\left[\bar{f}_1(p)\;;\;p \to t\,\right] + c_2\;L^{-1}\left[\bar{f}_2(p)\;;\;p \to t\,\right]$$
$$= c_1\;f_1(t) + c_2\;f_2(t).$$

Proof. Since, $L^{-1}\left[\bar{f}_1(p)\;;\;p \to t\,\right]\;=f_1(t)$ and

$$L^{-1}\left[\bar{f}_2(p)\;;\;p \to t\,\right]\;=f_2(t)$$

under the given condition, we have by the linearity property of the Laplace transform

$$L\,[c_1\;f_1(t) + c_2\;f_2(t); t \to\;p] = c_1 L\,[f_1(t);\;t \to p] + c_2 L\,[f_2(t)\;;\;t \to p]$$
$$= c_1 \bar{f}_1(p) + c_2 \bar{f}_2(p).$$

Hence, $L^{-1} L [c_1 f_1(t) + c_2 f_2(t) ; t \rightarrow p]$

$$= L^{-1} [c_1 \bar{f}_1(p) + c_2 \bar{f}_2(p) ; p \rightarrow t]$$

implying, $L^{-1} [c_1 \bar{f}_1(p) + c_2 \bar{f}_2(p) ; p \rightarrow t] = c_1 f_1(t) + c_2 f_2(t)$

$$= c_1 L^{-1} [\bar{f}_1(p) ; p \rightarrow t] + c_2 L^{-1} [\bar{f}_2(p) ; p \rightarrow t]$$

Thus, the theorem is proved.

4.2 Calculation of Laplace inversion of some elementary functions.

Before finding Laplace inversion of elementary functions which are a bit complex in nature, we state below some results obtained directly from the previous articles in 3.5 and 3.9. These results may be considered as formulae for further applications.

(i) $L[H(t) ; t \rightarrow p] = \frac{1}{p}$ implies $L^{-1} \left[\frac{1}{p} ; p \rightarrow t \right] = H(t)$

(ii) $L[t^\nu ; t \rightarrow p] = \frac{\Gamma(\nu+1)}{p^{\nu+1}}$ implies $L^{-1} \left[\frac{\Gamma(\nu+1)}{p^{\nu+1}} ; p \rightarrow t \right] = t^\nu, Re\ \nu > -1$

(iii) $L[e^{at} ; t \rightarrow p] = \frac{1}{p-a}$ implies $L^{-1} \left[\frac{1}{p-a} ; p \rightarrow t \right] = e^{at}, Re\ p > a$

(iv) $L[\sin at ; t \rightarrow p] = \frac{a}{p^2+a^2}$ implies $L^{-1} \left[\frac{a}{p^2+a^2} ; p \rightarrow t \right] = \frac{\sin at}{a}, p > 0$

(v) $L[\cos at ; t \rightarrow p] = \frac{p}{p^2+a^2}$ implies $L^{-1} \left[\frac{p}{p^2+a^2} ; p \rightarrow t \right] = \cos at, p > 0$

(vi) $L[sh\ at ; t \rightarrow p] = \frac{a}{p^2-a^2}$ implies $L^{-1} \left[\frac{1}{p^2-a^2} ; p \rightarrow t \right] - \frac{sh\ at}{a},$
$p > a > 0$

(vii) $L[ch\ at ; t \rightarrow p] = \frac{p}{p^2-a^2}$; implies $L^{-1} \left[\frac{p}{p^2-a^2} ; p \rightarrow t \right] = ch\ at,$
$p > a > 0$

(viii) $L[t^\nu e^{-at} ; t \rightarrow p] = \frac{\Gamma(\nu+1)}{(p+a)^{\nu+1}}$ implies
$L^{-1} \left[\frac{\Gamma(\nu+1)}{(p+a)^{\nu+1}} ; p \rightarrow t \right] = t^\nu e^{-at}, Re\ \nu + 1 > 0$

(ix) $L\left[\sin(t-a)H(t-a); t \to p\right] = \frac{e^{-pa}}{1+p^2}$ implies

$L^{-1}\left[\frac{e^{-pa}}{1+p^2} \; ; \; p \to t\right] = \sin(t-a)\,H(t-a)$

(x) $L\left[\cos(t-a)H(t-a); t \to p\right] = \frac{pe^{-pa}}{1+p^2}$ implies

$L^{-1}\left[\frac{pe^{-pa}}{1+p^2}; p \to t\right] = \cos(t-a)\,H(t-a)$

(xi) $L[sh(t-a)H(t-a); t \to p] = \frac{e^{-pa}}{p^2-1}$ implies

$L^{-1}[\frac{e^{-pa}}{p^2-1} \; ; \; p \to t] = sh(t-a)H(t-a)$

(xii) $L\left[ch(t-a)H(t-a); t \to p\right] = \frac{pe^{-pa}}{p^2-1}$ implies

$L^{-1}\left[\frac{pe^{-pa}}{p^2-1}; p \to t\right] = ch(t-a)\,H(t-a)$

We now turn our attention in discussing some rules of manipulation of Laplace inversion of some combinations of elementary functions of p through the following examples.

Example 4.1. Evaluate $L^{-1}\left[\frac{1}{(p+1)(p^2+1)} \; ; \; p \to t\right]$

Solution. Resolving into partial fractions under usual method we have

$$\frac{1}{(p^2+1)(p+1)} \equiv \frac{1}{2}\frac{1}{p+1} - \frac{1}{2}\frac{p}{p^2+1} + \frac{1}{2}\frac{1}{p^2+1}$$

Therefore,

$$L^{-1}\left[\frac{1}{(p+1)(p^2+1)}\right] = \frac{1}{2}L^{-1}\left[\frac{1}{p+1}\right] - \frac{1}{2}L^{-1}\left[\frac{p}{p^2+1}\right] + \frac{1}{2}L^{-1}\left[\frac{1}{p^2+1}\right]$$

$$= \frac{1}{2}\left[e^{-t} - \cos t + \sin t\right]$$

Example 4.2. Evaluate $L^{-1}\left[\frac{6p^2+22p+18}{p^3+6p^2+11p+6}\right]$

Solution.

$$L^{-1}\left[\frac{6p^2 + 22p + 18}{p^3 + 6p^2 + 11p + 6}\right]$$

$$= L^{-1}\left[\frac{6p^2 + 22p + 18}{(p+1)(p+2)(p+3)}\right]$$

$$= L^{-1}\left[\frac{1}{p+1} + \frac{2}{p+2} + \frac{3}{p+3}\right]$$

$$= e^{-t} + 2e^{-2t} + 3e^{-3t}$$

Example 4.3. Evaluate $L^{-1}\left[\left(\frac{\sqrt{p}-1}{p}\right)^2\right]$

Solution.

$$L^{-1}\left[\left(\frac{\sqrt{p}-1}{p}\right)^2\right]$$

$$= L^{-1}\left[\frac{p-2\sqrt{p}+1}{p^2}\right]$$

$$= L^{-1}\left[\frac{1}{p}-\frac{2}{p^{3/2}}+\frac{1}{p^2}\right]$$

$$= H(t)-4\sqrt{\frac{t}{\pi}}+t$$

4.3 Method of expansion into partial fractions of the ratio of two polynomials

Let $f(p)$ and $g(p)$ be two polynomials in p such that the degree of $f(p)$ is less than that of $g(p)$. Then $f(p)/g(p)$ can be expressed as

$$\frac{f(p)}{g(p)}=\sum_{r=1}^{n}\frac{A_r}{p-a_r}\ ,\ \text{where } p-a_r \text{ are the factors of } g(p)$$

with $\quad A_r=\lim_{p\to a_r}\frac{(p-a_r)\,f(p)}{g(p)}=\frac{f(a_r)}{g'(a_r)},\ \text{for } r=1,2,\cdots n.$

To prove the above result, we write

$$\frac{f(p)}{g(p)}\equiv\frac{A_1}{p-a_1}+\frac{A_2}{p-a_2}+\cdots\frac{A_n}{p-a_n} \tag{4.3}$$

Then, $\lim_{p\to a_r}\dfrac{(p-a_r)\,f(p))}{g(p)}=A_r,$ the other terms being zero in the limit.

Therefore

$$A_r=\lim_{p\to a_r}\frac{f(p)}{g(p)/p-a_r}=\frac{f(a_r)}{g'(a_r)},\ \text{provided } g'(a_r)\neq 0$$

meaning that $p-a_r$ is a non-repeated factor of $g(p)$. Thus, the result stated above is proved.

Example 4.4. Evaluate $L^{-1}\left[\frac{p+5}{(p+1)(p^2+1)}\right]$

Solution. We express by partial fraction that

$$\left[\frac{p+5}{(p+1)(p^2+1)}\right] \equiv \frac{A_1}{p+1} + \frac{A_2}{p+i} + \frac{A_3}{p-i}$$

$$= \frac{2}{p-1} + \frac{\frac{i-5}{2(1+i)}}{p+i} + \frac{\frac{i+5}{2(i-1)}}{p-i}$$

$$= \frac{2}{p+1} + \frac{-2p}{p^2+1} + \frac{3}{p^2+1} \ ,$$

Therefore, $L^{-1}\left[\frac{p+5}{(p+1)(p^2+1)}; p \to t\right] = 2e^{-t} - 2\cos\ t + 3\sin t$

Example 4.5. Evaluate $L^{-1}\left[\frac{p^2-6}{p^3+4p^2+3p}\right]$

Solution. We express, by partial fraction that

$$\frac{p^2-6}{p^3+4p^2+3p} \equiv \frac{A_1}{p} + \frac{A_2}{p+1} + \frac{A_3}{p+3} = \frac{f(p)}{g(p)} \ , \ \text{say}$$

where $f(p) = p^2 - 6, g(p) = p^3 + 4p^2 + 3p, g'(p) = 3p^2 + 8p + 3$

Then we have

$$A_1 = \frac{f(0)}{g'(0)} = -2, \ A_2 = \frac{f(-1)}{g'(-1)} = \frac{5}{2}, \ A_3 = \frac{f(-3)}{g'(-3)} = \frac{1}{2}$$

Hence,

$$L^{-1}\left[\frac{p^2-6}{p^3+4p^2+3p}\right] = \frac{5}{2}\ e^{-t} + \frac{1}{2}\ e^{-3t} - 2H(t)$$

Example 4.6. Evaluate $L^{-1}\left[\frac{4p+5}{(p-1)^2(p+2)}\right]$

Solution. By method of partial fraction we express

$$\frac{4p+5}{(p-1)^2(p+2)} \equiv \frac{A}{p-1} + \frac{B}{(p-1)^2} + \frac{C}{p+2}$$

Then, $4p + 5 \equiv A(p-1)(p+2) + B(p+2) + C(p-1)^2$

From this identity we get

$$9 \ = \ 3B, \ (\text{ by putting } p = 1) \Longrightarrow B = 3$$

$$-3 \ = \ 9C, \ (\text{ by putting } p = -2) \Longrightarrow C = -\frac{1}{3}$$

and $5 = -2A + 2B + C$, (by equating term independent of p) $\Longrightarrow A = \frac{1}{3}$

Therefore , $L^{-1}\left[\dfrac{4p+5}{(p-1)^2(p+2)}\right] = \dfrac{1}{3}e^t + 3L^{-1}\left[\dfrac{1}{(p-1)^2}\right] - \dfrac{1}{3}e^{-2t}$

$$= \dfrac{1}{3}e^t - \dfrac{1}{3}e^{-2t} + 3\ t\ e^t,\ \text{by article 4.2}$$

Example 4.7. Evaluate $L^{-1}\left[\dfrac{1}{\sqrt{2p+3}}\right]$,   after proving

$L^{-1}\left[\bar{f}\ (ap+b)\right] = \left(\dfrac{1}{a}\right)\ e^{-\frac{bt}{a}}f\left(\dfrac{t}{a}\right)$, where $L^{-1}\left[\ \bar{f}(p)\ \right] = f(t)$

Solution. Given that

$$\bar{f}(p) = \int_0^\infty e^{-pt}f(t)\ dt\ ,\ \text{by definition.}$$

So, $\bar{f}(ap+b) = \displaystyle\int_0^\infty e^{-pat}\ .\ e^{-bt}\ f(t)\ dt$

$$= \dfrac{1}{a}\int_0^\infty e^{-px}\ e^{-\frac{bx}{a}}\ f\left(\dfrac{x}{a}\right)\ dx\ ,\ \text{putting}\ at = x$$

$$= L\ \left[\ \dfrac{1}{a}\ e^{-\frac{bt}{a}}\ f\left(\dfrac{t}{a}\right)\ \right]$$

$$\Rightarrow L^{-1}\ \left[\ \bar{f}(ap+b)\right] = \dfrac{1}{a}\ e^{-\frac{bt}{a}}\ f\left(\dfrac{t}{a}\right).$$

Now putting $a = 2,\ b = 3$ we get.

$$\bar{f}(p) = \dfrac{1}{\sqrt{p}} = \dfrac{\sqrt{\pi}}{\sqrt{\pi}\sqrt{p}} = \dfrac{1}{\sqrt{\pi}}\ \dfrac{\Gamma\left(\frac{1}{2}\right)}{p^{\frac{1}{2}}}$$

so that $f(t) = \dfrac{1}{\sqrt{\pi t}}$.   Therefore, $f\ \left(\dfrac{t}{a}\right) = f\ \left(\dfrac{t}{2}\right) = \dfrac{\sqrt{2}}{\sqrt{\pi\ t}}$

These results give,

$$L^{-1}\left[\dfrac{1}{\sqrt{2p+3}}\right] = \dfrac{1}{2}\ e^{-\frac{3}{2}t}\dfrac{\sqrt{2}}{\sqrt{\pi\ t}} = \dfrac{1}{\sqrt{2\pi t}}\ e^{-\frac{3t}{2}}$$

Example. 4.8. Evaluate

$$L^{-1}\left[\dfrac{(p+1)\ e^{-\pi p}}{p^2+p+1}\right]$$

Solution.   We have

$$L^{-1}\left[\dfrac{p+1}{p^2+p+1}\right]$$

$$= L^{-1} \left[\frac{p + \frac{1}{2} + \frac{1}{2}}{(p + \frac{1}{2})^2 + \left(\frac{\sqrt{3}}{2} \right)^2} \right]$$

$$= L^{-1} \left[\frac{p + \frac{1}{2}}{(p + \frac{1}{2})^2 + \left(\frac{\sqrt{3}}{2} \right)^2} \right] + \frac{1}{\sqrt{3}} L^{-1} \left[\frac{\frac{\sqrt{3}}{2}}{(p + \frac{1}{2})^2 + \left(\frac{\sqrt{3}}{2} \right)^2} \right]$$

$$= e^{-\frac{t}{2}} L^{-1} \left[\frac{p}{p^2 + \left(\frac{\sqrt{3}}{2} \right)^2} \right] + \frac{e^{-\frac{t}{2}}}{\sqrt{3}} L^{-1} \left[\frac{\frac{\sqrt{3}}{2}}{p^2 + \left(\frac{\sqrt{3}}{2} \right)^2} \right]$$

$$= e^{-\frac{t}{2}} \cos \left(\frac{\sqrt{3}}{2} t \right) + e^{-\frac{t}{2}} \frac{1}{\sqrt{3}} \sin \left(\frac{\sqrt{3}}{2} t \right)$$

$$\therefore \quad L^{-1} \left[\frac{(p + 1)e^{-\pi p}}{p^2 + p + 1} \right]$$

$$= e^{-\frac{(t - \pi)}{2}} \left[\cos \left\{ \frac{\sqrt{3}}{2}(t - \pi) \right\} + \frac{1}{\sqrt{3}} \sin \left\{ \frac{\sqrt{3}}{2}(t - \pi) \right\} \right] \cdot H(t - \pi),$$

by the second shift theorem of Laplace transform

Example. 4.9. Evaluate $\quad L^{-1} \left[\frac{3(1 + e^{-p\pi})}{p^2 + 9} \right]$

Solution. We have

$$L^{-1} \left[\frac{3(1 + e^{-p\pi})}{p^2 + 9} \right]$$

$$= L^{-1} \left[\frac{3}{p^2 + 3^2} + e^{-p\pi} \frac{3}{p^2 + 3^2} \right]$$

$$= L^{-1} \left[\frac{3}{p^2 + 3^2} \right] + L^{-1} \left[e^{-p\pi} \frac{3}{p^2 + 3^2} \right]$$

$$= \sin \, 3t + H(t - \pi) \, \sin 3 \, (t - \pi)$$

$$= \sin \, 3t - \sin 3 \, t. \, H(t - \pi)$$

Example 4.10. Using the result of inverse Laplace transform evaluate
$L^{-1} \left[\frac{p+1}{(p^2 + 2p + 1)^2} \right]$.

Solution. We know that $L\left[t^n \, f(t) \right] = (-1)^n \frac{d^n}{dp^n} \, \bar{f}(p)$, for $n = 1, 2, 3, \cdots$.
This result implies

$$L^{-1} \left[\frac{d^n}{dp^n} \bar{f}(p) \right] = (-1)t^n f(t) \, , \quad \text{where} \quad L[f(t)] = \bar{f}(p)$$

Now, $\quad L^{-1}\left[\dfrac{p+1}{\{\,(p+1)^2+1\,\}^2}\right]$

$$= e^{-t}\,L^{-1}\left[\dfrac{p}{(p^2+1)^2}\right]$$

$$= -e^{-t}\,L^{-1}\left[\dfrac{1}{2}\dfrac{d}{dp}\cdot\left(\dfrac{1}{(p^2+1)}\right)\right]$$

$$= -e^{-t}\cdot(-1)^1\dfrac{t}{2}\ \sin\,t$$

$$= e^{-t}\dfrac{t}{2}\ \sin\,t$$

Example 4.11. Evaluate $L^{-1}\left[\,\log\,\frac{p+3}{p+2}\,\right]$

Solution. Let $\bar{f}(p)=\left[\,\log\,\frac{p+3}{p+2}\,\right]$. We wish to calculate $f(t)$.

Now, $\qquad \dfrac{d}{dp}\bar{f}(p)=\dfrac{1}{p+3}-\dfrac{1}{p+2}$

$$L^{-1}\left[\dfrac{d}{dp}\bar{f}(p)\right]=e^{-3t}-e^{-2t}$$

Therefore, $\quad -L^{-1}\left[\dfrac{d}{dp}\,\bar{f}(p)\right]=tf(t)=e^{-2t}\ -e^{-3t}$

implying, $\qquad f(t)=\dfrac{e^{-2t}-e^{-3t}}{t}$

Example 4.12. Evaluate $L^{-1}\left[\,\log\left(1-\frac{1}{p^2}\right)\,\right]$

Solution. Let

$$\bar{f}(p)=\log\left(1-\dfrac{1}{p^2}\right)\Rightarrow-\dfrac{d}{dp}\left[\,\bar{f}(p)\,\right]=\dfrac{2}{p}-\dfrac{2p}{p^2-1}$$

$$\therefore\ -L^{-1}\left[\dfrac{d}{dp}\bar{f}(p)\right]=L^{-1}\left[\dfrac{2}{p}-\dfrac{2p}{p^2-1}\right]$$

$$\Rightarrow\ tf(t)=2H(t)-2\cosh t$$

Oı, $\qquad f(t)=2\,[\,H(t)-\cosh t\,]\,/t.$

Example 4.13. Assuming that $\lim\limits_{t\to\infty}\dfrac{f(t)}{t}$ do exist, prove that $L^{-1}\left[\dfrac{\bar{f}(p)}{p}\right]=\int_0^t f(x)\,dx$. Hence, evaluate $L^{-1}\left[\dfrac{1}{p^3(p^2+1)}\right]$

Solution. First part is discussed in article 2.10.

By the above result, $L^{-1}\left[\dfrac{1}{p}\left(\dfrac{1}{p^2+1}\right)\right]=\int_0^t\sin x\ dx=1-\cos t$

$$L^{-1}\left[\dfrac{1}{p}\left\{\dfrac{1}{p(p^2+1)}\right\}\right]=\int_0^t(1-\cos x)\ dx=t-\sin t.$$

Hence, $L^{-1}\left[\dfrac{1}{p}\left\{\dfrac{1}{p^2(1+p^2)}\right\}\right] = \displaystyle\int_0^t (x - \sin x)\,dx = \dfrac{t^2}{2} + \cos t - 1.$

Example 4.14. Evaluate

$$L^{-1}\left[\dfrac{1}{p(p+1)^3}\right]$$

Solution. We have

$$\left[\dfrac{1}{p(p+1)^3}\right] = L^{-1}\left[\dfrac{1}{\{(p+1) - 1\,\}(p+1)^3}\right]$$

$$= e^{-t}L^{-1}\left[\dfrac{1}{(p-1)p^3}\right]$$

Now, $L^{-1}\left[\dfrac{1}{p-1}\right] = e^t$

$\therefore$ $L^{-1}\left[\dfrac{1}{p^3(p-1)}\right] = \displaystyle\int_0^t\left[\int_0^t\left\{\int_0^t e^t\,dt\right\}dt\right]dt$

$$= \int_0^t\left[\int_0^t (e^t - 1)\,dt\right]dt = \int_0^t (e^t - 1 - t)\,dt$$

$$= e^t - 1 - t - \dfrac{t^2}{2}$$

$\therefore$ $L^{-1}\left[\dfrac{1}{p(p+1)^3}\right] = e^{-t}\left[e^t - \left(1 + t + \dfrac{t^2}{2}\right)\right]$

$$= 1 - e^{-t}\left(1 + t + \dfrac{t^2}{2}\right)$$

Example 4.15. If $L^{-1}\left[\dfrac{p}{(p^2+1)^2}\right] = \frac{1}{2}\,t\sin t$, evaluate $L^{-1}\left[\dfrac{1}{(p^2+1)^2}\right]$

Solution. We have

$$L^{-1}\left[\dfrac{1}{(p^2+1)^2}\right] = L^{-1}\left[\dfrac{1}{p}\dfrac{p}{(p^2+1)^2}\right]$$

$$= \int_0^t\left(\dfrac{1}{2}\,x\sin x\right)dx$$

$$= \dfrac{1}{2}\,[\sin t - t\cos t]$$

Example 4.16. By using the convolution theorem evaluate $L^{-1}\left[\dfrac{1}{p^2(p+1)^2}\right]$
to show that its value is $t\,e^{-t} + 2\,e^{-t} + t - 2.$

Solution. We have by convolution theorem

$$L^{-1}[\bar{f}(p)\,\cdot\,\bar{g}(p)] = f * g = \int_0^t f(\tau)g(t - \tau)\,d\tau$$

Now, $\quad L^{-1}\left[\dfrac{1}{p^2}\right] = t$ and $L^{-1}\left[\dfrac{1}{(p+1)^2}\right] = (t)e^{-t}$.

Therefore, $\quad L^{-1}\left[\dfrac{1}{p^2} \cdot \dfrac{1}{(p+1)^2}\right] = \displaystyle\int_0^t e^{-u} \cdot u(t-u)du$

$$= te^{-t} + 2e^{-t} + t - 2 \text{ , on simplification.}$$

Example 4.17. Evaluate $L^{-1}\left[\dfrac{p}{(p^2+4)^3}\right]$ by using covolution theorem.

Solution. We know that $L^{-1}\left[\dfrac{1}{(p^2+4)}\right] = \dfrac{\sin 2t}{2}$

and $\quad L^{-1}\left[\dfrac{p}{(p^2+4)^2}\right] = L^{-1}\left[-\dfrac{1}{2}\dfrac{d}{dp}\dfrac{1}{(p^2+4)}\right]$

$$= -\dfrac{1}{2}.t(-1)^1\ L^{-1}\left[\dfrac{1}{p^2+4}\right]$$

$$= \dfrac{1}{2}t\,\dfrac{\sin 2t}{2}. = \dfrac{1}{4}\,t\,\sin\,2t$$

Therefore, $\quad L^{-1}\left[\dfrac{p}{(p^2+4)^2} \cdot \dfrac{1}{p^2+4}\right] = \displaystyle\int_0^t \dfrac{u\sin\,2u}{4}\cdot\dfrac{1}{2}\,\sin\,2(t-u)du$

$$= \dfrac{1}{8}\displaystyle\int_0^t u\sin\,2u(\,\sin\,2t\,\cos 2u - \cos 2t\sin\,2u)\,du$$

$$= \dfrac{\sin\,2t}{8}\displaystyle\int_0^t u\,\sin\,2u\cos\,2udu - \dfrac{\cos\,2t}{8}\displaystyle\int_0^t u\sin^2\,2u\,du$$

$$= \dfrac{1}{64}\,[\,\sin\,2t - 2t\cos\,2t\,],$$

after evalution of the integrals and on simplification.

Example 4.18. Prove that $L^{-1}\left[\dfrac{1}{p\sqrt{p+4}}\right] = \tfrac{1}{2}\,erf\,(2\sqrt{t})$.

Solution. We know that

$$L^{-1}\left[\dfrac{1}{p}\right] = H(t)$$

and $\quad L^{-1}\left[\dfrac{1}{\sqrt{p+4}}\right] = e^{-4t}\ L^{-1}\left[\dfrac{1}{\sqrt{p}}\right] = \dfrac{e^{-4t}}{\sqrt{\pi t}}$

Therefore, $\quad L^{-1}\left[\dfrac{1}{p}\dfrac{1}{\sqrt{p+4}}\right] = \displaystyle\int_0^t \dfrac{e^{-4\tau}}{\sqrt{\pi\tau}}\,H(t-\tau)d\tau$

$$= \displaystyle\int_0^t \dfrac{e^{-4\tau}}{\sqrt{4\tau}}\,d\tau = \dfrac{1}{\sqrt{\pi}}\displaystyle\int_0^{2\sqrt{t}} e^{-u^2}\,du, \quad \text{on putting } 4\tau = u^2$$

$$= \dfrac{1}{2}\,erf\,(2\sqrt{t})\,, \text{ by definition of } erf(x) = \dfrac{2}{\sqrt{\pi}}\displaystyle\int_0^x e^{-\alpha^2}d\alpha$$

Example 4.19. Evaluate $L^{-1}\left[\frac{1}{\sqrt{p}(p-a)}\right]$ by the convolution theorem.
Hence deduce the value of $L^{-1}\left[\frac{1}{p\sqrt{p+a}}\right]$

Solution. We know that $L^{-1}\left[\frac{1}{\sqrt{p}}\right] = \frac{1}{\sqrt{\pi t}}$ and $L^{-1}\left[\frac{1}{(p-a)}\right] = e^{at}$
Therefore, by the convolution theorem

$$L^{-1}\left[\frac{1}{\sqrt{p}} \cdot \frac{1}{(p-a)}\right] = \int_0^t \frac{1}{\sqrt{\pi u}} e^{a(t-u)} du$$

$$= \frac{e^{at}}{\sqrt{\pi}} \int_0^t \frac{1}{\sqrt{u}} e^{-au} du \qquad , \text{ putting } au = \varsigma^2$$

$$= \frac{e^{at}}{\sqrt{a}} \cdot \frac{2}{\sqrt{\pi}} \int_0^{\sqrt{at}} e^{-\varsigma^2} d\varsigma$$

$$= \frac{e^{at}}{\sqrt{a}} erf(\sqrt{at}).$$

We have now

$$L^{-1}\left[\frac{1}{p\sqrt{p+a}}\right] = L^{-1}\left[\frac{1}{(p+a)-a} \cdot \frac{1}{\sqrt{p+a}}\right]$$

$$e^{-at} L^{-1}\left[\frac{1}{(p-a)} \cdot \frac{1}{\sqrt{p}}\right]$$

$$= e^{-at} \frac{e^{at}}{\sqrt{a}} erf(\sqrt{at}) = \frac{1}{\sqrt{a}} erf(\sqrt{at}).$$

Example 4.20. A periodic function $f(t)$ of period 2π having a finite discontinuity at $t = \pi$ is given by

$$f(t) = \begin{cases} \sin t \ , \ 0 \leqslant t < \pi \\ \cos t \ , \ \pi < t \leqslant 2\pi \end{cases}$$

Evaluate its Laplace transform

Solution The Laplace transform of $f(t)$ is given by

$$\bar{f}(p) = \frac{1}{1 - e^{-2\pi p}} \left[\int_0^{2\pi} f(t)e^{-pt} dt\right]$$

$$= \frac{1}{1 - e^{-2\pi p}} \left[\int_0^{\pi-0} e^{-pt} \sin t \ dt + \int_{\pi+0}^{2\pi} e^{-pt} \sin t \ dt\right]$$

$$= \frac{1}{1 - e^{-2p\pi}} \left[\frac{1 + e^{-p\pi}}{1 + p^2} - \frac{pe^{-p\pi}(1 + e^{-p\pi})}{1 + p^2}\right]$$

$$= \frac{1}{1 - e^{-p\pi}} \left[\frac{1 - pe^{-p\pi}}{1 + p^2}\right]$$

4.4 The general evaluation technique of inverse Laplace Transform.

The inverse Laplace transform has already been discussed in articles 3.3 and 3.4 and is given by

$$f(t) = L^{-1}[\bar{f}(p) \ ; \ p \to t] = \frac{1}{2\pi i} \ \text{P.V.} \int_{c-i\infty}^{c+i\infty} e^{pt} \bar{f}(p) dp \ , \ t > 0 \quad (4.4)$$

where $Re \ p = c > 0$ and an arbitrary constant, provided $f(t) = 0(e^{\gamma t})$, $Re \ p > \gamma$ with

$$\bar{f}(p) = L[\ f(t); \ t \to p \ ] = \int_0^\infty f(t) e^{-pt} \ dt \quad (4.5)$$

where $\bar{f}(p)$ is being the function of a complex variable p in the half-plane $Re \ p > \gamma$. The integral in (4.4) is called a Bromwich integral.

The working rule for determining the inverse Laplace transform $f(t)$ of $\bar{f}(p)$ with the help of the contour integral of a complex variable is shown below:

$$\text{Since,} \quad f(t) = \frac{1}{2\pi i} \text{P.V.} \int_{c-i\infty}^{c+i\infty} e^{pt} \bar{f}(p) \ dp, \quad t > 0,$$

to evaluate $f(t)$ by (4.4), we evaluate the contour integral

$$\frac{1}{2\pi i} \int_C e^{pt} \bar{f}(p) \ dp, \quad (4.6)$$

where C is the contour $ABCDEA$ described in the counter-clockwise sense and consists of the line $Re \ p = c > \gamma$ and the sectorial part of the circle of large radius R with centre at the origin of the complex p-plane. All the singularities of $\bar{f}(p)$ are lying to the left of the line segment from $c - iR$ to $c + iR$ and therefore they are all enclosed inside the contour C. This contour C is known as Bromwich contour in turn and is shown in FIG 1 below.

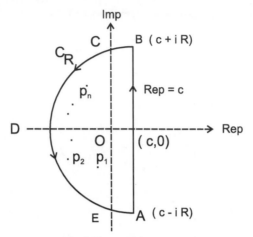

Fig 1. Bromwich contour

For actual evaluation of the integral in (4.6) let us assume that $p_n, (n = 1, 2, \cdots, N)$ denote the singularities of $\bar{f}(p)$ inside C. Let R_0 denotes the largest of the moduli of all these singular points. The parametric representation of the semicircular part, say, C_R is given by

$$p = c + Re^{i\theta} \ , \quad \frac{\pi}{2} \leqslant \theta \leqslant \frac{3\pi}{2} \tag{4.7}$$

where $R > R_0 + c$. Note that for each p_n,

$$|p_n - c| \leqslant |p_n| + c \leqslant R_0 + c < R.$$

Then by Cauchy's residue theorem, theorem we get

$$\int_\Gamma e^{pt} \bar{f}(p) \ dp = 2\pi i \sum_{n=1}^{N} \underset{p=p_n}{Res} \left[\ e^{pt} \bar{f}(p) \ \right] - \int_{C_R} e^{pt} \bar{f}(p) \ dp \tag{4.8}$$

It may be noted at this point that if $\bar{f}(p)$ has any branch point singularity, corresponding branch-cut must be given suitably for making $\bar{f}(p)$ analytic except for the singular points inside the modified contour C.

Suppose now that, for all point p on C_R, there is a positive constant M_R such that $|\bar{f}(p)| \leqslant M_R$, where $M_R \to 0$ as $R \to \infty$. We may use the parametric representation (4.7) for C_R to write

$$\int_{C_R} e^{pt} \ \bar{f}(p) \ dp = \int_{\frac{\pi}{2}}^{\frac{3\pi}{2}} exp \left(ct + Rte^{i\theta} \right) \bar{f} \left(c + Re^{i\theta} \right) \cdot Ri \ e^{i\theta} d\theta$$

Then since, $\left| \ exp \left(ct + Rte^{i\theta} \right) \right| = e^{ct} \cdot e^{Rt \cos \theta}$

and $\left| \bar{f}(c + Re^{i\theta}) \right| \leqslant M_R$, we find that

$$\left| \int_{C_R} e^{pt}\bar{f})(p)dp \right| \leqslant e^{ct}\, M_R\, R \int_{\pi/2}^{3\pi/2} e^{Rt\cos\theta}\, d\theta \qquad (4.9)$$

But the substitution $\phi = \theta - \frac{\pi}{2}$ together with Jordan's inequality reveals that

$$\int_{\pi/2}^{3\pi/2} e^{Rt\cos\theta}\, d\theta = \int_0^\pi e^{-Rt\sin\phi}\, d\phi < \frac{\pi}{Rt}$$

Thus inequality (4.9) becomes

$$\left| \int_{C_R} e^{pt}\bar{f}(p)\, dp \right| \leqslant e^{ct}\, \frac{M_R\, \pi}{t} \qquad (4.10)$$

and hence $\displaystyle\lim_{R\to\infty} \int_{C_R} e^{pt}\bar{f}(p)\, dp = 0 \qquad (4.11)$

Therefore, letting $R \to \infty$ in eqn. (4.4), we see that $f(t)$, defined by the equation

$$f(t) = \frac{1}{2\pi i} \lim_{R\to\infty} \int_{c-i\infty}^{c+i\infty} e^{pt}\bar{f}(p)\, dp \;, t > 0 \qquad (4.12)$$

can be written as

$$f(t) = \sum_{n=1}^{N} \operatorname*{Res}_{p=p_n} \left[e^{pt}\bar{f}(p) \right] \;, t > 0. \qquad (4.13)$$

Further it is to be noted in case that when $\bar{f}(p)$ has countable infinite number of singular points, the evaluation of $f(t)$ in (4.13) may be replaced by

$$f(t) = \sum_{n=1}^{\infty} \operatorname*{Res}_{p=p_n} \left[e^{pt}\bar{f}(p) \right] \;, t > 0. \qquad (4.14)$$

whenever it is possible to show that

$$\lim_{R,N\to\infty} \int_{C_R} e^{pt}\bar{f}(p)\, dp = 0 \qquad (4.15)$$

In many applications of Laplace transforms, such as the solution of ODE and PDE arising in initial value problem or boundary-value problem, the form of $\bar{f}(p)$ obtained there may not be that much simple to put it in a combination of the forms discussed in the heuristic approach with elementary functions. Thus, in those cases the necessity of this general evaluation technique of the inverse Laplace transform may be found useful.

It may also be noted that both heuristic method and the general contour integration method inversion of Laplace transform lead to identical results.

Example 4.21. Using heuristic as well as complex Inverse formula evaluate the inverse Laplace transform of the image function $\bar{f}(p) = \frac{p}{(p^2+1)^2}$.

Solution. By heuristic approach, we get

$$
\begin{aligned}
L^{-1}\left[\,\bar{f}(p)\,\right] &= L^{-1}\left[\frac{p}{(p^2+1)^2}\right] = L^{-1}\left[-\frac{1}{2}\frac{d}{dp}\left\{\frac{1}{(p^2+1)}\right\}\right] \\
&= \frac{1}{2}\,t\,L^{-1}\left[\frac{1}{p^2+1}\right] \\
&= \frac{1}{2}\,t\,\sin t = f(t)
\end{aligned}
$$

Also, by complex inversion rule we have

$$
f(t) = \frac{1}{2\pi i}\cdot\int_{c-i\infty}^{c+i\infty} e^{pt}\bar{f}(p)\,dp
$$

with $\bar{f}(p) = \frac{p}{(p^2+1)^2}$, which has singularities only at points $\pm i$, both of them being double poles. Therefore, we choose the arbitrary constant $c > 0$ and the corresponding Bromwhich contour C as shown below.

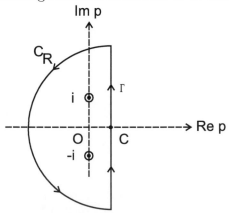

Fig 1. Contour C

In this case, $f(t) =$ Sum of the residues of $e^{pt}\frac{p}{(p^2+1)^2}$ at $p = \pm i$

$$= \lim_{p \to i} \frac{d}{dp}\left[(p-i)^2 \frac{e^{pt}p}{(p+i)^2(p-i)^2}\right] + \lim_{p \to -i} \frac{d}{dp}\left[(p+i)^2 \frac{e^{pt}p}{(p+i)^2(p-i)^2}\right]$$

$$= -\frac{it}{4} e^{it} + \frac{1}{4}it\, e^{-it}$$

$$= \frac{t}{2}\, \sin t \text{ , after simplification}$$

Example 4.22. Use complex inversion formula to evaluate inverse Laplace transform of $\bar{f}(p) = \frac{2p^2-4}{(p+1)(p-2)(p-3)}$

Solution. If inverse transform of $\bar{f}(p)$ be $f(t)$, then by complex Inversion formula for Laplace transform it is given by

$$f(t) = \frac{1}{2\pi i}\cdot \int_{c-i\infty}^{c+i\infty} \bar{f}(p)e^{pt}\, dp$$

$$= \frac{1}{2\pi i} \int_{c-i\infty}^{c+i\infty} \frac{(2p^2-4)\, e^{pt}\, dp}{(p+1)(p-2)(p-3)} = \frac{1}{2\pi i} \int_C \frac{(2p^2-4)\, e^{pt}\, dp}{(p+1)(p-2)(p-3)}$$

where C is a positively oriented closed contour with the Bramwich line from $c-i\infty$ to $c+i\infty$ as a part and enclosing all the poles of $\bar{f}(p)$.

Since, $\bar{f}(p)$ has only three simple poles at points $p = -1, 2, 3$ we choose here the arbitrary constant $c > 3$ for the Bromwich integral. Thus by article 2.21

$\quad f(t) =$ sum of the residues at the poles $p = -1, p = 2$ and $p = 3$ of the integrand $\left[\, \bar{f}(p)e^{pt}\, \right]$.

Now, residue of $\bar{f}(p)e^{pt}$ at $p = -1$

$$= \lim_{p \to -1}\left[\frac{(p \not{+} 1)\, (2p^2-4)e^{pt}}{(p \not{+} 1)\, (p+2)(p-3)}\right] = \frac{1}{6}\, e^{-t}$$

Residue of $\bar{f}(p)\, e^{pt}$ at $p = 2 = \lim_{p \to 2}\left[\frac{(p-2)(2p^2-4)e^{pt}}{(p+1)(p-2)(p-3)}\right]$

$$= -\frac{4}{3}e^{2t}$$

and residue of $\bar{f}(p)e^{pt}$ at $p = 3 = \lim_{p \to 3}\left[\frac{(p-3)(2p^2-4)e^{pt}}{(p+1)(p-2)(p-3)}\right]$

$$= \frac{7}{2}e^{3t}$$

Therefore, $f(t) = -\frac{1}{6}\, e^{-t} - \frac{4}{3}\, e^{2t} + \frac{7}{2}\, e^{3t}$

4.5   Inversion Formula from a different stand point : The Tricomi's method.

The Laguerre polynomial $L_n(t)$ is defined by

$$L_n(t) = \sum_{r=o}^{n} {}^{n}c_r \frac{(-t)^r}{r!} \tag{4.16}$$

Therefore, $L\left[L_n(t); t \to p\right] = \sum_{r=0}^{n}(-1)^r \, {}^{n}c_r \frac{1}{p^{r+1}} = \frac{1}{p}\left(1 - \frac{1}{p}\right)^2$

$$\tag{4.17}$$

Making use of change of scale and the first shift theorem of Laplace transform, we get

$$L\left[\,e^{-bt}\,L_n(2bt);\; t \to p\,\right] = \frac{1}{p+b}\left(\frac{p-b}{p+b}\right)^n \tag{4.18}$$

Therefore, if $f(t) = \sum_{n=0}^{\infty} a_n\, e^{-bt}\, L_n(2bt)$ $\tag{4.19}$

then $L[\,f(t); t \to p\,] = \bar{f}(p)$

$$= \sum_{n=0}^{\infty} a_n \frac{1}{p+b}\left(\frac{p-b}{p+b}\right)^n \tag{4.20}$$

by Lerch's theorem.

Thus the equation (4.19) defines the Tricomi's method of determining Laplace inversion $L^{-1}\left[\,\bar{f}(p); p \to t\,\right]$, when $\bar{f}(p)$ is expressed in the form (4.20). For this purpose we substitute $p = b\left(\frac{1+\sigma}{1-\sigma}\right)$ and then express the given $\bar{f}(p)$ in the form

$$\bar{f}(p) = \frac{1-\sigma}{2b} \sum_{n=0}^{\infty} a_n \sigma^n$$

Thus, $f(t) = L^{-1}\left[\,\bar{f}(p); p \to t\,\right]$ is interpreted as

$$f(t) = \sum_{n=0}^{\infty} a_n\, B_n(bt) \,, \tag{4.21}$$

where $B_n(bt) = e^{-bt} L_n(2bt)$ $\tag{4.22}$

To verify the accuracy of the method we discuss the following special case as discussed by Ward [(1954) Proc. camb. Phil. Soc.,50,49] , where

$$\bar{f}(p) = \frac{1}{(p+1)(p-2)} \text{ and } f(t) = e^{-t} - e^{-2t}$$

To evaluate $f(t)$ by the Tricomi's method, we substitute $p = (1 + \sigma)/(1 - \sigma)$, with $b = 1$. Then we get

$$\bar{f}(p) = \frac{1}{2}(1 - \sigma) \cdot \frac{1}{3}(1 - \frac{\sigma}{3})^{-1}(1 - \sigma)$$

$$= \frac{1 - \sigma}{2}\left[\frac{1}{3} - \frac{2}{3}\sum_{n=1}^{\infty}\left(\frac{\sigma}{3}\right)^n\right]$$

So, $\quad f(t) = \frac{1}{3} B_0(t) - \frac{2}{3}\sum_{n=1}^{\infty} 3^{-n} B_n(t),$ \hfill (4.23)

where $B_n(t)$ is defined in (4.22) above.

Estimated accuracy can be verified by keeping seven terms in (4.23), for fixed values of t.

A similar result can be derived involving Legendre Polynomials too.

Example 4.23. If $f(t) = L^{-1}\left[\bar{f}(s); s \to t\right]$, then prove that

$$L^{-1}\left[\frac{1}{s}\bar{f}(s) \; ; \; s \to t\right] = \int_0^t f(x) \, dx$$

Solution. Consider $g(t) = 1 \Rightarrow \bar{g}(s) = \frac{1}{s}$. Therefore, by convolution theorem $L^{-1}\left[\frac{1}{s}\bar{f}(s); s \to t\right] = \int_0^t f(t - \tau) \, d\tau \quad$ if $L[\,f(t); t \to s\,] = \bar{f}(s)$

This equation implies,

$$L^{-1}\left[\frac{1}{s}\bar{f}(s); s \to t\right] = \int_0^t f(x) dx \; , \; \text{where } t - \tau = x$$

Example 4.24 Evaluate

$$L^{-1}\left[\frac{ch \; x\sqrt{\frac{s}{a}}}{s \; ch\left(\sqrt{\frac{s}{a}}\, l\right)} \; ; \; s \to t\right]$$

Solution. By complex inversion formula

$$f(t) = \frac{1}{2\pi i}\int_{c-i\infty}^{c+i\infty} e^{st} \frac{ch(\alpha x)}{ch(\alpha l)}\frac{ds}{s} \; , \quad \text{where } \alpha = \sqrt{\frac{s}{a}}$$

Clearly, the integrand has poles at $s = 0, s = s_n = -(2n + 1)^2\frac{a\pi^2}{4l^2}$, $n = 0$ or integer.

Residue at $\quad s = 0$ is 1.

Residue at $s = s_n$ is $\dfrac{exp(-s_n t) ch\left\{ i(2n+1)\frac{\pi x}{2l} \right\}}{\left[s\frac{d}{ds}\left\{ ch\ l\ \sqrt{\frac{s}{a}} \right\} \right]_{s=s_n}}$

$$= \frac{4(-1)^{n+1}}{(2n+1)\ \pi}\ exp\left[-\left\{ \frac{(2n+1)\pi}{2l} \right\}^2 at \right]\ ch\left\{ (2n+1)\frac{\pi x}{2l} \right\}$$

Thus, $f(t) =$ Sum of the residues at the poles of the integrand

$$= 1 + \frac{4}{\pi}\sum_{n=0}^{\infty}\frac{(-1)^{n+1}}{(2n+1)}\ exp\left[-(2n+1)^2\frac{\pi^2 at}{4l^2} \right]\ ch\left\{ (2n+1)\frac{\pi x}{2l} \right\}.$$

Example 4.25. Using power series expansion in s due to Heaviside evaluate the Laplace inversion of $\bar{f}(s) = \frac{1}{s}\frac{sh\ x\sqrt{s}}{sh\sqrt{s}}$, $0 < x < 1$, $s > 0$.

Solution. We have

$$\bar{f}(s) = \frac{1}{s}\frac{e^{x\sqrt{s}} - e^{-x\sqrt{s}}}{e^{\sqrt{s}} - e^{-\sqrt{s}}}$$

$$= \frac{1}{s}\frac{e^{-(1-x)\sqrt{s}} - e^{-(1+x)\sqrt{s}}}{1 - e^{-2\sqrt{s}}}$$

$$= \frac{1}{s}\left[e^{-(1-x)\sqrt{s}} - e^{-(1+x)\sqrt{s}} \right]\cdot\sum_{n=0}^{\infty} exp(-2n\sqrt{s})$$

Or $\bar{f}(s) = \dfrac{1}{s}\displaystyle\sum_{n=0}^{\infty}\left[exp\left\{ -(1-x+2n)\sqrt{s} \right\} - exp\left\{ -(1+x+2n)\sqrt{s} \right\} \right]$

Therefore, $f(t) = \displaystyle\sum_{n=0}^{\infty}\left[erfc\left(\frac{1-x+2n}{2\sqrt{t}} \right) - erfc\left(\frac{1+x+2n}{2\sqrt{t}} \right) \right]$

Example 4.26. Evaluate $L^{-1}\left[\frac{1}{(s^2+a^2)^2}\ ;\ s \to t \right]$

Solution. We know that $L^{-1}\left[\frac{1}{(s^2+a^2)}\ ;\ s \to t \right] = \frac{sin\ at}{a}$

Therefore, $L^{-1}\left[\dfrac{1}{(s^2+a^2)^2}\ ;\ s \to t \right] = \dfrac{sin\ at}{a} * \dfrac{sin\ at}{a}$

$$= \frac{1}{a^2}\int_0^t sin\ a\ \tau\ sin\ a(t-\tau)\ d\tau,\ \text{by convolution theorem}$$

$$= \frac{1}{2a^3}(sin\ at - at\ cos\ at).$$

Example 4.27. Prove that $L^{-1}\left[\frac{1}{(\sqrt{s}+a)}; s \to t \right] = \frac{1}{\sqrt{\pi\ t}} - ae^{a^2 t}\ erfc\ (a\sqrt{t})$

Solution.

$$\frac{1}{(\sqrt{s}+a)} = \frac{1}{\sqrt{s}} - \frac{a}{\sqrt{s}(\sqrt{s}+a)}$$

$$\therefore \ L^{-1}\left[\frac{1}{(\sqrt{s}+a)} \ ; s \to t\right]$$

$$= \ L^{-1}\left[\frac{1}{\sqrt{s}} \ ; s \to t\right] - a \ L^{-1}\left[\frac{(\sqrt{s}-a)}{(s-a^2)\sqrt{s}} \ ; s \to t\right]$$

$$= \ L^{-1}\left[\frac{1}{\sqrt{s}} \ ; \ s \to t\right] - a \ L^{-1}\left[\frac{1}{s-a^2} \ ; \ s \to t\right]$$

$$+a^2 L^{-1}\left[\frac{1}{\sqrt{s}(s-a^2)} \ ; \ s \to t\right]$$

$$= \ \frac{1}{\sqrt{\pi t}} - a \ \exp\ (a^2 t) + a \ \exp\ (a^2 t) \ erf \ (a\sqrt{t})$$

$$= \ \frac{1}{\sqrt{\pi t}} - a \ \exp\ (a^2 t) \ erfc \ (a\sqrt{t}).$$

4.6   The Double Laplace Transform

The double Laplace transform of a function of two variables $f(x, y)$ in the positive quadrant of the xy- plane is defined by the equation

$$\bar{f}(p, q) = L_2\left[f(x, y) \ ; \ (x, y) \to (p, q)\right]$$
$$= L\left[L\left\{f(x, y) \ ; \ x \to p\right\} ; y \to q\right] \qquad (4.24)$$

and hence by the double integral

$$\bar{f}(p, q) = \int_0^\infty \int_0^\infty f(x, y) \ e^{-(px+qy)} \ dx \ dy \qquad (4.25)$$

whenever the integral exists .

It therefore, follows that

$$L_2\left[f(ax, by); (x, y) \to (p, q)\right] = \frac{1}{ab}\bar{f}\left(\frac{p}{a}, \frac{q}{b}\right), a > 0, b > 0 \qquad (4.26)$$

and $\quad L_2\left[f(ax) \ g(by); (x, y) \to (p, q) \ \right] = \frac{1}{ab} \ \bar{f}\left(\frac{p}{a}\right) \bar{g}\left(\frac{q}{b}\right), a > 0, b > 0 \qquad (4.27)$

In particular, we have therefore

$$L_2\left[\ f(x); (x, y) \to (p, q)\right] = \frac{1}{q}\bar{f}(p) \qquad (4.28)$$

The formulae (4.26)- (4.28) follow from the definitions (4.24) and (4.25) above.

To evaluate double Laplace transform of a function $f(x+y)$, we change the independent variables (x, y) to (ξ, η) where we define $x = \frac{1}{2}(\xi - \eta)$, $y = \frac{1}{2}(\xi + \eta)$ implying $dxdy = \frac{1}{2} d\xi d\eta$

Thus,

$$
\begin{aligned}
L_2\left[f(x+y); (x,y) \to (p,q)\right] &= \frac{1}{2} \int_0^\infty f(\xi) d\xi \left[\int_{-\xi}^{\xi} e^{-\frac{1}{2}(p+q)\xi - \frac{1}{2}(p-q)\eta} d\eta\right] \\
&= \frac{1}{2} \int_0^\infty f(\xi) \left[\frac{2}{p-q}(e^{-q\xi} - e^{-p\xi})\right] d\xi \\
&= \frac{1}{p-q} \left[\bar{f}(q) - \bar{f}(p)\right] \qquad (4.29)
\end{aligned}
$$

after introducing the new region of double integration in the $\xi \eta$ - plane and evaluating the double integral.

To evaluate the double Laplace transform of $f(x-y)$ no such simple formula as above exists. For the case when $f(t)$ is an even function of t, we have from definition that

$$
\begin{aligned}
L_2\left[f(x-y); (x,y) \to (p,q)\right] &= \int\int_{Q_1} f(x-y) \, e^{-px-qy} \, dx \, dy \\
&+ \int\int_{Q_2} f(y-x) \, e^{-px-qy} \, dx \, dy
\end{aligned}
$$

$$(4.30)$$

while if $f(t)$ is an odd function of t

$$
\begin{aligned}
L_2\left[f(x-y); (x,y) \to (p,q)\right] &= \int\int_{Q_1} f(x-y) \, e^{-px-qy} \, dx \, dy \\
&- \int\int_{Q_2} f(y-x) \, e^{-px-qy} \, dx \, dy \qquad (4.31)
\end{aligned}
$$

where Q_1 and Q_2 are infinite regions shown below in Fig. 1.4.4

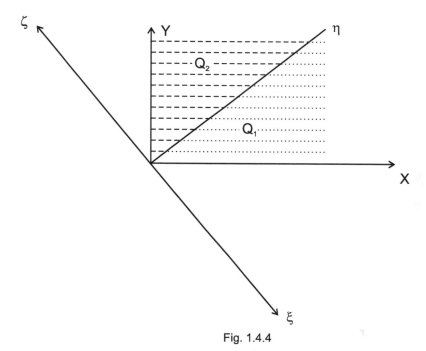

Fig. 1.4.4

After making the transformation $x = \frac{1}{2}(\xi + \eta), y = \frac{1}{2}(\eta - \xi)$ we find that

$$\int\int_{Q_1} f(x-y)e^{-px-qy}\,dx\,dy = \frac{1}{2}\int_0^\infty f(\xi)d\xi \left[\int_\xi^\infty e^{-\frac{1}{2}(p-q)\xi - \frac{1}{2}(p+q)\eta}d\eta\right]$$

$$= \bar{f}(p)/(p+q) \tag{4.32}$$

On the other hand making the transformation

$$x = \frac{1}{2}(\eta - \varsigma), \ y = \frac{1}{2}(\eta + \varsigma)$$

we get $\displaystyle\int\int_{Q_2} f(y-x)\, e^{-px-qy}\, dx\, dy$

$$= \frac{1}{2}\int_0^\infty f(\varsigma)d\varsigma \left[\int_\varsigma^\infty e^{-\frac{1}{2}(q-p)\varsigma - \frac{1}{2}(p+q)\eta}\, d\eta\right]$$

$$= \frac{\bar{f}(q)}{p+q} \tag{4.33}$$

Therefore, from above when $f(t)$ is an even function of t

$$L_2[\, f(x-y); (x,y) \to (p,q)] = \frac{1}{p+q}\left[\bar{f}(p) + \bar{f}(q)\,\right]. \tag{4.34}$$

But, if $f(t)$ is an odd function of t, we have

$$L_2\left[\,f(x-y);(x,y)\to(p,q)\right]=\frac{1}{p+q}\left[\bar{f}(p)-\bar{f}(q)\,\right].\qquad(4.35)$$

Also, $L_2\left[\,f(x)\,H\,(x-y);(x,y)\to(p,q)\,\right]$

$$=\int_0^\infty f(x)e^{-px}\left(\int_0^x e^{-qy}dy\right)dx$$

$$=\frac{1}{q}\left[\,\bar{f}(p)-\bar{f}(p+q)\,\right]\,,\qquad(4.36)$$

$$L_2\left[\,f(x)\,H\,(y-x);(x,y)\to(p,q)\,\right]$$

$$=\int_0^\infty f(x)e^{-px}\left[\int_x^\infty e^{-qy}dy\right]dx$$

$$=\frac{1}{q}\left[\bar{f}(p+q)\,\right]\,,\qquad(4.37)$$

and $L_2\left[\,f(x)\,H\,(x+y)\,;\,(x,y)\to(p,q)\,\right]=\dfrac{1}{q}\bar{f}(p)\quad(4.38)$

As a special case if $f(x)=1,\bar{f}(p)=\frac{1}{p}$ so that

$$L_2\left[H\,(x-y);(x,y)\to(p,q)\,\right]=\frac{1}{p(p+q)}\qquad(4.39)$$

The formulae for double Laplace transform of the partial derivatives of a function of two variables can be obtained from the definition as given below.

$$L_2\left[\frac{\partial u(x,y)}{\partial x};(x,y)\to(p,q)\,\right]=\int_0^\infty e^{-qy}dy\left[\int_0^\infty\frac{\partial u}{\partial x}e^{-px}dx\right]$$

$$=-\int_0^\infty e^{-qy}dy\left[u(0,y)-p\int_0^\infty e^{-px}u(x,y)dx\right]$$

$$=p\,\bar{u}\,(p,q)-\bar{u}_1(q),$$

where $\bar{u}_1(q)=L\left[u(0,y);y\to q\right]$

and similarly,

$$L_2\left[\frac{\partial u(x,y)}{\partial y}\,;\,(x,y)\to(p,q)\right]=q\,\bar{u}\,(p,q)-\bar{u}_2(p),\qquad(4.40)$$

where $\bar{u}_2(p)=L\left[u(x,0);x\to p\right]$

Laplace transform of the higher order partial derivatives of a function $u(x, y)$. of the variables x, y can similarly be derived as

$$L_2 \left[\frac{\partial^2 u}{\partial x \, \partial y} \; ; \; (x, y) \to (p, q) \right] = pq\bar{u}(p, q) - q\bar{u}_1(q) - p\bar{u}_2(p) + u(0, 0)$$

(4.41)

$$L_2 \left[\frac{\partial^2 u}{\partial x^2} \; ; \; (x, y) \to (p, q) \right] = p^2\bar{u}(p, q) - p\bar{u}_1(q) - \bar{u}_3(q)$$

(4.42)

$$L_2 \left[\frac{\partial^2 u}{\partial y^2} \; ; \; (x, y) \to (p, q) \right] = q^2\bar{u}(p, q) - q\bar{u}_2(p) - \bar{u}_4(p)$$

(4.43)

where, $\bar{u}_3(q) = L \left[\dfrac{\partial u}{\partial x}(0, y); y \to q \right]$ and $\bar{u}_4(p) = L \left[\dfrac{\partial u}{\partial y}(x, 0); x \to p \right]$

The convolution theorem in this case is generalised as below.

The convolution of two double variable functions $f(x, y)$ and $g(x, y)$ is defined by

$$f ** g = \int_0^x ds \int_0^y f(s, t) \, g\,(x - s, \; y - t)dt$$

(4.44)

and its double Laplace transform is given by

$$\mathcal{L}_2[\, f \, * * g \; ; \; (x, y) \to (p, q)\,] = \bar{f}(p, q)\,\bar{g}(p, q)$$

(4.45)

The proof of (4.45) follows directly from the definition of double Laplace transform.

From (4.45) it therefore follows that

$$L_2^{-1}\left[\bar{f}(p, q)\,\bar{g}\,(p, q)\;;\;(p, q)\;\to(x, y)\right] = f ** g.$$

Example 4.28 Solve the PDE $\frac{\partial u}{\partial x} = \frac{\partial u}{\partial y}$ satisfying the boundary conditions $u(x, 0) = a(x), \; u(0, y) = b(y), \; \text{for} \; x > 0, y > 0.$

Solution. Taking double Laplace transform, the PDE gives

$$L_2 \left[\frac{\partial u}{\partial x}; (x, y) \to (p, q) \right] = L_2 \left[\frac{\partial u}{\partial y}; (x, y) \to (p, q) \right]$$

implying, $\quad p\bar{u}(p, q) - \bar{b}(q) = q\bar{u}(p, q) - \bar{a}(p).$

Thus, $\quad \bar{u}(p, q) = \dfrac{\bar{b}(q) - \bar{a}(p)}{p - q}$

Or, $\quad u(x, y) = L_2^{-1}\left[\dfrac{\bar{b}(q) - \bar{a}(p)}{p - q} \; ; \; (p, q) \to (x, y) \right]$

4.7   The iterative Laplace Transform.

In case of double Laplace transform of a function of two variables and
those for its partial derivatives if the transformed variables p and q are
idential, then such a double Laplace transform, in particular, is called
an iterated Laplace transform which is denoted and defined as

$$I[\, f(x,y)\, ;\, p\,] \;=\; L_2[\, f(x,y); (x,y) \to (p,p)\,]$$

$$=\; \int_0^\infty \int_0^\infty f(x,y) e^{-p(x+y)}\, dx\, dy \qquad (4.46)$$

Thus, by substituting $q = p$ in the last section 4.4 all the properties
of iterated Laplace transform can be derived.

In this case the generalised convolution of a function $f(x,y)$ is denoted and defined by

$$f^{(2)}(x) = \int_0^x f(x - y, y)\, dy \qquad (4.47)$$

and hence

$$L\left[f^{(2)}(x); \; x \to p \right] = \int_0^\infty e^{-px}\, dx \left[\int_0^x (x - y, y)\, dy \right] \qquad (4.48)$$

Using the variable transformation $x = u + v$, $y = v \Rightarrow \frac{\partial(x,y)}{\partial(u,v)} = 1$

$$L\left[f^{(2)}(x); \; x \to p \right] = \int_0^\infty \int_0^\infty f(u,v) e^{-p(u+v)}\, du\, dv , \qquad \text{from (4.48)}$$

$$=\; I[\, f(x,y)\, ;\, p\,] \quad , \quad [\text{ by (4.46) }] \qquad (4.49)$$

4.8   The Bilateral Laplace Transform.

Here, we shall now define a transform on functions which are defined
for all real values of the independent variables with the kernel e^{-px} as
denoted and defined by the equation

$$\bar{f}_+(p) = L_+[\, f(x)\, ;\, x \to p\,] = \int_{-\infty}^{+\infty} f(x) e^{-px} dx \qquad (4.50)$$

provided that $f(x)$ is such that integral on the right hand side of eqn.
(4.50) is convergent for some values of p. $f_+(p)$ is sometimes called the
two-sided Laplace transform of $f(x)$ by some authors.

Writing (4.50) in the form

$$L_+[\,f(x)\,;\,x \to p\,] = \int_0^\infty f(x)e^{-px}dx + \int_0^\infty f(-x)e^{px}dx$$

we have

$$L_+[f(x)\,;\,x \to p] = L[\,f(x)\,;\,x \to p\,] + L[\,f(-x)\,;\,x \to -p\,]$$

$$(4.51)$$

Let the first integral on the r.h.s converges for $Re\;p > \sigma_1$. and the second for $Re(-p) > -\sigma_2 \Rightarrow Re\;p < \sigma_2$. Therefore, for the bilateral Laplace transform of $f(x)$ is to exist these two half-planes must overlap implying that $\sigma_2 > \sigma_1$. Hence, the condition for existance of the bilateral Laplace transform defined above in eqn. (4.50) is when Re p satisfies

$$\sigma_1 < Re p < \sigma_2 \qquad (4.52)$$

If $\sigma_1 = \sigma_2$, the strip contracts to a line $Re\;p = \sigma_1$ and the region of convergence is the line $Re\;p = \sigma_1$. But, if $\sigma_2 < \sigma_1$, the bilateral Laplace transform for $f(x)$ as defined above does not exist.

For example, if $f(x) = e^{-x^2}$, the bilateral Laplace transform is defined for all $Re\;p$. While if $f(x) = e^{-a|x|}$, the bilateral transform exists for $-a < Re\;p < +a$. But the function $f(x) = e^{-ax}$ does not have bilateral Laplace transform.

Thus , if $\bar{f}_+(p) = \int_{-\infty}^{+\infty} f(x)e^{-px}dx,\;\;\sigma_1 < Re\;p < \sigma_2,$

its inversion formula is given by

$$f(x) = \frac{1}{2\pi i}\int_{c-i\infty}^{c+i\infty} \bar{f}_+(p)e^{px}dp,\;\;\sigma_1 < c < \sigma_2 \qquad (4.53)$$

Here of course the region of convergence must always be specified since two distinct object functions may have the same image function. For example, $L_+[\,H(t);t \to p\,] = 1,\;\;Re\,p > 0$ and $L_+[\,-H(-t)\,;\,t \to p\,] = 1,\;\;Re\;p < 0$

4.9   Application of Laplace Transforms.

A. Solution of ODE with constant coeffiecients.

Suppose that we wish to solve the nth order linear ODE

$$\frac{d^n y}{dt^n} + c_1 \frac{d^{n-1} y}{dt^{n-1}} + \cdots + c_{n-1} \frac{dy}{dt} + c_n \, y = f(t) \qquad (4.54)$$

where $c_i, i = 1, 2, \cdots n$ are given constants, subject to the initial conditions $y(0) = k_1, y'(0) = k_2, \cdots y^{n=1}(0) = k_n$. Taking Laplace transform of both sides of the ODE using the results of article 3.9, and the initial conditions the ODE gives,

$$\left[\bar{y}(p)p^n - p^{n-1}k_1 - \cdots - p \, k_{n-1} - k_n \right] + c_1 \left[\bar{y}(p)p^{n-1} - p^{n-2}k_1 - \cdots k_{n-1} \right]$$
$$+ \cdots + c_{n-1} \left[p\bar{y}(p) - k_1 \right] + c_n.\bar{y}(p) = \bar{f}(p).$$

Thus,

$$\bar{y}(p) \left[p^n + c_1 p^{n-1} + \cdots + c_n \right] = \bar{f}(p) + k_1 \left[p^{n-1} + c_1 p^{n-2} \cdots + c_{n-1} \, p \right]$$
$$+ k_2 \left[p^{n-2} + c_1 p^{n-3} + \ldots \right] + \cdots + k_{n-1} \left[p + c_1 \right] + k_n$$
$$\implies \quad \bar{y}(p) = \left[\bar{f}(p) + \bar{g}(p) \right] / \left[p^n + c_1 p^{n-1} + \cdots + c_n \right]$$
$$\bar{y}(p) = \frac{\bar{f}(p)}{h(p)} + \frac{\bar{g}(p)}{h(p)}, \quad \text{where } \bar{f}(p) = L\left[f(t) \right] , \qquad (4.55)$$

$\bar{g}(p)$ is a polynomial of degree $n - 1$ in p and $h(p)$ is also a polynomial of degree n in p.

Thus, the solution of the ODE under the given initial conditions can be obtained after taking inverse Laplace transform of the last equation in(4.55). Since, the right side is a known function, we have

$$y(t) = L^{-1} \left[\frac{\bar{f}(p)}{h(p)} \right] + L^{-1} \left[\frac{\bar{g}(p)}{h(p)} \right] \qquad (4.56)$$

If eqn.(4.54) is a homogeneous one, then $f(t) \equiv 0$ and so, the required solution of the homogeneous ODE under given initial conditions is

$$y(t) = L^{-1} \left[\, \bar{g}(p)/h(p) \, \right]. \qquad (4.57)$$

Example 4.29   Solve by using Laplace transform the initial -value problem $y'''(t) + 2y''(t) - y'(t) - 2y(t) = 0$ with initial conditions $y(0) = y'(0) = 0$ and $y''(0) = 6$.

Solution. Taking Laplace transform of both sides of the given ODE under the initial conditions, we get

$$\left[p^3\bar{y}(p) - p^2y(0) - py'(0) - y''(0)\right] + 2\left[p^2\bar{y}(p) - py(0) - y'(0)\right]$$
$$- \left[p\bar{y}(p) - y(0)\right] - 2\bar{y}(p) = 0$$
$$\implies \bar{y}(p) = 6/\left[p^3 - 2p^2 - p - 2\right]$$
$$\implies \bar{y}(p) = \frac{6}{(p-1)(p+1)(p+2)} \equiv \frac{1}{p-1} - \frac{3}{p+1} + \frac{2}{p+2}$$

Thus inverting both sides , we get

$$y(t) = e^t - 3e^{-t} + 2e^{-2t} \ .$$

Example 4.30. Solve the initial-value problem defined by

$$\left[\frac{d^2}{dt^2} + n^2\right] x(t) = a \ \sin \ (nt + \alpha) \ ; \ x(0) = x'(o) = 0 \ .$$

Solution. Taking Laplace transform of both sides of the given ODE under the initial conditions, we get

$$\left(p^2 + n^2\right)\bar{x}(p) = a\cos\ \alpha\frac{n}{p^2 + n^2} + a\ \sin\ \alpha\ \frac{p}{p^2 + n^2}$$
$$\Rightarrow \quad \bar{x}(p) = an\cos\ \alpha\ \frac{1}{(p^2 + n^2)^2} + a\ \sin\ \alpha\ \frac{p}{(p^2 + n^2)^2}$$

On inversion, the above eqn. gives

$$\begin{aligned}x(t) &= an\cos\alpha\ \frac{1}{2n^3}\ [\ \sin nt - nt\cos\ nt\ ] + a\sin\alpha\ \frac{t}{2n}\ \sin\ nt \\ &= a\,[\ \sin nt \cos\alpha - nt\cos(nt+\alpha)\ ]/2n^2\end{aligned}$$

Example 4.31. Solve the initial-value problem

$$\frac{dy}{dt} + 2y + \int_0^t ydt = \sin t \ , \ y(0) = 1.$$

Solution. Taking Laplace transform of the ODE we get under the given initial condition that

$$p\bar{y}(p) - 1 + 2\bar{y}(p) + \frac{1}{p}\bar{y}(p) = \frac{1}{p^2 + 1}$$
$$\Rightarrow \quad \bar{y}(p) = \frac{p(2 + p^2)}{(p^2 + 1)(p + 1)^2}$$

By partial fraction method we can express

$$\bar{y}(p) = \frac{1}{(p+1)} - \frac{3}{2}\frac{1}{(p+1)^2} + \frac{1}{2}.\frac{1}{p^2+1}$$

On inverting, we get

$$y(t) = e^{-t} - \frac{3}{2}t\,e^{-t} + \frac{1}{2}\,\sin t$$

as the required solution of the given problem.

Example 4.32. Solve the BVP $\frac{d^2x}{dt^2} + 9x = \cos 2t$, under the boundary conditions $x(0) = 1$, $x\left(\frac{\pi}{2}\right) = -1$

Solution. Taking Laplace transform, the given ODE leads to

$$(p^2 + 9)\,\bar{x}(p) = \frac{p}{p^2+4} + p + A \ ; \ \text{where } A = x'(0), \text{ say}$$

Therefore,

$$\begin{aligned}
\bar{x}(p) &= \frac{p}{(p^2+3^2)(p^2+2^2)} + \frac{p}{p^2+3^2} + \frac{3A}{3(p^2+3^2)} \\
&= \frac{1}{5}\left[\frac{p}{p^2+2^2} - \frac{p}{p^2+3^2}\right] + \frac{p}{p^2+3^2} + \frac{A}{3}.\frac{3}{p^2+3^2}
\end{aligned}$$

Inverting,

$$\begin{aligned}
x(t) &= \frac{1}{5}\left[\cos 2t - \cos 3t\right] + \cos 3t + \frac{A}{3}\cdot\sin 3t \\
&= \frac{1}{5}\cos 2t + \frac{4}{5}\cos 3t + \frac{A}{3}\sin 3t
\end{aligned}$$

Since, $x\left(\frac{\pi}{2}\right) = -1$ is being given, the above equation gives $A = \frac{4}{5}$

Thus the solution of the BVP is

$$x(t) = \frac{1}{5}\left[\cos 2t + 4\cos 3t + 4\sin 3t\right]$$

Example 4.33. A voltage Ee^{-at} is applied at $t = 0$ to a circuit of inductance L and resistance R connected in series where a, E, L and R are constants. Using Laplace transform show that the current $i(t)$ at any time t is given by

$$i(t) = \frac{E}{R - aL}\left[e^{-at} - e^{\frac{-Rt}{L}}\right]$$

Solution. In an electrical network with given voltage $E(t)$, resistance R and inductance L, the current $i(t)$ builds at the rate given by

$$L\frac{di}{dt} + Ri(t) = E(t)$$

Here it is given that $E(t) = Ee^{-at}$ and $i(0) = 0$. Therefore, taking Laplace transform of the above equation under given conditions, we get

$$L\left[\, p\,\bar{i}(p) - 0 \,\right] + R\,\bar{i}(p) = E\frac{1}{p+a}$$

$$\Rightarrow \quad \bar{i}(p) = \frac{E}{(Lp+R)(p+a)} \equiv \frac{E}{L\left(a - \frac{R}{L}\right)}\left[\frac{1}{p+\frac{R}{L}} - \frac{1}{p+a}\right]$$

$$= \frac{E}{R - aL}\left[\frac{1}{p+a} - \frac{1}{p+\frac{R}{L}}\right]$$

Inverting, $i(t) = \dfrac{E}{R - aL}\left[\, e^{-at} - e^{-\frac{Rt}{L}} \,\right].$

B. Solution of simultaneous ODE with constant co-efficeents

The ODE involving more than one dependent variables but with a single independent variable give rise to simultaneous equations. The procedure for solving such simultaneous equations is almost same as that discussed in section A. Here also, we have to take Laplace transform of the simultaneous equations to reduce them to corresponding number of algebraic equations which can then be solved for the Laplace transformed dependent variables. Finally inverting these relations we can recover the dependent variables forming the required solutions. The method is illustrated through the following examples.

Example 4.34. Solve the following initial value problem defined by the simultaneous ODE

$$\frac{dx}{dt} - y = e^t \quad \frac{dy}{dt} + x = \sin t, \text{ given that } x(0) = 1, y(0) = 0$$

Solution. Taking Laplace transform the two given ODE under assigned initial conditions, we get

$$p\bar{x}(p) - 1 - \bar{y}(p) = \frac{1}{p-1}$$

$$p\bar{y}(p) + \bar{x}(p) = \frac{1}{p^2 + 1}$$

Solving these two equations for $\bar{x}(p)$ and $\bar{y}(p)$, we get

$$\bar{x}(p) = \frac{p}{p^2 + 1} + \frac{p}{(p-1)(p^2+1)} + \frac{1}{(p^2+1)^2}$$

$\implies x(t) = \frac{1}{2} \left[\cos t + 2\sin t + e^t - t\cos t \right]$, on inversion of Laplace transform.

Also, $\bar{y}(p) = -1 - \frac{1}{p-1} + p\bar{x}(p)$

This expression after Laplace inversion and on simplification gives rise to

$$y(t) = \frac{1}{2} \left[t\sin t - e^t + \cos t - \sin t \right]$$

Example 4.35. Solve the IVP defined by

$x'(t) + y'(t) = t,\ x''(t) - y(t) = e^{-t}$ if $x(0) = 3, x'(0) = -2$ and $y(0) = 0$.

Solution. Taking Laplace transform of the ODE under given initial conditions we get,

$$p\bar{x}(p) - 3 + p\bar{y}(p) = \frac{1}{p^2}$$

$$\text{and} \quad p^2\bar{x}(p) - 3p + 2 - \bar{y}(p) = \frac{1}{p+1}$$

Solving these equations for $\bar{x}(p)$ and $\bar{y}(p)$, we get

$$\bar{x}(p) = \frac{3p^2 + 1}{p^3(p^2 + 1)} + \frac{3p^2 + p - 1}{(p + 1)(p^2 + 1)}$$

$$\text{and} \quad \bar{y}(p) = \frac{3p^2 + 1}{p^2(p^2 + 1)} - \frac{3p^2 + p - 1}{(p + 1)(p^2 + 1)}$$

Expressing into partial fractions, we get

$$\bar{x}(p) = \frac{2}{p} + \frac{1}{p^3} - \frac{3}{2(p^2 + 1)} + \frac{p}{2(p^2 + 1)} + \frac{1}{2(p + 1)}$$

$$\text{and} \quad \bar{y}(p) = \frac{1}{p} + \frac{1}{2(p + 1)} + \frac{3}{2(p^2 + 1)} - \frac{p}{2(p^2 + 1)}$$

Taking inverse Laplace transform, it is found that

$$x(t) = 2 + \frac{1}{2}t^2 + \frac{1}{2}e^t - \frac{3}{2}\sin t + \frac{1}{2}\cos t$$

$$y(t) = 1 - \frac{1}{2}e^{-t} - \frac{1}{2}\cos t + \frac{3}{2}\sin t,$$

as solutions of the given problem.

C. Solution of ODE with variable co-efficients.

In this case let the differenial equation be a combination of terms of the form $t^n y^n(t)$ and the Laplace transform of which are given by the formulae

$$L\left[t^m y^n(t)\right] = (-1)^m \frac{d^m}{dp^m} \left[\mathrm{L}\left\{y^{(n)}(t)\right\} \right], m = 1, 2, \ldots$$

Then the usual method of solution, as in the case A, may be employed for its solution.

Example 4.36. Solve the ODE with variable coefficients

$$ty''(t) + y'(t) + ty(t) = 0$$

under the initial condition $y(0) = 1$

Solution. Taking Laplace transform, the differential equation gives

$$-\frac{d}{dp}\left[\, p^2\,\bar{y}(p) - p.1 - y'(0)\,\right] + \left[\, p\,\bar{y}(p) - 1\,\right] - \frac{d}{dp}\bar{y}(p) = 0$$

under the given initial condition. On simplification, it gives

$$-\frac{d}{dp}\left[\,(p^2 + 1)\,\bar{y}\,(p)\,\right] + p\bar{y}(p) = -1 + 1 = 0$$

$$\Rightarrow \quad -(p^2 + 1)\,\frac{d\,\bar{y}(p)}{dp} - p\bar{y}(p) = 0$$

or, $\quad \dfrac{d\bar{y}(p)}{\bar{y}(p)} = -\dfrac{p\,dp}{p^2 + 1}$

Integrating, $\quad \log\,\bar{y}(p) = -\dfrac{1}{2}\log(p^2 + 1) + \log\,c$

or, $\quad \bar{y}(p) = \dfrac{c}{\sqrt{1 + p^2}}$

or, $\quad y(t) = c\,J_0(t)$

But, by given condition $y(0) = 1 \Rightarrow 1 = c.1 \Rightarrow c = 1$. Therefore, the required particular solution is given by

$$y(t) = J_0(t).$$

Example 4.37. Solve the initial value problem defined by the ODE with variable coefficients given by

$$ty''(t) + 2y'(t) + ty(t) = 0, \ y(0) = 1$$

Solution. Taking Laplace transform under the given initial condition, the ODE leads to

$$-\frac{d}{dp}\left[\,p^2\bar{y}(p) - py(0) - y'(0)\,\right] + 2\left[\,p\bar{y}(p) - y(0)\,\right] - \frac{d}{dp}\,\bar{y}(p) = 0$$

$$\Rightarrow \qquad (p^2+1)\frac{d\bar{y}(p)}{dp} = -1$$

Or, $\qquad \displaystyle\int d\bar{y}(p) = -\int \frac{dp}{1+p^2}$

Or, $\qquad \bar{y}(p) = \cot^{-1} p + \left(c - \dfrac{\pi}{2}\right)$, c is being an arbitrary constant.

After inverting, the above relation gives

$$y(t) = L^{-1}\left(\,-\frac{\pi}{2} + c\,\right) + \frac{\sin\,t}{t}$$

As $\qquad t \to 0$, $y(0) = 1$ and therefore we get

$$1 = L^{-1}\left[\,c - \frac{\pi}{2}\,\right] + 1 \Rightarrow L^{-1}\left[\,c - \frac{\pi}{2}\,\right] = 0$$

Hence, $y(t) = \frac{\sin t}{t}$ is the required solution of the IVP.

[Note here that this last result can also be obtained by considering the limit of $\bar{y}(p)$ as $t \to 0$ or as $p \to \infty$ leading to $c = \frac{\pi}{2}$].

Example 4.38. Solve the ODE

$$ty''(t) + (t-1)y'(t) - y(t) = 0 \text{ under the conditions } y(0) = 5 \text{ and}$$
$y'(0) = -5$, if $y(\infty)$ is finite.

Solution. Taking Laplace transform, the ODE under given initial conditions leads to

$$-\frac{d}{dp}\left[\,(p^2+p)\bar{y}(p) - 5p - y'(0) - 5\,\right] = p\bar{y}(p) + \bar{y}(p) - 5$$

or, $\qquad \dfrac{d\bar{y}(p)}{dp} + \dfrac{3p+2}{p(p+1)}\,\bar{y}(p) = \dfrac{10}{p(p+1)}$

The I.F of this linear ODE in $\bar{y}(p)$ is $e^{\int \frac{(3p+2)dp}{p(p+1)}} = p^2(p+1)$. Hence its solution is given by

$$\bar{y}(p) \cdot p^2(p+1) = \int\left[\,I.F \times \frac{10}{p(p+1)}\,\right] dp + A, \ A = \text{arbitrary constant}$$

$$\Rightarrow \qquad (p^3 + p^2)\bar{y}\,(p) = 5p^2 + A$$

Now by the final value theorem of Laplace transform as $t \to \infty$ implying $p \to 0$ we get

$$\lim_{p \to 0}(p + p^2) \cdot \lim_{p \to 0} p\ \bar{y}(p)\ =\ A$$
$$\Rightarrow\ 0 \cdot y(\infty) = A$$

$\therefore$ $A = 0$ and hence $\bar{y}(p)\ =\ \dfrac{5}{p+1} \Longrightarrow y(t) = 5e^{-t}$, as the required solution.

D. Ordinary differential equations in application to Electrical networks.

We consider an electrical network consisting of (i) a power source or supplier of electromotive force E(volts) (ii) a resistor having rasistance R(ohms) (iii) an inductor having inductance L(henrys) and (iv) a capacitor having capacitance C(farads). When the circuit is switched on at time $t = 0$ by Kirchhoff law the circuit usually known as an L-C-R circuit satisfies the ODE

$$L\frac{d^2Q}{dt^2} + R\frac{dQ}{dt} + \frac{Q(t)}{C} = E$$

with the current $i(t)$ at time t is given by $i(t) = \frac{dQ}{dt}$, where Q is the charge measured in coulombs. To discuss such a problem we consider the following examples.

Example 4.39. At time $t = 0$, a constant voltage E is applied to a L-C-R circuit. The initial current i and the charge Q are zero. Find the current at any time $t > 0$, when $R^2\ <$ or $=$ or $> \frac{4L}{C}$.

Solution. By Kirchhoff law, the ODE in Q at time t for the circuit is given by $L\frac{d^2Q}{dt^2} + R\frac{dQ}{dt} + \frac{Q}{C} = E$, $i(t) = \frac{dQ}{dt}$ under $i(0) = 0, Q(0) = 0$. The above pair of simultaneous ODE can also expressed as

$$L\frac{di}{dt} + Ri + \frac{Q}{C} = E\ ,\quad \frac{dQ}{dt} = i$$

Taking Laplace transform, these equations under given initial conditions become

$$L\,[\,p\,\bar{i}\,(p) - i(0)\,] + R\,\bar{i}\,(p) + \frac{1}{C}\,\bar{Q}\,(p) = \frac{E}{p}$$

and $$p\,\bar{Q}\,(p) - Q(0) = \bar{i}(p)$$

Since $i(0) = 0$, $Q(0) = 0$, these equations become

$$\bar{i}(p) [Lp + R] + \frac{1}{C} \bar{Q} (p) = \frac{E}{p} , \; \bar{i} (p) = p \bar{Q}(p) \; \cdots (i)$$

Solving for $\bar{i}(p)$, we get

$$\left[Lp + R + \frac{1}{Cp} \right] \bar{i} (p) = \frac{E}{p} \Rightarrow \bar{i} (p) = \frac{E}{L \left[(p + \frac{R}{2L})^2 + n^2 \right]} \; \cdots (ii)$$

where we assume $\dfrac{1}{CL} - \dfrac{R^2}{4L^2} = n^2$

We shall now discuss three different cases individually.

Case I. If $R^2 < \frac{4L}{C}, \frac{R^2}{4L^2} < \frac{1}{CL} \Rightarrow n^2 > 0$.

Then inverting, the equation (ii) gives

$$
\begin{aligned}
i(t) & = L^{-1} \left[\frac{E}{L \left\{ (p + \frac{R}{2L})^2 + n^2 \right\}} \right] \\
& = \frac{E}{L} e^{-\frac{Rt}{2L}} L^{-1} \left\{ \frac{1}{p^2 + n^2} \right\} = \frac{E}{Ln} e^{-\frac{Rt}{2L}} \sin nt
\end{aligned}
$$

Case II. If $R^2 = \frac{4L}{C}, \frac{R^2}{4L^2} = \frac{1}{CL} \Rightarrow n = 0$

Then after inversion of Laplace transform, the eqn.(ii) gives

$$i(t) = \frac{E}{L} L^{-1} \left[\frac{1}{(p + \frac{R}{2L})^2} \right] = \frac{E}{L} e^{-\frac{Rt}{2L}} L^{-1} \left(\frac{1}{p^2} \right) = \frac{Et}{L} e^{-\frac{Rt}{2L}}.$$

Case III. If $R^2 > \frac{4L}{C}, n^2 < 0 \Rightarrow n^2 = -m^2$, say

Then after inversion of Laplace transform, the equation(ii) gives

$$
\begin{aligned}
i(t) & = \frac{E}{L} L^{-1} \left[\frac{1}{\left(p + \frac{R}{2L} \right)^2 - m^2} \right] \\
& = \frac{E}{L} e^{-\frac{Rt}{2L}} L^{-1} \left(\frac{1}{p^2 - m^2} \right) \\
& = \frac{E}{Lm} e^{-\frac{Rt}{2L}} \sinh mt
\end{aligned}
$$

Example 4.40. In an electrical network with e.m.f $E(t)$, rasistance R and inductance L in series, the current i builds up at the rate given by $L\frac{di}{dt} + Ri = E(t)$. If the switch of the circuit is connected at $t = 0$ and disconnected at $t = a$, find the current i at any time.

Solution. In this case, it is given that $i = 0$ at $t = 0$ and

$$E(t) = \begin{cases} E, & \text{for} \quad 0 < t < a \\ 0, & \text{for} \quad t > a \,. \end{cases}$$

Taking Laplace transform, the given ODE in $i(t)$ gives

$$(Lp + R)\, \bar{i}(p) \; = \; \int_0^\infty e^{-pt} E(t)\, dt = \int_0^a e^{-pt}\, E\, dt = \frac{E}{p} \left(1 - e^{-ap} \right)$$

This gives $\; \bar{i}(p) \; = \; \dfrac{E}{p\,(Lp + R)} - \dfrac{E e^{-ap}}{p\,(Lp + R)}$

To find $i(t)$ we have to invert the above equation.

$$\text{Now,} \quad L^{-1}\left[\frac{E}{p\,(Lp + R)} \right] = L^{-1}\left[\frac{E}{R} \left\{ \frac{1}{p} - \frac{1}{p + \frac{R}{L}} \right\} \right]$$

$$= \frac{E}{R} \left(1 - e^{-\frac{Rt}{L}} \right)$$

$$\text{and} \quad L^{-1}\left[\frac{E\, e^{-ap}}{p(Lp + R)} \right] = L^{-1}\left[\frac{E\, e^{-ap}}{R} \left\{ \frac{1}{p} - \frac{1}{p + \frac{R}{L}} \right\} \right]$$

$$= \frac{E}{R} \left(1 - e^{-\frac{R(t-a)}{L}} \right)\; H(t \quad a), \text{ by 2nd shift theorem.}$$

Therfore, we get

$$i(t) = \frac{E}{R} \left(1 - e^{-\frac{Rt}{L}} \right) - \frac{E}{R} \left(1 - e^{\frac{-R(t-a)}{L}} \right) H(t - a).$$

$$\text{where} \quad H(t - a) = \begin{cases} 0, & 0 < t < a \\ 1, & t > a \end{cases}$$

E. Application to Mechanics.

In the vibration problem the displacement of a particle measured from given fixed point satisfies some ODE. In case of loaded beams the transverse deflection of the beam also satisfies some ODE. Under initial and boundary conditions for these respective cases these displacements will be evaluated in the following examples. It may be noted that in case of beam problems we assume the boundary conditions as (i) $y = y' = 0$ for clamped or built in fixed end (ii) $y = y'' = 0$ for hinged or simply supported ends and (iii) $y'' = y''' = 0$ for free ends, y being the deflection of the beam.

Example 4.41 A mass m moves along the x-axis under the influence of a force which varies to its instantaneous speed and acts in a direction opposite to the direction of motion of the mass. Assuming that the mass is located at $x = a$ initially and is moving to the right side or along positive x-axis with the speed v_0, find the position where the mass comes to rest if the position of the mass at any time is x and it satisfies the ODE $m \frac{d^2 x}{dt^2} = -\mu \frac{dx}{dt}$.

Solution. The initial conditions of the problem are $x(0) = a$, $\frac{dx}{dt}(0) = v_0$. Now taking Laplace transform of the ODE under the given initial conditions, we get

$$m\left[p^2 \, \bar{x}(p) - pa - v_0 \right] = -\mu\left[\, p\bar{x}(p) - a\right]$$

This implies , $\bar{x}(p)\left[\, mp^2 + \mu p \, \right] = mpa + mv_0 + \mu a$

or
$$\bar{x}(p) = \frac{mv_0 + \mu a}{\mu p} - \frac{mv_0}{\mu(p + \frac{\mu}{m})}$$

Inverting Laplace transform, the above equation gives the position $x(t)$ of the mass as

$$x(t) = \frac{mv_0 + \mu a}{\mu} - \frac{mv_0}{\mu} \, e^{-\frac{\mu}{m} \, t}$$

Thus the velocity of the mass at any time is given by

$$\frac{dx}{dt} = v_0 \, e^{-\frac{\mu t}{m}}$$

When the mass comes to rest, $\frac{dx}{dt} = 0 \Rightarrow e^{-\frac{\mu t}{m}} = 0$

Thus for such a time t when $e^{\frac{-\mu t}{m}} = 0$, we get

$$x = \frac{mv_0 + \mu a}{\mu}.$$

Hence, the mass comes to rest when it is at a distance $a + \frac{mv_0}{\mu}$ from the origin.

Example 4.42. Obtain the deflection y at a distance x from one end of a weightless beam of length l and freely supported at both ends, when a concentrated load W acts at $x = a$. The differential equation for deflection y is being

$$EI\frac{d^4 y}{dx^4} = W\delta(x - a),$$

where E is the Young's modulus of elasticity of the beam and I is the moment of inertia of a cross-section of the beam about its axis. It may be noted that for freely supported ends, the following conditions $y(0) = y''(0) = 0$ and $y(l) = y''(l) = 0$ are satisfied

Solution. Taking Laplace transform in variable x the given ODE after using the initial conditions $y(0) = y''(0) = 0$ and assuming $y'(0) = c_1$, $y'''(0) = c_2$, we get

$$p^4 \bar{y}(p) = c_1 p^2 + c_2 + \frac{W}{EI} e^{-ap}$$

Or, $\bar{y}(p) = \frac{c_1}{p^2} + \frac{c_2}{p^4} + \frac{W}{EI} \frac{e^{-ap}}{p^4}$

Taking inverse Laplace transform the last equation gives

$$y(x) = c_1 x + c_2 \frac{x^3}{3!} + \frac{W}{EI} \frac{(x-a)^3}{3!} H(x-a)$$

Therefore, $y'(x) = c_1 + \frac{1}{2} c_2 x^2 + \frac{W}{2EI}(x-a)^2 H(x-a)$

$$y''(x) = c_2 x + \frac{W}{EI}(x-a) H(x-a)$$

Using the conditions $y(l) = 0,\ y''(l) = 0$ we get

$$c_2 = \frac{-W}{EI} \frac{l-a}{l}$$

$$c_1 = \left[\frac{W a(l-a)}{6EIl} \right] [\, 2l - a \,]$$

Therefore, the required deflection of the beam at a distance x from the end $x = 0$ is

$$y(x) = \frac{W}{6EI} \left[\frac{a(l-a)(2l-a)}{l} x - \frac{l-a}{l} x^3 + (x-a)^3 H (x-a) \right]$$

In particular, if $0 < x < a$, $H(x-a) = 0$

$$\therefore \quad y(x) = \frac{W}{6(EI)} \left[\frac{a(l-a)(2l-a)}{l} x - \frac{(l-a)}{l} x^3 \right]$$

Also, if $a < x < l$, $H(x-a) = 1$

$$\therefore \quad y(x) = \frac{W}{6EI} \left[\frac{a(l-a)(2l-a)}{l} x - \frac{l-a}{l} x^3 + (x-a)^3 \right]$$

From these above two results we have at $x = a$

$$y(a) = \frac{1}{3} \frac{W}{EI} \frac{a^2(l-a)^2}{l}.$$

F. Application to partial differential equation : Boundary value Problem.

It is known that in case of ordinary differential equation the Laplace transform reduces it to an algebraic equation of the transformed function. But it may now be noted that by the use of Laplace transform one of the independent variable of the equation will be removed by the transformed parameter p or s. Thus the PDE in two independent variables will be reduced to an ODE and for PDE in more than two independent variables, one of the variable will be removed by the transformed parameter p , after taking account of the boundary conditions on that variable.

To start with the matter under discussion, let $u(x,t)$ be a function of two independent variables defined for $a \leqslant x \leqslant b$ and $t \geqslant 0$. Also let $u(x,t)$, regarded as a function t , be of some exponential order for large t and is a piecewise continuous function over $t \geqslant 0$. Then Laplace transform over the variable t of $u(x,t)$ is defined as

$$L[\, u(x,t);\; t \to p \,] = \int_0^\infty e^{-pt} u(x,t)\, dt = \bar{u}(x,p) \tag{4.58}$$

Therefore , $L\left[\dfrac{\partial u}{\partial t}\right] = p\bar{u}(x,p) - u(x,0)$

$$L\left[\frac{\partial^2 u}{\partial t^2}\right] = p^2\bar{u}(x,p) - pu(x,0) - \frac{\partial u}{\partial t}(x,0) \ , \ \text{etc.}$$

The proof of the above results will follow from the definition in (4.58). Similarly, for functions of more than two independent variables of which one of them is t, say $u(x,y,t)$, or $u(x,y,z,t)$ etc, let it be defined for $t \geqslant 0$ and is of some exponential order for large t and is piecewise continuous function in $t \geqslant 0$. Then Laplace transform of such function and of their partial derivatives can be defined as an extension of the above case of two or more variables.

Example 4.43. Find the solution of the PDE $\frac{\partial^2 y}{\partial x^2} - \frac{\partial^2 y}{\partial t^2} = xt$ subject to $y(x,0) = \frac{\partial y}{\partial t}(x,0) = 0$ and $y(0,t) = 0$. It is given that the required solution is bounded both for $x \geqslant 0$ and $t \geqslant 0$.

Solution. Taking Laplace transform on the variable t the given PDE under first two initial conditions gives

$$\frac{d^2\bar{y}(x,p)}{dx^2} - \left[\, p^2\, \bar{y}(x,p) - p.0 - 0\,\right] = \frac{x}{p^2}$$

Or, $\quad \dfrac{d^2\bar{y}(x,p)}{dx^2} - p^2\, \bar{y}(x,p) = \dfrac{x}{p^2}$

The general solution of the last equation is

$$\bar{y}\,(x,p) = Ae^{px} + Be^{-px} - \frac{x}{p^4} \tag{i}$$

Since $y\,(x,t)$ is bounded for all x and t, so also $\bar{y}\,(x,p)$ for all x and p. Hence for large x, $\bar{y}\,(x,p)$ is bounded and therefore, the arbitrary constant $A = 0$ in the above form (i) of $\bar{y}\,(x,p)$. So,

$$\bar{y}\,(x,p) = B\, e^{-px} + \frac{-x}{p^4} \tag{ii}$$

But, by the given condition $y(0,t) = 0$ we get

$$\bar{y}(0,p) = 0 \tag{iii}$$

Therefore (ii) implies under (iii) that $B = 0$. Then eqn. (ii) implies

$$\bar{y}\,(x,p) = -\frac{x}{p^4}$$

Taking inverse Laplace transform the above solution reduces to

$$y\,(x,t) = -\frac{x\, t^3}{6}$$

as the required solution of the PDE under given conditions.

Example 4.44. A bar of length a is at zero temperature. At $t = 0$, the end $x = a$ is suddenly raised to a temperature u_0 and the end $x = 0$ is insulated. Find the temperature at any point x of the bar at any time $t > 0$ assuming that the surface of bar is insulated.

Solution. Let $u(x,t)$ be the temperature at any point x and at any time t of the bar so that it satisfies the heat conduction equation

$$\frac{\partial u}{\partial t} = c^2 \frac{\partial^2 u}{\partial x^2}, \quad (0 \leqslant x \leqslant a, t \geqslant 0) \tag{i}$$

subject to the conditions

$$u(x, 0) = 0 \hspace{6cm} \text{(ii)}$$

$u_x(0, t) = 0$ (insulated end means heat flow from that end is nil)   (iii)

and $u(a, t) = u_0$ \hspace{5cm} (iv)

Taking Laplace transform of eqn. (i) above over the variable t we get

$$p\bar{u}(x, p) - u(x, 0) = c^2 \frac{d^2}{dx^2} \bar{u}(x, p) \hspace{2cm} \text{(v)}$$

where $\quad \bar{u}\,(x, p) = L\,[\,u(x, t)\,;\,t \rightarrow p\,] = \int_0^\infty e^{-pt} u(x, t)\ dt$

Using condition given in eqn. (ii) , the eqn (v) becomes

$$\frac{d^2\bar{u}(x, p)}{dx^2} - \frac{p}{c^2}\bar{u}(x, p) = 0 \hspace{2cm} \text{(vi)}$$

Also, Laplace transform of eqns. (iii) and (iv) become

$$\bar{u}_x(0, p) = 0 \hspace{4cm} \text{(vii)}$$

$$\text{and} \quad \bar{u}(a, p) = \frac{u_0}{p} \hspace{3.5cm} \text{(viii)}$$

Solving eqn. (vi), we have

$$\bar{u}(x, p) = A\ e^{\frac{\sqrt{p}x}{c}} + B\ e^{-\frac{\sqrt{p}x}{c}}$$

Under the condition (vii) we get $A = B$ and therefore,

$$\bar{u}(x, p) = 2A\ \cosh \frac{\sqrt{p}x}{c}$$

Now, using condition in eqn. (viii) we have finally

$$\bar{u}(x, p) = \frac{u_o\ \cosh\left(\frac{\sqrt{p}x}{c}\right)}{p\ \cosh\left(\frac{\sqrt{p}a}{c}\right)} \hspace{3cm} \text{(ix)}$$

To obtain the temperature $u(x, t)$ we have to invert Laplace transform of $\bar{u}(x, p)$ given in eqn. (ix). But the inversion can not be obtained by heuristic method. Therefore, we take recourse to general inversion formula of Laplace transform as discussed in article 4.4. By such a formula we have here,

$$u(x,t) = \frac{1}{2\pi i} \cdot P.V \int_{\lambda - i\infty}^{\lambda + i\infty} e^{pt} \bar{u}(x,p) \, dp , \qquad \text{(x)}$$

where the Bromwich line contour Γ from $\lambda - i\infty$ to $\lambda + i\infty$, (λ is a real constant) is such that all the singularities of $\bar{u}(x,p)$ in p-plane lie to the left of $Re \; p = \lambda$. Then if C as shown in the figure below is a positively oriented closed contour consisting of Γ and a large (infinite) circle C_R enclosing all the singularities of $\bar{u}(x,p)$ after giving branch-cut in the p-plane required to make the integrand single valued eqn. (x) can be written equivalently as

$$u(x,t) = \frac{1}{2\pi i} \int_C e^{pt} \bar{u}(x,p) \, dp \qquad \text{(xi)}$$

$$= \sum_{n=1}^{N} \operatorname*{Res}_{p=p_n} \left[\, e^{pt} \bar{u}(x,p) \, \right] , \qquad \text{(xii)}$$

where N= number of poles of $\bar{u}(x,p)$ inside C. Here also

$$\lim_{R \to \infty} \int_{C_R} e^{pt} \bar{u}(x,p) \, dp = 0$$

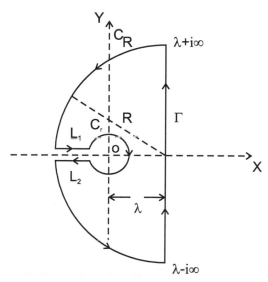

Fig. The contour C. (Γ+C_R+L_1+C_r+L_2)

Also, $\displaystyle\int_{L_1+L_2} e^{pt}\,\bar{u}\,(x,p)\,dp = 0$,

$$\lim_{r\to 0}\int_{C_r} e^{pt}\,\bar{u}\,(x,p)\,dp = 0$$

Thus eqn. (xii) follows from eqns. (x) and (xi). Now the poles of the integrand in (xi) inside C are given by

$$p = p_n = -\frac{(2n-1)^2}{4a^2}c^2\pi^2 \ , \ n = 0,1,2,\cdots$$

Residue at $p = p_n$

$$= \lim_{p\to p_n}\left[\frac{u_o(p-p_n)\,e^{pt}\,\cosh\left(\frac{\sqrt{p}x}{c}\right)}{p\,\cosh\left(\sqrt{p}\,\frac{a}{c}\right)}\right]$$

$$= u_0\lim_{p\to p_n}\left[\left\{\frac{p-p_n}{ch\sqrt{p}\,\frac{a}{c}}\right\}\right]\cdot\lim_{p\to p_n}\left[\frac{e^{pt}ch(\sqrt{p}\,\frac{x}{c})}{p}\right]$$

$$= u_0\lim_{p\to p_n}\left\{\frac{1}{sh(\sqrt{p}\,\frac{a}{c})}\cdot\frac{2\sqrt{p}c}{a}\right\}\cdot\lim_{p\to p_n}\left[\frac{e^{pt}ch(\sqrt{p}\,\frac{x}{c})}{p}\right]$$

$$= \frac{4u_0(-1)^n}{(2n-1)\pi}\cdot e^{\frac{-(2n-1)^2\pi^2\,c^2\,t}{4a^2}}\cdot\cos\frac{(2n-1)\pi x}{2a}$$

Thus, we get

$$u(x,t) = \frac{4u_0}{\pi}\sum_{n=1}^{\infty}\frac{(-1)^n}{(2n-1)}e^{\frac{-(2n-1)^2\pi^2\,c^2\,t}{4a^2}}\cos\frac{(2n-1)\pi x}{2a},$$

as the required solution of the given heat conduction problem.

Example 4.45. A semi-infinite solid $x > 0$ has its initial temperature zero. A constant heat flux H is applied at the face $x = 0$ and hence $-ku_x(0,t) = A$, where k is the thermal diffusivity of the material of the solid and $u(x,t)$ is the temperature of the solid at any point of it at time t. If the heat conduction equation of the solid is given by

$$\frac{\partial u}{\partial t} = k\frac{\partial^2 u}{\partial x^2} \ , \ \ 0 < x < \infty,\ t > 0 \tag{i}$$

with the boundary condition $\frac{\partial u(0,t)}{\partial x} = -\frac{H}{\tau}$ and the initial condition $u(x,0) = 0$, find the temperature of the solid at any point $x > 0$ at any instant, given that

$$L^{-1}\left[\frac{e^{-x\sqrt{p}}}{p^{\frac{3}{2}}}\ ;\ p\to t\right] = \sqrt{\frac{t}{\pi}}\ e^{\frac{-x^2}{4t^2}} - \frac{x}{2}\ erfc\left\{\frac{x}{(2\sqrt{t}\ )}\right\}$$

Solution. Taking Laplace transform over the variable t the eqn. (i) under the initial condition takes the form

$$p\bar{u}(x,p) = k\frac{d^2}{dx^2}\ \bar{u}\ (x,p)$$

Its general solution is given by

$$\bar{u}(x,p) = Ae^{x\sqrt{p/k}} + Be^{-x\sqrt{p/k}} \tag{ii}$$

Since the temperature $u(x,t)$ is bounded for large x, so is $\bar{u}(x,p)$ and therefore eqn. (ii) gives $A = 0$. Thus eqn.(ii) becomes

$$\bar{u}(x,p) = Be^{-x\sqrt{\frac{p}{k}}} \tag{iii}$$

Also, Laplace transform of the boundary condition becomes

$$\frac{\partial\bar{u}}{\partial x}(0,p) = -\frac{H}{\tau p}\ ,\quad (\tau = \text{thermal conductivity }) \tag{iv}$$

Using (iv) in (iii), we get $B = \frac{H}{\tau\sqrt{k}p^{\frac{3}{2}}}$ and hence eqn. (iii) implies

$$\bar{u}(x,p) = \frac{\sqrt{k}\ H\ e^{-x\sqrt{\frac{p}{k}}}}{\tau\ p^{\frac{3}{2}}}$$

Now, inverting Laplace transform the required solution of the problem is given by

$$u(x,t) = L^{-1}\left[\frac{H\sqrt{k}}{\tau}\ \frac{e^{-x\sqrt{\frac{p}{k}}}}{p^{\frac{3}{2}}};\quad p\to t\right]$$

Therefore, by given condition

$$u(x,t) = \frac{H\sqrt{k}}{\tau}\ \left[\sqrt{\frac{t}{\pi}}\ e^{-\frac{x^2}{4t^2 k}} - \frac{x}{2\sqrt{k}}\ erfc\left\{\frac{x}{2\sqrt{kt}}\right\}\right]$$

Example 4.46. The temperature $u(x,t)$ of a slab of material medium satisfies the heat conduction equation $\frac{\partial u}{\partial t} = k\frac{\partial^2 u}{\partial x^2}, 0 < x < l,\ t > 0$. If the initial temperature at the bounding planes $x = 0$ and $x = l$ of the slab be u_0, find the temperature in this solid after the face at $x = 0$ is insulated and the temperature of the face at $x = l$ is reduced to zero.

Solution. The temperature distribution $u(x, t)$ of the slab satisfies the PDE

$$\frac{\partial u}{\partial t} = k \frac{\partial^2 u}{dx^2} \tag{i}$$

subject to the boundary conditions $\frac{du(o,t)}{dx} = 0$ and $u(l, t) = 0$ and the initial condition $u(x, 0) = u_0, 0 < x < l$. Taking Laplace transform equation (i) under the initial condition gives

$$p\bar{u}(x, p) - u_0 = k \frac{d^2}{dx^2} \bar{u}(x, p)$$

Solving the above equation we get

$$\bar{u}(x, p) = A \cosh\left(\sqrt{\frac{p}{k}}\, x\right) + B \sinh\left(\sqrt{\frac{p}{k}}\, x\right) + \frac{u_0}{p} \tag{ii}$$

The Laplace transform of the given boundary conditions become

$$\frac{d}{dx} \bar{u}(0, p) = 0 \quad \text{and} \quad \bar{u}(l, p) = 0 \tag{iii}$$

Using (iii) in (ii) we get $B = 0$ and $A = -\dfrac{u_0}{p \cosh\left(\sqrt{\frac{p}{k}}\, l\right)}$

Therefore, eqn. (ii) becomes

$$\bar{u}(x, p) = \frac{u_0}{p} - \frac{u_0}{p} \frac{\cosh\left(\sqrt{\frac{p}{k}}\, x\right)}{\cosh\left(\sqrt{\frac{p}{k}}\, l\right)}$$

Inverting Laplace transform the above relation gives

$$u(x, t) = u_0 - u_0 L^{-1}\left[\frac{\cosh\left(\sqrt{\frac{p}{k}}\, x\right)}{p \cosh\left(\sqrt{\frac{p}{k}}\, l\right)}\right] \tag{iv}$$

The second term in the above is almost equivalent to eqn. (ix) of example-4.44. Hence, following the same technique (iv) can be evaluated to get the solution $u(x, t)$ of the problem in terms of x and t.

Note: As an alternative procedure for evaluation of Laplace inversion the expression $\frac{1}{p} \cdot \text{ch}\left(\sqrt{\frac{p}{k}}\,x\right)/\text{ch}\left(\sqrt{\frac{p}{k}}\,l\right)$ may be expanded in an infinite series and then term by term inversion may be obtained for solution, noting that $L^{-1}\left[\frac{1}{p}\,e^{-c\sqrt{p}}\right] = erfc\left(\frac{c}{2\sqrt{t}}\right)$.

Example 4.47. Consider the diffusion equation in a semi-infinite line $0 < x < \infty$ defined by $\frac{\partial^2\theta}{\partial x^2} = \frac{1}{k}\frac{\partial\theta}{\partial t}$, $t \geqslant 0$ and subject to the boundary condition $\theta(0,t) = f(t)$ and the initial condition $\theta(x,0) = 0$ together with the limiting condition $\theta(x,t) \to 0$ as $x \to \infty$. Obtain the temperature $\theta(x,t)$ of the line under this general case and of the particular cases (*i*) $\theta(0,t) = \theta_0\,(T/t)^{\frac{1}{2}}$, θ_0 and T are positive constants

(ii) $\theta(0,t) = \theta_0$, a constant and (iii) $\frac{\partial}{\partial x}\theta(0,t) = f(t)$ instead of the condition $\theta(0,t) = f(t)$ by using Laplace transform.

Solution. Taking Laplace transform the *PDE* under the given initial condition reduces to

$$\frac{d^2\bar\theta(x,p)}{dx^2} = \frac{1}{k}\left[\,p\,\bar\theta(x,p)\,\right]$$

We solve the above second order *ODE* to obtain

$$\bar\theta(x,p) = Ae^{-\sqrt{\frac{p}{k}}x} + Be^{\sqrt{\frac{p}{k}}x} \qquad (i)$$

Now, taking Laplace transform to the given boundary condition $\theta(0,t) = f(t)$ we get

$$\bar\theta(0,p) = \bar f(p)$$

Under Laplace transformed the limiting condition becomes $\bar\theta(x,p) \to 0$ as $x \to \infty$. Using these two results, we get

$$\bar\theta(x,p) = \bar f(p)\,e^{-\sqrt{\frac{p}{k}}x}, \quad \text{since } B = 0 \qquad (ii)$$

Therefore, inverting this equation we have by the convolution theorem that

$$\theta(x,t) = \int_0^t f(\tau)\,g(x,t-\tau)\,d\tau \qquad (iii)$$

where $\quad g(x,t) = L^{-1}\left[e^{-\sqrt{\frac{p}{k}}x}\,;\,p \to t\right] \qquad (iv)$

But since, $L\left[t^{-\frac{1}{2}}e^{-c/t}\,;\,t \to p\right] = \sqrt{\frac{\pi}{p}}e^{-2\sqrt{cp}}$, differentiating both sides with respect to c we obtain

$$L\left[t^{-\frac{3}{2}}\ e^{-\ c/t}\ ;\ t \to p \right] = \sqrt{\frac{\pi}{c}}\ e^{-2\sqrt{cp}} \qquad (v)$$

Therefore, $g\,(x,t) = \dfrac{1}{\sqrt{4k\pi t^3}}\ e^{-x^2/4kt}$

and hence the solution of this general problem is given by

$$\theta\,(x,t) = \frac{1}{\sqrt{4\pi k}} \int_0^t f(\tau) exp\left[\frac{x^2}{4k\,(t-\tau)} \right] \frac{d\tau}{(t-\tau)^{\frac{3}{2}}}$$

We shall now take up the particular cases shown below.

Case-I

$$\theta(0,t) = \theta_0 \left(\frac{T}{t} \right)^{\frac{1}{2}}$$

So, $\bar{f}(p) = \theta_0\ T^{\frac{1}{2}}\ L\left[t^{-\frac{1}{2}}\ ;\ t \to p \right] = \theta_0 \left(\dfrac{\pi T}{p} \right)^{\frac{1}{2}}$ and hence

$$\bar{\theta}(x,p) = \theta_0\ (\pi T)^{\frac{1}{2}}\ p^{-\frac{1}{2}}\ e^{-x\sqrt{\frac{p}{k}}}$$

Inverting , $\theta(x,t) = \theta_0 \left(\dfrac{T}{t} \right)^{\frac{1}{2}} e^{-\frac{x^2}{4kt}}$

Case-II

$$\theta(0,t) = \theta_0 \ ,\ \text{a constant} .$$

In this case , $\bar{f}(p) = \dfrac{\theta_0}{p}$ and   so $\bar{\theta}(x,p) = \dfrac{\theta_0}{p}\ e^{-x\sqrt{\frac{p}{k}}}$

After Laplace inversion we get

$$\begin{aligned}
\theta\,(x,t) &= \theta_0\ L^{-1}\left[\frac{1}{p}\ e^{-x\sqrt{\frac{p}{k}}}\ ;\ p \to t \right] \\
&= \theta_0\ Erfc\left(\frac{1}{2} \frac{x}{\sqrt{kt}} \right) .
\end{aligned}$$

Case - III

In this case $\dfrac{\partial}{\partial x}\left[\bar{\theta}(0,p) \right] = \bar{f}(p)\ ,\ \bar{\theta}(x,p) \to 0$ as $x \to \infty$

Therefore, $\bar{\theta}(x,p) = -\sqrt{k}\bar{f}(p)\ \dfrac{1}{\sqrt{p}}\ e^{-x\sqrt{\frac{p}{k}}}$

Hence, $\quad \theta\,(x,t) = \displaystyle\int_0^t f\,(\tau)\,g\,(x, t - \tau)\,d\tau$

where $\quad g(x, t) = -k^{\frac{1}{2}}\,L^{-1}\left[\dfrac{1}{\sqrt{p}}\,e^{-x\sqrt{\frac{p}{k}}}\;;\; p \to t\right]$

$$= -\left(\dfrac{k}{\pi t}\right)^{\frac{1}{2}} e^{-\frac{x^2}{4kt}}$$

Therefore, the required solution for this general case is given by

$$u(x, t) = -\sqrt{\dfrac{k}{\pi}}\int_0^t f(\tau)\,exp\left[-\dfrac{x^2}{4k(t - \tau)}\right]\dfrac{d\,\tau}{\sqrt{t - \tau}}\,.$$

Example 4.48. An infinitely long string of negligible mass having one end at $x = 0$, is initially at rest along the x-axis. The end $x = 0$ is given a transverse displacement $f(t)$, $t > 0$. Find the displacement of any point of the string at any time by using Laplace transform.

Solution. Let $y(x, t)$ be the transverse displacement of any point x of the string at any time t. Then, $y(x, t)$ satisfies the wave equation $\frac{\partial^2 y}{\partial t^2} = c^2\,\frac{\partial^2 y}{\partial x^2}$, $x > 0, t > 0$; subject to the initial conditions $y(x, 0) = 0$, $\frac{\partial y}{\partial t}(x, 0) = 0$ and the boundary condition $y(0, t) = f(t)$ and $y(x, t)$ is bounded. Laplace transform of the PDE over the variable t gives

$$p^2\bar{y}\,(x, p) - p\,y\,(x, 0) - \dfrac{\partial y\,(x, 0)}{\partial t} = c^2\dfrac{d^2\bar{y}(x, p)}{dx^2}$$

Using the initial conditions we rewrite the above equation as

$$\dfrac{d^2\bar{y}}{dx^2} = \left(\dfrac{p}{c}\right)^2 \bar{y}\,(x, p) \qquad\qquad \text{(i)}$$

Again, Laplace transform of the given boundary conditions are

$$\bar{y}(0, p) = \bar{f}(p) \quad \text{and} \quad \bar{y}(x, p) \text{ is bounded.}$$

Solving (i), we get

$$\bar{y}\,(x, p) = A\,e^{\frac{px}{c}} + B\,e^{-\frac{px}{c}}$$

Since $y(x, t)$ is bounded so is $\bar{y}(x, p)$ for all x. Then we have

$$\bar{y}\,(x, p) = Be^{-\frac{px}{c}}, \quad \text{because } A = 0.$$

By use of Laplace transformed boundary condition $\bar{y}(0,p) = \bar{f}(p)$ we have

$$\bar{y}\,(x,p) = \bar{f}(p)\,.\,e^{-\frac{px}{c}}$$

Therefore, by complex inversion formula of Laplace transform we get

$$y(x,t) \;=\; \frac{1}{2\pi i}\int_{\lambda-i\infty}^{\lambda+i\infty} e^{(t-\frac{x}{c})p}\,\bar{f}(p)\,dp$$

$$\;=\; f\left(t-\frac{x}{c}\right).$$

Example 4.49 A light flexible string of length a is stretched between two fixed points at $x = 0$ and at $x = a$. If the string is displaced into the curve $y = b\sin\frac{\pi\,x}{a}$ and released from rest in that position initially, find its displacement at any time t and at any point x.

Solution. The transverse vibration $y(x,t)$ of the string satisfies the PDE $\frac{\partial^2 y}{\partial t^2} = c^2\frac{\partial^2 y}{\partial x^2}$ subject to boundary conditions $y(0,t) = 0$, $y(a,t) = 0$ and the initial conditions $y(x,0) = b\sin\frac{\pi x}{a}$, $\frac{\partial y}{\partial t}(x,0) = 0$. Taking Laplace transform of the PDE over the variable t under the initial conditions we get

$$\frac{d^2\bar{y}(x,p)}{dx^2} - \frac{p^2}{c^2}\,\bar{y}\,(x,p) = -\frac{bp}{c^2}\,\sin\frac{\pi x}{a}$$

The general solution of this second order ODE is

$$\bar{y}(x,p) = A\,e^{\frac{px}{c}} + B\,e^{-\frac{px}{c}} + \frac{bp\,\sin\left(\frac{\pi x}{a}\right)}{p^2 + \left(\frac{\pi c}{a}\right)^2} \qquad (i)$$

Now, Laplace transform of the given boundary conditions are

$$\bar{u}(0,p) = 0 \quad and \quad \bar{u}(a,p) \;=\; 0$$

Using these results in (i) we get

$$A + B \;=\; 0 \quad and \quad A\,e^{\frac{pa}{c}} + B\,e^{-\frac{pa}{c}} \;=\; 0.$$

Solving these equations for A and B, we get $A = B = 0$, then

$$\bar{y}(x,p) \;=\; b\,\sin\frac{\pi x}{a}\,.\,\frac{p}{p^2 + \left(\frac{\pi c}{a}\right)^2}$$

Therefore, inverting this equation we get the required solution of the given problem as

$$y(x,t) = b \sin \frac{\pi x}{a} L^{-1} \left[\frac{p}{p^2 + \left(\frac{\pi c}{a} \right)^2} ; p \to t \right]$$

$$= b \sin \frac{\pi x}{a} \cos \frac{\pi c t}{a}$$

Example 4.50. A tightly stretched flexible string has its end points at $x = 0$ and at $x = l$. At time $t = 0$ the string is given the shape defined by $y = \mu \sum_{n=1}^{\infty} \sin \frac{n\pi}{l} x$, where μ is a constant, and then released from rest. Find the displacement $y = (x,t)$ of the string.

Solution. The partial differential equation satisfying the transverse displacement $y(x,t)$ of the string and satisfying the initial conditions $y(x,0) = \mu \sum_{n=1}^{\infty} \sin \frac{n\pi x}{l}$ and $\frac{\partial y \ (x,0)}{\partial t} = 0$ and boundary conditions $y(0,t) = y(l,t) = 0$ is given by

$$\frac{\partial^2 y}{\partial t^2} = c^2 \frac{\partial^2 y}{\partial x^2} ,$$

c being a material constant of the string.

Taking Laplace transform of the PDE under the initial conditions, we have

$$p^2 \bar{y}(x,p) - py(x,0) - \frac{\partial y(x,0)}{\partial t} = c^2 \frac{d^2}{dx^2} \bar{y}(x,p)$$

Or $\qquad \left[\frac{d^2}{dx^2} - \frac{p^2}{c^2} \right] \bar{y} \ (x,p) = -\frac{p\mu}{c^2} \sum_{n=1}^{\infty} \sin \frac{n\pi x}{l}$

This equation has the solution

$$\bar{y}(x,p) = A \ ch \ \frac{p}{c}(l-x) + B \ sh\frac{p}{c}(l-x) + \mu \sum_{n=1}^{\infty} \frac{p \sin \frac{n\pi x}{l}}{p^2 + \left(\frac{n\pi c}{l} \right)^2}$$

Laplace transform of the given boundary conditions are

$$\bar{y}(0,p) = \bar{y}(l,p) = 0$$

Using these results, the above form of $\bar{y}(x,p)$ gives

$$A \ ch \ \frac{pl}{c} + B \ sh\frac{pl}{c} = 0$$

and $\qquad A = 0$. Thus , $B = 0$ and hence

$$\bar{y}(x,p) = \mu \sum_{n=1}^{\infty} \frac{p \ \sin \frac{n\pi x}{l}}{p^2 + \left(\frac{n\pi c}{l} \right)^2}$$

Inverting Laplace transform we get the solution of the problem as

$$y(x,t) = \mu \sum_{n=1}^{\infty} \cos \frac{n\pi ct}{l} \sin \frac{n\pi x}{l}.$$

Example 4.51. An infinite string lying along the x-axis is vibrating by giving a displacement $f(t)$ to the end $x = 0$. The other end of the string at large distance from the origin remains fixed. If initially the displacement and the velocity of every point of the string were zero, find the displacement of every point of the string at any time, when the displacement satisfies the PDE

$$\frac{\partial^2 y}{\partial t^2} = c^2 \frac{\partial^2 y}{\partial x^2} \qquad\qquad \text{(i)}$$

Solution. As per statement of the problem the initial and the boundary conditions are

$$y(x,0) = 0, \frac{\partial y\,(x,0)}{dt} = 0 \text{ and } y(0,t) = f(t), \ \lim_{x \to \infty} y(x,t) = 0$$

respectively. Taking Laplace transform over t, eqn. (i) becomes

$$p^2 \bar{y}(x,p) - py(x,0) - \frac{\partial y}{\partial t}\,(x,0) = c^2 \frac{d^2}{dx^2}\,\bar{y}(x,p)$$

Using the initial conditions, we have then

$$\frac{d^2\bar{y}}{dx^2}(x,p) - \frac{p^2}{c^2}\,\bar{y}(x,p) = 0 \qquad , \ p > 0.$$

The general solution of this ODE is

$$\bar{y}(x,p) = Ae^{\frac{p}{c}\,x} + Be^{-\frac{p}{c}\,x} \qquad \text{(ii)}$$

Since, $\lim_{x \to \infty} y(x,0) = 0$, we must have from the above solution that $A = 0$. Therefore, the solution (ii) becomes

$$\bar{y}(x,p) = Be^{-\frac{p}{c}\,x} \qquad \text{(iii)}$$

But, again from the boundary condition $y(0,t) = f(t)$ under Laplace transform, we get $\bar{y}(0,p) = \bar{f}(p) = L[\,f(t)\,]$. Hence, by (iii) we get

$$\bar{y}(0,p) \ = \ B \ = \ \bar{f}(p)$$

Therefore, $\bar{y}(x,p) = \bar{f}(p)\, e^{-\frac{p}{c}x}$. Inverting the above transformed solution $\bar{y}(x,p)$, we have then finally

$$L^{-1}\left[\,\bar{y}(x,p)\,;\,p\to t\,\right] = L^{-1}\left[\,\bar{f}(p)\,.\,e^{-\frac{p}{c}x}\,;\,p\to t\,\right]$$

Or $$y(x,t) = f\left(t - \frac{x}{c}\right)\,H\left(t - \frac{x}{c}\right).$$

by the second shift theorem of Laplace transform. Thus the required displacement is

$$
\begin{aligned}
y(x,t) &= 0, \qquad t < \frac{x}{c} \\
&= f\left(t - \frac{x}{c}\right), \qquad t > \frac{x}{c}
\end{aligned}
$$

G. Application to solution of Integral equation and Integro-differntial equation.

An integral equation is an equation in which the unknown function appears under integral sign. For example,

$$f(s) = \int_a^b k\,(s,t)\,g(t)\,dt\,,$$

$$g(s) = f(s) + \int_a^b k\,(s,t)\,g\,(t)\,dt$$

and $$g(s) = \int_a^b k(s,t)\,[g\,(t)]^2\,dt\,,$$

where b, the upper limit of the integral may be either a variable or a constant. The function $g(s)$ is the unknown function while all other functions are known. These functions may be complex valued functions of the real values of s and t. If the limits a and b are both constants, the integral equation is known as Fredholm integral equation. If the upper limit b is a variable and a is a constant, the corresponding integral equation is called Volterra equation. The kernel $k(s,t)$ of the integral equation is a known function. If $k(s,t) = K(s-t)$, a function of the difference $s - t$ only, the corresponding integral equation is called a convolution type integral equation. Many interesting problems of mechanics and physics lead to integral equation in which the kernel $k(s,t)$ is a function of the difference $(s-t)$ only. Thus, when $k(s,t) = K(s-t)$, where K is a function of one variable, though $k(s,t)$ is a function of two variables, Laplace transform is one of the necessary tools towards the solution of such convolution type Volterra integral equation

$$g(s) = f(s) + \int_0^s K(s-t)g(t) \, dt \tag{4.59}$$

as shown below.

On applying Laplace transform to both sides of the eqn.(4.59) and using the convolution formula, we have

$$\bar{g}(p) = \bar{f}(p) + \bar{K}(p) \cdot \bar{g}\,(p)$$

Or $$\bar{g}(p) = \bar{f}(p)/\left[1 - \bar{K}(p)\,\right], \tag{4.60}$$

and then the inversion of eqn. (4.60) yields the solution of (4.59)

A special type convolution integral equation

$$\int_0^s \frac{f(u)}{(s-u)^\alpha} \, du = g(s), \qquad 0 < \alpha < 1 \tag{4.61}$$

is called an Abel's integral equation. Here $f(u)$ is an unknown and $g(s)$ is a known function.

Laplace transform method may also be effectively used to solve such an equation in (4.61) too.

The following examples may be considered for much more details in applications of the method.

Example 4.52. Solve the integral equation

$$s = \int_0^s e^{s-t} \, g(t) \, dt$$

Solution. Taking Laplace transform of both sides, we obtain

$$\frac{1}{p^2} = \bar{K}(p) \, \bar{g}\,(p) = \frac{1}{p-1} \, \bar{g}(p)$$

Or , $$\bar{g}\,(p) = \frac{1}{p} - \frac{1}{p^2}$$

Inverting , $g\,(s) = 1 - s$,   as the solution of the integral equation

Example 4.53. Solve the integral equation

$$f(s) = s^2 + \int_0^s f(u) \, \sin\,(s-u) \, du$$

Solution. Taking Laplace transform on both sides, we get

$$\bar{f}(p) \;=\; \frac{2}{p^3} + L[\,f(s) * \sin s\,] = \frac{2}{p^3} + \bar{f}(p)/(p^2+1)$$

$$\Rightarrow \bar{f}(p) \;=\; \frac{2(p^2+1)}{p^5} = \frac{2}{p^3} + \frac{2}{p^5}\;.$$

Inverting , $f(s) \;=\; t^2 + \dfrac{t^4}{12}\;.$

Example 4.54. Solve the integral equation

$$g(t) = 1 - \int_o^t (\,t - \tau)\; g\;(\tau)\; d\tau$$

Solution. Taking Laplace transform, the given integral equation becomes

$$\bar{g}(p) \;=\; \frac{1}{p} - L\left[\int_o^t (\,t - \tau)\; g\;(\tau)\; d\tau\;;\; t \to p\right]$$

$$=\; \frac{1}{p} - \frac{1}{p^2}\cdot \bar{g}\;(p)\;,\;\; \text{by convolution theorem .}$$

Therefore , $\qquad \bar{g}(p)\left[\dfrac{1+p^2}{p^2}\right] = \dfrac{1}{p} \;\Rightarrow\; \bar{g}(p) = \dfrac{p}{p^2+1}$

Inverting the last equation, the required solution of the integral equation is

$$g(t) \;=\; L^{-1}\left[\frac{p}{p^2+1}\;;\; p \to t\right]$$

$$=\; \cos t$$

Example 4.55. Solve the integral equation

$$\sin t = \int_0^t J_0\;(t - \tau\,)\; g\;(\tau)\; d\tau$$

Solution. Taking Laplace transform, the given integral equation becomes

$$\frac{1}{p^2+1} \;=\; L[\,J_o\;(t)\;;\; t \to p]\cdot \bar{g}\;(p) = \frac{1}{\sqrt{p^2+1}}\;\bar{g}\;(p)$$

Thus , $\quad \bar{g}\;(p) \;=\; \dfrac{1}{\sqrt{p^2+1}}\;.$

Inverting, we get the solution of the problem as

$$g(t) = L^{-1}\left[\frac{1}{\sqrt{p^2 + 1}}\right] = J_0(t).$$

Example 4.56. Solve the integral equation

$$\cos t = \int_0^t J_0(t - \tau) \, g(\tau) \, d\tau$$

Solution. Taking Laplace transform the integral equation gives

$$\frac{p}{p^2 + 1} = \frac{1}{\sqrt{p^2 + 1}} \cdot \bar{g}(p) \Rightarrow \bar{g}(p) = \frac{p}{\sqrt{p^2 + 1}}$$

Inverting, $g(t) = L^{-1}\left[p.\dfrac{1}{\sqrt{p^2 + 1}}\right] = \displaystyle\int_0^t H(\tau) J_0(t - \tau) d\tau$

$$= \int_0^t J_0(t - \tau) \, d\tau \, , \, t > \tau > 0$$

since it is known that , $L\left[\displaystyle\int_0^t f(\tau) \, g(t - \tau) d\tau\right] = \bar{f}(p) \cdot \bar{g}(p).$

Example 4.57. Solve the Abel's integral equation

$$\int_0^t \frac{f(\tau) d\tau}{(t - \tau)^{\frac{1}{3}}} = t(1 + t)$$

Solution.   It is known that $f(t) * t^{-\frac{1}{3}} = t + t^2$. Taking the Laplace transform of both sides we get

$$L[f(t) \, ; \, t \to p] \cdot L\left[t^{-\frac{1}{3}}\right] = L[t] + L[t^2]$$

Or , $\bar{f}(p) \cdot \dfrac{\Gamma\left(\frac{2}{3}\right)}{p^{\frac{2}{3}}} = \dfrac{1}{p^2} + \dfrac{2}{p^3}$

Or , $\bar{f}(p) = \dfrac{1}{\Gamma\left(\frac{2}{3}\right)}\left[\dfrac{1}{p^{\frac{4}{3}}} + \dfrac{2}{p^{\frac{7}{3}}}\right]$

Taking inverse Laplace transform, the above equation gives the solution of the Abel's integral equation as

$$f(t) = \frac{1}{\Gamma\left(\frac{2}{3}\right)}\left[\frac{t^{\frac{4}{3} - 1}}{\Gamma\left(\frac{4}{3}\right)} + 2\frac{t^{\frac{7}{3} - 1}}{\Gamma\left(\frac{7}{3}\right)}\right]$$

$$= \frac{1}{\Gamma\left(\frac{2}{3}\right)} \left[\frac{t^{\frac{1}{3}}}{\frac{1}{3}\,\Gamma\left(\frac{1}{3}\right)} + \frac{2\,t^{1+\frac{1}{3}}}{\frac{4}{3}\cdot\frac{1}{3}\,\Gamma\left(\frac{1}{3}\right)} \right]$$

$$= \frac{t^{\frac{1}{3}}}{\Gamma\left(\frac{1}{3}\right)\Gamma\left(1-\frac{1}{3}\right)} \left[3 + \frac{9}{2}t \right]$$

$$= \frac{3t^{\frac{1}{3}} \sin\frac{\pi}{3}}{\pi} \left[1 + \frac{3}{2}\,t \right]$$

$$= \frac{3\sqrt{3}\,t^{\frac{1}{3}}}{4\pi}\,(2 + 3t)$$

Example 4.58. Solve the integro-differential equation

$$f'(t) = t + \int_0^t f\,(t - \tau)\,\cos\tau\,\,d\tau\ ,\ \text{if } f(0) = 4.$$

Solution. Taking Laplace transform of both sides the given equation becomes

$$p\bar{f}(p) - 4 = \frac{1}{p^2} + \bar{f}(p)\cdot\frac{p}{p^2+1}$$

$$\Rightarrow p\bar{f}(p) \left[1 - \frac{1}{1+p^2} \right] = \frac{1}{p^2} + 4$$

Or, $$\frac{p^3}{1+p^2}\,\bar{f}(p) = \frac{1}{p^2} + 4 = \frac{1+4p^2}{p^2}$$

Or, $$\bar{f}(p) = \frac{1}{p^5} + \frac{5}{p^3} + \frac{4}{p}\ .$$

Inverting Laplace transform, the solution of the equation is

$$f(t) = \frac{t^4}{24} + \frac{5t^2}{2} + 4.$$

G. Application to solution of Difference and Differential-difference eqations.

Difference equation arise in a natural way in physics and engineering. Let $a, a + 1, a + 2, \ldots$ be consequitive finite values of the argument x of a function denoted by $u(x)$ or u_x at discrete points. Here the constant 1 is called the interval of difference. The nth order finite difference of u_r is defined by

$$\Delta^n u_r \equiv \Delta^{n-1}\left[\Delta u_r\right] = \Delta^{n-1}\left[u_{r+1} - u_r\right] = \sum_{k=0}^n (-1)^k \binom{n}{k} u_{r+n-k}$$

$$(4.62)$$

Any equation containing the differences of different orders of the function u_r is called a difference equation. The highest order finite difference appearing in a difference equation is called its order. Again any difference equation containing the derivatives of the unknown function is called a differential-difference equation. Thus a differential - difference equation possesses two orders - one due to the largest order difference and the other associated with the highest order derivatives. For example,

$$\Delta^2 u_n - \Delta u_n + u_n = 0 \tag{4.63}$$

and $\quad u'(t) + u(t-1) = f(t) \tag{4.64}$

are second order difference equation and first order differential-difference equation respectively. To solve difference equation it facilitates to introduce a step function $S_n(t)$ defined by

$$S_n(t) = H(t-n) - H(t-n-1), \quad n \leqslant t \leqslant n+1 \tag{4.65}$$

where $H(t)$ is the Heaviside unit step function

$$H(t) = \begin{cases} 1, & t > 0 \\ 0, & \text{otherwise} \end{cases} \tag{4.66}$$

Then Laplace transform of $S_n(t)$ is given by

$$L[S_n(t) ; t \to p] = \int_0^\infty e^{-pt} [H(t-n) - H(t-n-1)] \, dt \tag{4.67}$$

$$= \int_n^{n+1} e^{-pt} \, dt = \frac{1}{p}(1 - e^{-p}) e^{-np}$$

$$= \bar{S}_0(p) e^{-np} \tag{4.68}$$

where $\quad \bar{S}_0(p) = \frac{1}{p}(1 - e^{-p})$

Let us now define a function $u(t)$ by a series

$$u(t) = \sum_{n=0}^\infty u_n S_n(t) \tag{4.69}$$

where $\{u_n\}$ is a given sequence of finite numbers. Then it follows from (4.69) that

$$u(t) = u_n \quad \text{for } n \leqslant t < n+1$$

$$\text{and} \quad u(t+1) = \sum_{n=0}^{\infty} u_n \, S_n \, (t+1)$$

$$= \sum_{n=0}^{\infty} u_n \, [H(t+1-n) - H(t-n)]$$

$$= \sum_{n=1}^{\infty} u_n S_{n-1}(t)$$

$$= \sum_{n=0}^{\infty} u_{n+1} S_n(t) \tag{4.70}$$

$$\text{Similarly}, \ u \, (t+2) = \sum_{n=0}^{\infty} u_{n+2} \, S_n(t) \tag{4.71}$$

$$\text{and more generally}, \ u \, (t+j) = \sum_{n=0}^{\infty} u_{n+j} \, S_n(t) \tag{4.72}$$

when $\quad j = 1, 2, 3, \cdots\cdots\cdots$

Then from (4.69) Laplace transform of $u(t)$ is given by

$$L \, [u(t) \, ; \, t \to p \,] \quad = \quad \bar{u} \, (p) = \int_0^{\infty} e^{-pt} u \, (t) \, dt$$

$$= \int_0^{\infty} e^{-pt} \, [\, \sum_{n=0}^{\infty} u_n S_n \, (t) \,] \, dt$$

$$= \sum_{n=0}^{\infty} u_n \int_0^{\infty} e^{-pt} S_n \, (t) \, dt$$

$$= \sum_{n=0}^{\infty} u_n \, \bar{S}_n \, (p) \tag{4.73}$$

Now this eqn. (4.73) gives

$$\bar{u} \, (p) = \sum_{n=0}^{\infty} \bar{S}_0 \, (p) \, e^{-np}$$

$$= \frac{1}{p} \, (1 - e^{-p}) \sum_{n=0}^{\infty} u_n \, e^{-np}$$

$$= \bar{S} \, (p) \, \zeta \, (p) \, , \tag{4.74}$$

where $\qquad \zeta(p) = \sum_{n=0}^{\infty} u_n e^{-np}$ (4.75)

Then, $\qquad u(t) = L^{-1}[\bar{S}_0(p)\zeta(p) ; p \to t]$ (4.76)

In particular, if $u_n = a^n$ then in this case

$$\begin{aligned} \zeta(p) &= \sum_{n=0}^{\infty} a^n e^{-np} = \sum_{n=0}^{\infty}(a e^{-p})^n \\ &= \frac{1}{1 - ae^{-p}} = \frac{e^p}{e^p - a} \end{aligned}$$ (4.77)

$L[a^n] \equiv L[a^n(t)] = \bar{S}_0(p)\zeta(p)$ implying , $n \leqslant t < n+1$

$$L^{-1}[\bar{S}_0(p) \frac{e^p}{e^p - a}] = a^n, \text{ by eqn. (4.77)}$$ (4.78)

Also, if $u_n = (n+1) a^n$, then correspondingly

$$\begin{aligned} \zeta(p) &= \sum_{n=0}^{\infty}(n+1) a^n e^{-np} \\ &= \sum_{n=0}^{\infty}(n+1)(ae^{-p})^n = (1 - ae^{-p})^{-2} \end{aligned}$$ (4.79)

Hence,

$$\begin{aligned} L[(n+1) a^n] = L[(n+1) a^n(t)] &= \bar{S}_0(p)\zeta(p), \qquad n \leqslant t < n+1 \\ &= \frac{1}{(1 - ae^{-p})^2}\bar{S}_0(p) \\ &= \frac{e^{2p}}{(e^p - a)^2}\bar{S}_0(p) \end{aligned}$$ (4.80)

This equation therefore implies

$$L^{-1}\left[\frac{e^{2p}\bar{S}_0(p)}{(e^p - a)^2} ; p \to t\right] = (n+1) a^n$$ (4.81)

Again, $\qquad \sum_{n=0}^{\infty} na^n e^{-np} = \frac{ae^p}{(1 - ae^{-p})^2}$

and so

$$\begin{aligned} L[na^n] \equiv L[na^n(t)] &= \bar{S}_0(p)\zeta(p) , n \leqslant t < n+1 \\ &= \frac{\bar{S}_0(p) ae^p}{(e^p - a)^2} \end{aligned}$$

Therefore, $\qquad L^{-1}\left[\frac{a\bar{S}_0(p) e^p}{(e^p - a)^2}\right] = n a^n$ (4.82)

Theorem 4.3.

$$\text{If} \quad \bar{u}\,(p) \;=\; L[u\,(t)\,;\, t \to p]\,, \text{ then}$$

$$L\,[u\,(t+1)\,;\, t \to p] \;=\; e^p\,[\bar{u}\,(p) - u_0\,\bar{S}_0\,(p)] \qquad (4.83)$$

where $u_0 = u(0)$.

Proof. By definition

$$L[u\,(t+1)\,;\, t \to p] \;=\; \int_0^\infty e^{-pt}\,u\,(t+1)\,dt$$

$$= \int_1^\infty e^{-p\tau}\,e^p\,u(\tau)\,d\tau$$

$$= e^p\left[\int_0^\infty e^{-p\tau}u\,(\tau)\,d\tau - \int_0^1 e^{-p\tau}\,u(\tau)\,d\tau\right]$$

$$= e^p[\bar{u}(p) - \int_0^1 e^{-p\tau}u(0)\,d\tau]\,, \text{ since } u\,(\tau) = u_0 \text{ in } (0,1)$$

$$= e^p\,\bar{u}(p) - u_0\,e^p\,\bar{S}_0\,(p)$$

$$= e^p\left[L\{u(t)\,;\, t \to p\} - u_0\bar{S}_0(p)\right]$$

Also in view of the above theorem, we get

$$L[u\,(t+2)\,;\, t \to p] = e^p\,[\,L\{\,u\,(t+1)\,;\, t \to p\,\} - u(1)\bar{S}_0\,(p)\,]$$

$$= e^{2p}\,[\bar{u}\,(p) - u_0\,\bar{S}_0(p)\,] - e^p\,u(1)\bar{S}_0\,(p)$$

$$= e^{2p}\,[\,\bar{u}\,(p) - \{\,u_0 + u_1\,e^{-p}\,\}\,\bar{S}_0\,(p)\,] \qquad (4.84)$$

More generally,

$$L\,[\,u\,(t+j)\,;\, t \to p\,] = e^{jp}\,[\bar{u}(p) - \bar{S}_0\,(p)\sum_{i=0}^{j-1} u_i\,e^{-ip}\,] \qquad (4.85)$$

Example 4.59. Solve the difference equation

$$y_{k+2} - 5y_{k+1} + 6y_k = 0 \quad \text{when } y_0 = 0,\ y_1 = 1.$$

Solution. Taking Laplace transform the given equation results in

$$e^{2p}\,[\bar{y}\,(p) - \{\,y_0 + y_1e^{-p}\,\}\,\bar{S}_0\,(p)] - 5e^p\,[\,\bar{y}\,(p) - y_0\,\bar{S}_0\,(p)\,] + 6\bar{y}\,(p) = 0$$

We use the given initial condition to obtain

$$[\,e^{2p} - 5e^p + 6\,]\,\bar{y}\,(p) \;=\; e^p\,\bar{S}_0\,(p)$$

$$\text{Thus}\,, \quad \bar{y}(p) = \frac{e^p}{e^p - 3}\,\bar{S}_0\,(p) - \frac{e^p}{e^p - 2}\,\bar{S}_0\,(p)$$

$$\Rightarrow \qquad y_k \;=\; 3^k - 2^k$$

Example 4.60. Solve the difference equation

$$y_{n+2} - 2\lambda\, y_{n+1} + \lambda^2 y_n = 0 \,, \quad \text{when } y_0 = 0 \text{ and } y_1 = 1$$

Solution. Taking Laplace transform the given difference equation under the initial condition becomes

$$(e^p - \lambda)^2\, \bar{y}\,(p) = e^p\, \bar{S}_0\,(p)]$$

$$\Rightarrow \qquad \bar{y}\,(p) = \frac{e^p}{(e^p - \lambda)^2}\, \bar{S}_0(p)$$

$$= \frac{1}{\lambda}\, \frac{\lambda\, e^p}{(e^p - \lambda)^2}\, \bar{S}_0(p)$$

Inverting this Laplace transformed equation we get

$$y_n \;=\; \frac{1}{\lambda}\, n\, \lambda^n \,, \qquad \text{by (4.83)}$$

$$=\; n\, \lambda^{n-1}$$

Example 4.61. Solve the differential-difference equation $y'(t) = y(t-1)$ under the initial condition $y(0) = 1$

Solution. Laplace transform of the given equation we get

$$p\, \bar{y}\,(p) - 1 = e^{-p}\, [\, \bar{y}\,(p) - \bar{S}_0\,(p)]$$

$$\Rightarrow \qquad \bar{y}(p)\, [\, p - e^{-p}\,] = 1 + \frac{e^{-p} - 1}{p} \cdot e^{-p}$$

Therefore,
$$\bar{y}\,(p) \;=\; \frac{1}{p - e^{-p}} - \frac{e^{-p}}{p\,(p - e^{-p})} + \frac{e^{-2p}}{p(p - e^{-p})}$$

$$=\; \frac{1}{p} + \frac{e^{-2p}}{p^2\left(1 - \frac{e^{-p}}{p}\right)}$$

$$=\; \frac{1}{p} + \left(\frac{e^{-p}}{p}\right)^2 \left[\, 1 - \frac{e^{-p}}{p}\, \right]^{-1}$$

$$=\; \frac{1}{p} + \frac{e^{-2p}}{p^2} + \frac{e^{-3p}}{p^3} + \frac{e^{-4p}}{p^4} + \cdots\cdots$$

On inversion of Laplace transform, the above result gives

$$y(t) = 1 + \frac{t - 2}{1!} + \frac{(t - 3)^2}{2!} + \cdots\cdots + \frac{(t - n)^{n-1}}{(n - 1)!} \,, \quad t > n.$$

Example 4.62. Solve the non-homogeneous difference equation
$y_{k+1} - 3y_k = 1$ under the condition $y_0 = \frac{1}{2}$

Solution. Laplace transform of the given difference equation results in

$$e^p \left[\, \bar{y} \, (p) \, - y_0 \, \bar{S} \, (p) \, \right] - 3\bar{y} \, (p) = \frac{1}{p}$$

$$\Rightarrow \quad (e^p - 3) \, \bar{y} \, (p) = \frac{1}{2} \, e^p \, \bar{S}_0 \, (p) + \frac{1}{p}$$

$$\Rightarrow \quad \bar{y} \, (p) = \frac{1}{2} \frac{e^p}{e^p - 3} \, \bar{S}_0 \, (p) + \frac{1}{p(e^p - 3)}$$

$$= \frac{1}{2} \frac{e^p}{e^p - 3} \, \bar{S}_0 \, (p) + \frac{1}{p} \frac{1 - e^{-p}}{(1 - e^{-p})(e^p - 3)}$$

$$= \frac{1}{2} \frac{e^p}{e^p - 3} \, \bar{S}_0 \, (p) + \bar{S}_0 \, (p) \frac{e^p}{e^{2p} - 4e^p + 3}$$

$$= \left[\frac{1}{2} \frac{e^p}{e^p - 3} + \frac{e^p}{(e^p - 3) \, (e^p - 1)} \right] \bar{S}_0 \, (p)$$

$$= \left[\frac{1}{2} \frac{e^p}{e^p - 3} + \frac{1}{2} \frac{e^p}{e^p - 3} - \frac{1}{2} \frac{e^p}{e^p - 1} \right] \bar{S}_0 \, (p)$$

$$= \frac{e^p}{e^p - 3} \, \bar{S}_0 \, (p) - \frac{1}{2} \frac{e^p}{e^p - 1} \, \bar{S}_0 \, (p)$$

Inverting Laplace transform, we get

$$y_k = 3^k - \frac{1}{2} \cdot 1^k$$

Or, $$y_k = 3^k - \frac{1}{2}.$$

Example 4.63. Show that the solution of the difference equation

$$u_{n+2} + 4 \, u_{n+1} + u_n = 0 \quad \text{with} \quad u_0 = 0 \quad \text{and} \quad u_1 = 1 \text{ is given by}$$

$$u_n = \frac{1}{2\sqrt{3}} \left[(\sqrt{3} - 2)^n + (-1)^{n+1} \, (2 + \sqrt{3})^n \, \right]$$

Solution. Application of Laplace transform yields

$$e^{2p}[\bar{u} \, (p) - (e^{-p}u_1 \, + u_0) \, \bar{S}_0(p) \,] + 4e^p[\bar{u} \, (p) - u_0 \, \bar{S}_0(p)] + \bar{u} \, (p) = 0$$

Or $$[e^{2p} + 4 \, e^p + 1 \,]\bar{u} \, (p) = e^p \, \bar{S}_0(p)$$

Or, $$\bar{u}(p) = \frac{e^p}{(e^p - \alpha)(e^p - \beta)} \bar{S}_0(p), \text{where } \alpha = \sqrt{3} - 2 \text{ and } \beta = -(\sqrt{3} + 2)$$

$$= \frac{1}{-\beta + \alpha} \left[\frac{-e^p}{e^p - \beta} + \frac{e^p}{e^p - \alpha} \right] \bar{S}_0 \, (p)$$

$$= \frac{1}{2\sqrt{3}} \left[\frac{e^p}{e^p - \alpha} \, \bar{S}_0 \, (p) - \frac{e^p}{e^p - \beta} \, \bar{S}_0 \, (p) \right]$$

Inverting Laplace transform we get

$$u_n = \frac{1}{2\sqrt{3}}\left[(\sqrt{3}-2)^n - \{ -(\sqrt{3}+2)\}^n \right]$$

$$= \frac{1}{2\sqrt{3}}\left[(\sqrt{3}-2)^n + (-1)^{n+1}(\sqrt{3}+2)^n \right]$$

as the solution of the given difference equation.

Exercises

(1) Find inverse Laplace transforms of :

(a) $\dfrac{p}{(p+1)^2(p^2+1)}$ (b) $\dfrac{p^2+6}{(p^2+1)(p^2+4)}$

(c) $\dfrac{p+2}{(p^2+4p+5)^2}$ (d) $\dfrac{p}{p^4+p^2+1}$

(e) $\dfrac{p}{p^4+4a^4}$ (f) $\dfrac{1}{p(p+1)^3}$

$$\left[\text{Ans. (a) } \frac{1}{2}(\sin t - t\,e^{-t}) \text{ (b) } \frac{1}{3}(5\sin t - \sin 2\,t) \right.$$

$$\text{(c) } \frac{1}{2}t\,e^{-2t}\,\sin t \quad \text{(d) } \frac{2}{\sqrt{3}}\sin\frac{t}{2}\cdot\sin\frac{\sqrt{3}}{2}t$$

$$\left. \text{(e) } \frac{1}{2a^2}\sin at\,sh\,at \text{ (f) } 1 - e^{-t}\left(1+t+\frac{t^2}{2}\right) \right]$$

(2) Prove the following :

(a) $L\left[(1-2t)\,e^{-2t} \; ; \; t \to s \right] = s\,(s+2)^{-2}$

(b) $L\left[t\,H\,(t-a) \; ; \; t \to s \right] = (1+sa)\,s^{-2}\exp(-as)$

(c) $L\left[(1+2at)\,t^{-\frac{1}{2}}\exp(at) \; ; \; t \to s \right] = s\sqrt{\pi}\,(s-a)^{-\frac{3}{2}}$

(d) $L\left[\frac{1}{t}\left\{ e^{-at} - e^{-bt} \right\} \; ; \; t \to s \right] = \log\left(\dfrac{s+b}{s+a} \right)$

(3) Prove that

(i) $L^{-1}\left[\dfrac{1}{p}\sin\dfrac{1}{p} \right] = t - \dfrac{t^3}{(3!)^2} + \dfrac{t^5}{(5!)^2} - \dfrac{t^7}{(7!)^2} + \cdots$ and

(ii) $L^{-1}\left[\dfrac{1}{p}\cos\dfrac{1}{p} \right] = 1 - \dfrac{t^2}{(2!)^2} + \dfrac{t^4}{(4!)^2} - \dfrac{t^6}{(6!)^2} + \cdots$

(4) Find inverse Laplace transform of :

(a) $\log \dfrac{p+1}{p-1}$ (b) $\log \dfrac{p-1}{p}$ (c) $\dfrac{3p+1}{(p+1)^4}$

(d) $\dfrac{1}{2}\log \dfrac{p^2+b^2}{p^2+a^2}$ (e) $\cos^{-1}\left(\dfrac{p}{2}\right)$ (f) $\log\left(1+\dfrac{1}{p^2}\right)$

$$\left[\text{Ans. (a)}\ \dfrac{2\ sh\ t}{t}\ \text{(b)}\ \dfrac{1-e^t}{t}\ \text{(c)}\ e^{-t}\left(\dfrac{3}{2}t^2-\dfrac{1}{3}t^3\right)\right.$$

$$\left.\text{(d)}\ \dfrac{\cos\ at-\cos\ bt}{t}\ \text{(e)}\ \dfrac{\sin\ (2t)}{t}\ \text{(f)}\ \dfrac{2(1-\cos t)}{t}\right]$$

(5) Prove that

$$L^{-1}\left[\dfrac{1}{p}\bar{f}(p)\right]=\int_0^t f(x)dx \quad \text{and hence obtain}$$

(i) $L^{-1}\left[\dfrac{1}{p(p+1)}\right]$ and   (ii) $L^{-1}\left[\dfrac{1}{p^2(p+a)}\right]$

$$\left[\text{Ans. (i)}\ \dfrac{1}{a}\left(1-e^{-at}\right)\ \text{(ii)}\ \dfrac{1}{a^2}\left(at+e^{-at}-1\right)\right]$$

(6) Prove that

(a) $L^{-1}\left[\dfrac{e^{-as}}{s^2}\ ;\ s\to t\right]=(t-a)H(t-a)$

(b) $L^{-1}\left[\dfrac{w}{s^2+w^2}\ \bar{f}(s)\ ;\ s\to t\right]=\int_0^t f(t-\tau)\sin\ w\tau\ d\tau$

(c) $L^{-1}\left[\dfrac{1}{s}\ e^{-a\sqrt{s}}\ ;\ s\to t\right]=erfc\left(\dfrac{a}{2\sqrt{t}}\right)$

(d) $L^{-1}\left[\tan^{-1}\left(\dfrac{a}{s}\right)\ ;\ s\to t\right]=\dfrac{1}{t}\ sh(at)$

(e) $L^{-1}\left[\dfrac{\bar{f}(s)}{s^2}\ ;\ s\to t\right]=\int_0^t (t-\tau)\ f(\tau)\ d\tau$

(7) Show that

(i) $\displaystyle\int_0^\infty \cos x^2 dx=\dfrac{\sqrt{\pi}}{2\sqrt{2}}=\int_0^\infty \sin x^2 dx$ and

(ii) $\displaystyle\int_0^\infty e^{-x^2} dx=\dfrac{\sqrt{\pi}}{2},$

by using Laplace transform and its inversion and considering functions $f(t)=\int_0^\infty \cos(t\ x^2)dx$ and $f(t)=\int_0^\infty e^{-tx^2}dx$ in cases (i) and (ii) respectively.

(8) Evaluate inverse Laplace transform of :

(i) $\dfrac{3(1 + e^{-p\pi})}{(p^2 + 9)}$ (ii) $\dfrac{32p}{(16\ p^2 + 1)^2}$ (iii) $\dfrac{1}{p} \log \dfrac{p + 2}{p + 2}$

(iv) $\log(1 - 1/p^2)$

$\Bigg[$ Ans. (i) $- \sin\ 3t\ H(t - \pi) + \sin 3t$ (ii) $\dfrac{t}{4} \sin \dfrac{t}{4}$

(iii) $\displaystyle\int_0^t \dfrac{e^{-x} - e^{-2x}}{x}\ dx$ (iv) $\dfrac{2(1 - ch\ t)}{t}\Bigg]$

(9) Prove that

(i) $L^{-1} \left[\dfrac{1}{p} \log \left(1 + \dfrac{1}{p^2} \right) \right] = \displaystyle\int_\partial^x \dfrac{2}{x}(1 - \cos\ x) dx$ and

(ii) $L^{-1} \left[\dfrac{1}{p^2(1 + p^2)} \right] = t\ e^{-t} + 2\ e^{-t} + t - 2$

(10) Apply convolution theorem to evaluate

(a) $L^{-1} \left[\dfrac{p^2}{(p^2 + a^2)(p^2 + b^2)} \right]$ (b) $L^{-1} \left[\dfrac{1}{p(p^2 + 4)} \right]$

(c) $L^{-1} \left[\dfrac{p}{(p^2 + 4)^3} \right]$ (d) $L^{-1} \left[\dfrac{8}{(p^2 + 1)^3} \right]$

(e) $L^{-1} \left[\dfrac{1}{(p + 2)^2(p - 2)} \right]$

[Ans.   (a) $\dfrac{(a \sin at - b \sin bt)}{(a^2 - b^2)}$ (b) $\dfrac{1}{4}(1 - \cos 2t)$

(c) $\dfrac{t}{64}[\sin 2t - 2t \cos 2t]$ (d) $(3 - t^2) \sin\ t - 3t\ \cos t$

(e) $\dfrac{1}{16} \left[e^{2t} - (4t + 1)e^{-2t} \right]$

(11) Using Laplace transform properly prove that

$$\int_0^\infty \dfrac{e^{-t} - e^{-3t}}{t}\ dt = \log\ 3$$

(12) Find inverse Laplace transform of

(a) $\dfrac{p\ e^{-p/2} + \pi\ e^{-p}}{p^2 + 4}$ (b) $\dfrac{p\ e^{-ap}}{p^2 - w^2}$, $a > 0$

(c) $\dfrac{w}{(1 - e^{\frac{-\pi p}{w}})(p^2 + w^2)}$ (d) $\dfrac{e^{-p}}{(p - 1)(p - 2)}$

$$\left[\text{Ans.} \quad \text{(a)} \quad \sin \pi t \left[H\left(t - \frac{1}{2}\right) - H(t-1) \right] \right.$$

(b) $ch\ w(t-a)\ H(t-a)$

(c) $f(t)$, where $f(t) = \begin{cases} \sin wt, & 0 < t < \frac{\pi}{w} \\ 0, & \frac{\pi}{w} < t < \frac{2\pi}{w} \end{cases}$

and $f(t)$ is a periodic function of period $\dfrac{2\pi}{w}$

$$\left. \text{(d)} \quad [e^{2(t-1)} - e^{(t-1)} H(t-1)] \; \right]$$

(13) Using $L[J_0(t)]$ prove that $L[J_1(t)] = 1 - \dfrac{p}{\sqrt{p^2+1}}$. Also prove that
$L[t\ J_1(t)] = \dfrac{1}{(1+p^2)^{\frac{3}{2}}}$ and hence show that

$$L^{-1}\left[\frac{1}{\sqrt{p^2 + 2ap + 2a^2}} \right] = e^{-at} J_0(at)$$

(14) By expanding $\tan^{-1}\left(\frac{1}{p}\right)$ in ascending powers of p, and taking term by term inverse Laplace transform show that

$$L^{-1}\left[\tan^{-1}\left(\frac{1}{p}\right) \; ; \; p \to t \right] = \frac{\sin t}{t} \; .$$

(15) Evaluate $L[e^t\ erf\ \sqrt{t}]$ and hence evaluate $L^{-1}\left[\dfrac{3p+2}{2p^2(p+1)^{\frac{3}{2}}} \right]$

$$\left[\text{Ans.} \quad \frac{1}{(p-1)\sqrt{p}} \; , \quad t\ erf\ \sqrt{t} \right]$$

(16) Evaluate $L^{-1}\left[\frac{1}{p^3(1+p^2)} \right]$ by using

(a)   partial fractions,

(b)   the convolution theorem.

[Ans. $-1 + \frac{t^2}{2} - \cos t$]

(17) Apply convolution theorem to prove that

(a) $L^{-1}\left[\dfrac{1}{p^2(p^2 - a^2)} \right] = \dfrac{(sh\ at - at)}{a^3}$

(b) $L^{-1}\left[\dfrac{1}{(p-1)\sqrt{p}} \right] = e^t\ erf(\sqrt{t})$

(c) $L^{-1}\left[\dfrac{1}{p}\dfrac{p}{(p^2 + 1)^2} \right] = \dfrac{1}{2}(\sin t - t\cos t)$,

using $L^{-1}\left[\dfrac{p}{(p^2 + 1)^2} \right] = \dfrac{1}{2}\ t\sin t$

(18) Prove that

$$L^{-1}\left[e^{-\sqrt{p}}\right] = \frac{e^{-\frac{1}{4t}}}{2\sqrt{\pi}\, t^{\frac{3}{2}}}$$

(19) Use convolution theorem to prove that

$$\int_0^t J_0(\tau)\, J_0(t-\tau)\, d\tau = \sin t$$

(20) (a) If $f(t) = H\left(t - \frac{\pi}{2}\right)\sin t$, prove that $L[f(t)\,;\, t \to s] = \frac{s}{s^2+1} e^{-\frac{\pi s}{2}}$

(b)   Prove that $L[\,|\sin at|\,;\, t \to s] = \dfrac{a}{s^2+a^2}\coth\dfrac{\pi s}{2a}\,,\ s > 0$

(c)   Prove that $L\left[\dfrac{d}{dt}(f * g)\,;\, t \to s\right]$

$$= g(0)\bar{f}(s) + L[f * g'(t)\,;\, t \to s] = s\bar{f}(s)\,\bar{g}(s)$$

(d)   Show that $f(t) = \sin(a\sqrt{t})$ satisfies the ODE
$$4t\, f''(t) + 2f'(t) + a^2 f(t) = 0$$

(e)   Establish that $L\left[\displaystyle\int_t^\infty \dfrac{f(x)}{x}\, dx\,;\, t \to s\right] = \dfrac{1}{s}\displaystyle\int_0^s \bar{f}(x)\, dx$

(f)   Prove that $L\left[\dfrac{\cos at - \cos bt}{t}\,;\, t \to s\right] = \dfrac{1}{2}\log\left(\dfrac{s^2+a^2}{s^2+b^2}\right)$

(21) Find Laplace transform of the triangular wave function defined in $(0, 2a)$ by

$$f(t) = \begin{cases} t & ,\ 0 < t < a \\ 2a - t & ,\ a < t < 2a \end{cases}$$

$\left[\text{Ans. }\ s^{-2}\ \text{th}\ \left(\dfrac{as}{2}\right)\right]$

(22) Apply convolution theorem to show that

(a) $\displaystyle\int_0^t \sin u \cos(t-u)\,du = \dfrac{1}{2}\, t\sin t$

(b) $L^{-1}\left[\dfrac{1}{\sqrt{p}(p-a)}\right] = \dfrac{e^{at}}{\sqrt{a}}\, erf\ \sqrt{at}\ $ and deduce that

$$L^{-1}\left[\dfrac{1}{p\,\sqrt{p+a}}\right] = \dfrac{erf\sqrt{at}}{\sqrt{a}}$$

(23) Evaluate, by the method of residues that

(a) $L^{-1}\left[\dfrac{p}{(p+1)^3(p-1)^2}\right]$

(b) $L^{-1}\left[\dfrac{1}{(p^2+1)^2}\right]$

(c) $L^{-1}\left[\dfrac{2p^2-4}{(p+1)(p-2)(p-3)}\right]$

(d) $L^{-1}\left[e^{-\sqrt{p}}\right]$

(e) $L^{-1}\left[\dfrac{\operatorname{ch} x\sqrt{p}}{p\cosh\sqrt{p}}\right]$, $0<x<1$

(f) $L^{-1}\left[\dfrac{1}{p^2}\operatorname{th}\dfrac{\pi p}{2}\right]$

[Ans. (a) $\dfrac{e^{-t}}{16}(1-2t^2)+\dfrac{e^t}{16}(2t-1)$

(b) $\dfrac{1}{2}t\cos t+\dfrac{1}{2}\sin t$

(c) $\dfrac{7}{2}e^{3t}-\dfrac{4}{3}e^{2t}-\dfrac{1}{6}e^{-t}$

(d) $t^{-\frac{3}{2}}e^{-\frac{1}{4t}}/(2\sqrt{\pi})$

(e) $\dfrac{4}{\pi}\sum_{n=1}^{\infty}\dfrac{(-1)^n}{2n-1}e^{-(2n-1)^2\frac{\pi^2 t}{4}}\cos\left(n-\dfrac{1}{2}\right)\pi x+1$

(f) A periodic function $f(t)=\begin{cases} t, & 0<t<\pi \\ 2\pi-t, & \pi<t<2\pi \end{cases}$ of period 2π]

(24) Solve the following equations by using Laplace transform :

(a) $x''(t)-2x'(t)+x(t)=e^t$, when $x(0)=2$, $x'(0)=-1$

(b) $y''(t)+4y'(t)+3y(t)=e^{-t}$, when $y(0)=y'(0)=1$

(c) $y''(t)-3y'(t)+2y(t)=4t+e^{3t}$, when $y(0)=1$, $y'(0)=-1$

(d) $y'''(t)+2y''(t)-y'(t)-2y(t)=0$, when $y(0)=1, y'(0)=2,$
$$y''(0)=2$$

(e) $y''(t)+2y'(t)+5y(t)=e^t\sin t$, when $y(0)=0, y'(0)=1$

(f) $x''(t)+9x(t)=\cos 2t$, when $x(0)=1, x\left(\dfrac{\pi}{2}\right)=-1$

(g) $(D^3-3D^2+3D-1)y(t)=t^2e^{2t}$, where $y(0)=1, y'(0)=0,$
$$y''(0)=-2$$

(h) $(D^4 + 2D^2 + 1)y(t) = 0$, where $y(0) = 0, y'(0) = 1, y''(0) = 2,$
$$y'''(0) = -3$$

(i) $(D^3 - 2D^2 + 5D)x(t) = 0$, where $x(0) = 0, x'(0) = 1, x\left(\dfrac{\pi}{8}\right) = 1$

(j) $(D^2 - 1) x\ (t) = a\ ch\ nt$, where $x(0) = x'(0) = 0,$

(k) $(D^3 + 1) x\ (t) = 1$, $x(0) = x'(0) = x''(0) = 0$

(l) $x''(t) - k^2 x(t) = f(t)$, where $x(0) = x'(0) = 0, k \neq 0$

(m) $(D^2 + 1)x(t) = f(t)$, $x(0) = x'(0) = 0$

[Ans. (a) $x = 2e^t - 3t\ e^t + \dfrac{1}{2}\ t^2\ e^t$ (b) $y = \dfrac{7}{4}\ e^{-t} - \dfrac{3}{4}\ e^{-3t} - \dfrac{1}{2}t\ e^{-t}$

(c) $y = 2t + 3 + \dfrac{1}{3}\left(e^{3t} - e^t\right) - 2e^{2t}$ (d) $y = \dfrac{1}{3}(5e^t + e^{-2t}) - e^{-t}$

(e) $y = \dfrac{11}{3}e^{-t}(\sin t + \sin 2t)$ (f) $x = \dfrac{1}{5}(4\cos 3t + 4\sin 3t + \cos 2t)$

(g) $y = (t^2 - 6t + 12)\ e^{2t} - \left(\dfrac{3}{2}t^2 + 7t + 11\right)e^t$

(h) $y = t(\sin t + \cos t)$ (i) $x = 1 - e^t(\cos at - \sin 2t)$

(j) $x = \dfrac{a(ch\ nt - ch\ t)}{(n^2 - 1)}$ (k) $x = 1 - \dfrac{1}{3}\ e^{-t} - \dfrac{2}{3}\ e^{\frac{t}{2}}\cos\dfrac{\sqrt{3}t}{2}$

(l) $x = \dfrac{1}{2k}\left[e^{kt}\displaystyle\int_0^t e^{-ku}f(u)\ du - e^{-kt}\int_0^t e^{ku}f(u)\ du\right]$

(m) $x = \displaystyle\int_0^t \sin(t - u)f(u)\ du$]

(25) Solve the following simultaneous ODE by using Laplace transform:

(a) $x'(t) - 2x(t) + 3y(t) = 0$, $y'(t) + 2x(t) - y(t) = 0$,
$$\text{when}\quad x(0) = 8,\ y(0) = 3$$

(b) $(D^2 - D) x\ (t) + y(t) = 0$, $\left(D - \dfrac{1}{2}\right) x\ (t) + D\ y\ (t) = 0$,
$$\text{when}\ x(0) = 0,\ y(0) = 1,\ x'(0) = 0$$

(c) $D^2 x\ (t) + y(t) = -5\cos 2t$, $D^2 y\ (t) + x(t) = 5\cos 2t$,
$$\text{when}\ x(0) = x'(0) = y'(0) = y'(0) = 1,\ y(0) = -1$$

(d) $x'(t) - y'(t) - 2x(t) + 2y(t) = 1 - 2t$, $x''\ (t) + 2y'(t)$
$$+x(t) = 0,\ x(0) = x'(0) = y(0) = 0$$

(e) $x'(t) + 5x(t) - 2y(t) = t$, $y'(t) + 2x(t) + y(t) = 0$,

$$\text{when} \quad x(0) = y(0) = 0$$

(f) $(D^2 - 1)x(t) + 5Dy(t) = t, \ -2Dx(t) + (D^2 - 4)\, y\,(t) = 2,$
$$\text{if} \quad x(0) = y(0) = x'(0) = y'(0) = 0$$

(g) $Dx(t) + Dy(t) = t, \ D^2x(t) - y(t) = e^{-t},$
$$\text{if} \quad x(0) = 3, \ x'(0) = -2, \ y(0) = 0$$

(h) $(D^2 + 2)\, x\,(t) - Dy(t) = 1, \ Dx(t) + (D^2 + 2)\, y\,(t) = 0,$
$$\text{if} \quad x(0) = x'(0) = y(0) = y'(0) = 0$$

[Ans. (a) $x(t) = 5e^{-t} + 3e^{4t} , \ y(t) = 5e^{-t} - 2e^{4t}$

(b) $x(t) = \dfrac{1}{2}\left(sh\ t - te^t\right), y(t) = ch\ t$

(c) $x(t) = \sin t + \cos 2t, \ y(t) = \sin t - \cos 2t$

(d) $x(t) = 2\left(1 - e^{-t} - te^{-t}\right), \ y(t) = -t(1 + 2e^{-t})$
$$+2(1 - e^{-t})$$

(e) $x(t) = -\dfrac{1}{27}(1 + 6t)e^{-3t} + \dfrac{1}{27}(1 + 3t) ,$

$\quad y(t) = -\dfrac{1}{9}(2 + 3t)e^{-3t} + \dfrac{1}{27}(2 - 3t)$

(f) $x(t) = 5\sin t - 2\sin 2t - t, \ y(t) = 1 - 2\cos t + \cos 2t$

(g) $x(t) = 2 + \dfrac{t^2}{2} + \dfrac{e^t}{2} - \dfrac{3\sin t}{2} + \dfrac{\cos t}{2} ,$

$\quad y(t) = 1 - \dfrac{1}{2}e^{-t} - \dfrac{\cos t}{2} + \dfrac{3\sin t}{2}$

(h) $x(t) = \dfrac{1}{6}[3 - 2\cos t - \cos 2t], y(t) = \dfrac{1}{6}[-2\sin t + \sin 2t]$]

(26) Solve $(D^2 + tD - 1)x(t) = 0$, when $x(0) = 0$, $x'(0) = 1$ using Laplace transform to prove that $x(t) = t + cL^{-1}\left[\dfrac{1}{p^2}\,e^{\frac{p^2}{2}}\right]$. Also discuss for the particular solution when $x(0) = 0$.

[Ans. $x = t$]

(27) Using Laplace transform prove that $x(t) = 1 + 2t$ is the particular solution of the ODE $x''(t) - tx'(t) + x(t) = 1$, when $x(0) = 1$, $x'(0) = 2$. Assume here that $L^{-1}(p^n) = 0$, for $n = 0, 1, 2, 3, \cdots$

(28) Using Laplace transform prove that $x(t) = \sin t/t$ is the solution of the ODE with variable coefficients $tx''(t) + 2x'(t) + tx(t) = 0$, if $x(0) = 0$.

(29) Prove, by using Laplace transform that $y = (1 - t) e^{-t} + \frac{1}{2} \sin t$ is the solution of the integro-differential equation given by

$$\frac{dy}{dt} + 2y(t) + \int_0^t y(u)du = \sin t , \quad y(0) = 1$$

(30) Solve the following integral/integro-differential equation :

(a) $f(t) = e^{-t} - 2 \int_0^t \cos(t - u)f(u)du$

(b) $16 \sin 4t = \int_0^t f(u)f(t - u)du$

(c) $f'(t) = \sin t + \int_0^t f(t - u) \cos u \, du$, if $f(0) = 0$

(d) $f'(t) = t + \int_0^t f(t - u) \cos u \, du$, if $f(0) = 4$

(e) $\int_0^t f'(t)f(t - u)du = 24t^3$, if $f(0) = 0$

(f) $\int_0^t f(u) \cos(t - u)du = f'(t)$, if $f(0) = 1$

[Ans. (a) $f(t) = e^{-t}(1 - t)^2$ (b) $f(t) = \pm 8 \, J_0(4t)$

(c) $f(t) = \dfrac{t^2}{2}$ (d) $f(t) = 4 + \dfrac{5}{2} t^2 + \dfrac{t^4}{24}$

(e) $f(t) = \pm 16 \, t^{\frac{3}{2}}/\sqrt{\pi}$ (f) $f(t) = 1 + \dfrac{t^2}{2}$]

(31) A mechanical system, with two degrees of freedom, satisfies the equations $2x''(t) + 3y'(t) = 4$, $2y''(t) - 3x'(t) = 0$. Use Laplace transform to determine $x(t)$ and $y(t)$, given that $x(t), y(t), x'(t), y'(t)$ all vanish at $t = 0$.

$$\left[\text{Ans.} \quad x = \frac{8}{9}\left(1 - \cos\frac{3}{2} t\right) , \quad y = \frac{8}{9}\left(\frac{3}{2} t - \sin\frac{3}{2} t\right)\right]$$

(32) The co-ordinates (x, y) of a particle moving along a plane curve at any time t are given by

$$y'(t) + 2x(t) = \sin 2t , \quad x'(t) - 2y(t) = \cos 2t , \quad (t > 0)$$

If $x(t = 0) = 1, y(t = 0) = 0$, using Laplace transform show that the particle moves on the curve $4x^2 + 4xy + 5y^2 = 4$

(33) Solve the differential system

$$\frac{dx}{dt} = Ax \ , \ x(0) = \begin{bmatrix} 0 \\ 1 \end{bmatrix} \quad \text{where} \ \ x = \begin{bmatrix} x_1 \\ x_2 \end{bmatrix}, \ A = \begin{bmatrix} 0 & 1 \\ -2 & 3 \end{bmatrix}$$

$$\left[\text{Ans.} \ \ x(t) = \begin{bmatrix} e^{2t} - e^t \\ 2e^{2t} - e^t \end{bmatrix} \right]$$

(34) Solve the system

$$\frac{d^2 x_1}{dt^2} - 3x_1 - 4x_2 = 0 \ , \ \frac{d^2 x_2}{dt^2} + x_1 + x_2 = 0 \ , \ t > 0 \ \text{with}$$

$$x_1 = x_2 = 0 \ , \ \frac{dx_1}{dt} = 2 \ , \ \frac{dx_2}{dt} = 0 \ \text{at} \ \ t = 0$$

[Ans. $x_1 = 2t \, ch \, t, \ x_2 = sh \, t - t \, ch \, t$]

(35) Obtain the solution of $tx''(t) + x'(t) + a^2 x(t) = 0$, $x(0) = 1$
in the form

$$x(t) = AL^{-1} \left[\frac{1}{\sqrt{s^2 + a^2}} \ ; s \to t \right] = A \, J_0(at)$$

(36) Solve the following differential equations under the given initial conditions :

(a) $\dfrac{dX}{dt} = AX, X(0) = \begin{bmatrix} 1 \\ 0 \end{bmatrix}$, where $X = \begin{bmatrix} x \\ y \end{bmatrix}$, $A = \begin{bmatrix} 1 & -2 \\ -2 & 1 \end{bmatrix}$

(b) $\dfrac{dx_1}{dt} = x_1 + 2x_2 + t, \dfrac{dx_2}{dt} = x_2 + 2x_1 + t, x_1(0) = 2, x_2(0) = 4$

(c) $\dfrac{dx}{dt} = 2x - 3y, \dfrac{dy}{dt} = y - 2x$; $x(0) = 2, \ y(0) = 1$

[Ans. (a) $x(t) = \dfrac{1}{2} \left(e^{3t} + e^{-t} \right), \ y(t) = \dfrac{1}{2} \left(e^{3t} - e^{-t} \right)$

(b) $x_1 = \dfrac{28}{9} e^{3t} - e^{-t} - \dfrac{t}{3} - \dfrac{1}{9} , \ x_2 = \dfrac{28}{9} e^{3t} + e^{-t} - \dfrac{t}{3} - \dfrac{1}{9}$

(c) $x = \dfrac{1}{5} \left(7e^{-t} + 3e^{4t} \right), \ y = \dfrac{1}{5} \left(7e^{-t} - 2e^{4t} \right)$]

(37) The zero-order chemical reaction satisfies the initial value problem

$$\frac{dC(t)}{dt} + k_0 = 0 \ , \ t > 0 \quad \text{with} \ \ C(0) = c_0 \ \text{at} \ \ t = 0$$

where k_0 is a positive constant and $C(t)$ is the concentration of a reacting substance at time t. Show that

$$C(t) = c_0 - k_0 \, t$$

(38) Using Laplace transform investigate the motion of a particle governed by the equations of motion

$$\frac{d^2x}{dt^2} - w\frac{dy}{dt} = 0, \quad \frac{d^2y}{dt^2} + w\frac{dx}{dt} = w^2a \quad \text{under initial conditions}$$

$$x(0) = y(0) = \frac{d}{dt}(x(0)) = \frac{d}{dt}y(0) = 0$$

[Ans. $x(t) = a(wt - \sin\ wt)$, $y(t) = a(1 - \cos\ wt)$]

(39) A weightless beam of length l has its ends at $x = 0$ and $x = l$. The beam is hinged at these two points. If a concentrated load W acts at point $x = \frac{l}{3}$, show that the deflection of the beam at any point x of it is given by

$$EI\ y(x) = \frac{W}{81}\ x\ (5l^2 - 9x^2) + \frac{W}{6}\left(x - \frac{l}{3}\right)^3 H\left(x - \frac{l}{3}\right),$$

on the assumption that the deflection $y(x)$ satisfies the ODE

$$EI\ y^{(iv)}(x) = W\delta\left(x - \frac{l}{3}\right),\ E, I \text{ being the elastic constants of the}$$

beam.

(40) The deflection of a beam of length l, clamped horizontally at both ends and loaded at $x = \frac{l}{4}$ by a weight W is given by

$$EI\ y^{(iv)}(x) = W\delta\left(x - \frac{l}{4}\right). \text{ Find the deflection curve of the beam,}$$

given that $y(x) = y'(x) = 0$, when $x = 0, l$.

$$\left[\text{Ans. } EI\ y(x) = \frac{W}{6}\left(x - \frac{l}{4}\right)^3 H\left(x - \frac{l}{4}\right) + \frac{9}{128}WIx^2 - \frac{9}{64}W^2x^3.\right]$$

(41) Find Laplace transform of
(i) a square-wave periodic function of period a given by

$$f(t) = \begin{cases} 1, & 0 < t < \frac{a}{2} \\ -1, & \frac{a}{2} < t < a \end{cases}$$

(ii) a rectified semi-wave periodic function of period $\frac{2\pi}{w}$ given by

$$f(t) = \begin{cases} \sin\ wt, & 0 < t < \frac{\pi}{w} \\ 0, & \frac{\pi}{w} < t < \frac{2\pi}{w} \end{cases}$$

$$\left[\text{Ans. (i) } \frac{1}{p}\ th\left(\frac{ap}{4}\right) \quad \text{(ii) } \frac{w}{\left[\left(1 - e^{\frac{-\pi p}{w}}\right)(p^2 + w^2)\right]}\right]$$

(42) Using Laplace transform solve the initial value problem defined by

$$(D^2 + a^2)^2 \, x \, (t) = \cos at, \quad \text{if} \quad x(0), \, x'(0), \, x''(0), \, x'''(0) \text{ all vanish.}$$

$$\left[\text{Ans.} \quad x(t) = \frac{1}{8a^3} \left[t \sin at - at^2 \cos at \right] \right]$$

(43) Solve the following ODE :

(a) $ty''(t) + (2t + 3)y'(t) + (t + 3)y(t) = a \, e^{-t}$, $y(0) = 0 \;\; y'(0) = \dfrac{a}{3}$

(b) $y''(t) + at \, y'(t) - 2ay(t) = 1, y(0) = y'(0) = 0$, $a > 0$

$$\left[\text{Ans.} \quad (a) \;\; y(t) = \frac{at \, e^{-t}}{3} \qquad (b) \;\; y(t) = \frac{t^2}{2} \; \right]$$

(44) Solve the following initial value problems defined by

(a) $x''(t) + y'(t) + 3x(t) = 15e^{-t}, y''(t) - 4x'(t) + 3y(t) = 15\sin 2t$,

$\qquad x(0) = 35, x'(0) = -48, y(0) = 27$ and $y'(0) = -55$

(b) $x'(t) + 2y''(t) = e^{-t}$, $x'(t) + 2x(t) - y(t) = 1, x(0) = y(0) = 0$

$$\left[\text{Ans.} \quad (a) \;\; x(t) = 30\cos t - 15\sin 3t - 3e^{-t} + 2\cos 2t \right.$$

$$y(t) = 30\cos 3t - 60\sin t - 3e^{-t} + \sin 2t$$

$$(b) \;\; x(t) = 1 + e^{-t} - e^{-at} - e^{-bt}, y(t) = 1 + e^{-t} - be^{-at} - ae^{-bt}$$

$$\text{where} \;\; a = 1 - \frac{1}{\sqrt{2}} \;\; \text{and} \;\; b = 1 + \frac{1}{\sqrt{2}} \; \Big]$$

(45) An alternating e.m.f. $E \sin wt$ is applied to an electrial network consisting of a constant inductance L and capacitance C connected in series. If $Q(t)$ be the charge at any time t and $i(t)$ be the current at that instant and they satisfy the ODE

$L\frac{dQ}{dt} + \frac{Q}{C} = E \sin \, wt$, $i(t) = \frac{dQ}{dt}$ with initial charge $Q(0) = 0$ and initial current $i(0) = 0$, show that the current $i(t)$ is given by

$$i(t) = \frac{Ew(\cos wt - \cos nt)}{[\,(n^2 - w^2)\,L\,]} \quad \text{where} \;\; LC = \frac{1}{n^2}$$

(46) (a) A beam which is clamped at the ends $x = 0$ and $x = l$, carries a uniform load w_0 per unit length. Show that the deflection $y(x)$ at any point is given by

$$y(x) = \frac{[w_0 \, x^2(l - x)^2]}{(24 \, E \, I)}$$

(b) If the beam in (a) be clamped at $x = 0$ but is free at $x = l$ and carries a uniform load w_0 per unit length, show that its deflection at any point x is given by

$$y(x) = \left(\frac{w_0 \, x^2}{24 \, E \, I} \right) (x^2 - 4lx + 6l^2)$$

(47) An impulsive voltage $E\delta(t)$ is applied to a circuit consisting of L, R, C in series with zero initial conditions. If $I(t)$ be the current at any subsequent time t, find the limit $\lim_{t \to 0} I(t)$ and justify your answer.   [Ans. $\frac{E}{L}$]

(48) A semi-infinite transmission line of negligible inductance and leakage per unit length satisfy the PDE $\frac{\partial v}{\partial t} = -Ri$, $\frac{\partial i}{\partial x} = -C\frac{\partial v}{\partial t}$ where v is the voltage and i is the current, R is the constant resistance and C is the constant capacitance. If v_0, the constant voltage, is applied at the sending end $x = 0$ at $t = 0$, prove that the voltage and current at any point are given by $v(x,t) = v_0 \; erfc \left(\frac{x}{2} \sqrt{\frac{RC}{t}} \right)$

and $i(x,t) = v_0 \frac{\sqrt{x}}{2} \sqrt{\frac{C}{R}} \, t^{-3/2} \, e^{-(RCx^2/4t)}$.

(49) Using Laplace transform prove that the bounded solution of the PDE

(a) $\frac{\partial y}{\partial t} = \frac{\partial^2 y}{\partial x^2} - 4y$ under conditions $y(0,t) = y(\pi, t) = 0$ and
     $y(x,0) = 6 \sin x - 4 \sin 2x$ is $y(x,t) = 6e^{-5t} \sin x - 4e^{-8t} \sin 2x$

(b) $\frac{\partial y}{\partial x} = y + 2\frac{\partial y}{\partial t}$ under condition $y(x,0) = 6e^{-3x}$ is given by
     $y(x,t) = 6 \, e^{-(3x+2t)}$

(50) A semi-infinite solid $x > 0$ is initially at temperature zero. At time $t = 0$, a constant temperature $u_0 > 0$ is applied and maintained at the face $x = 0$. Prove that at any point of the solid at a later time is given by $u_0 \; erfc[x/2\sqrt{kt}]$ provided the heat conduction equation of the solid is given by $\frac{\partial u}{\partial t} = k\frac{\partial^2 u}{\partial x^2}$, $0 < x < \infty$, $t > 0$, k is the diffusivity of the material.

(51) The faces $x = 0$ and $x = l$ of a slab of material for which $k = 1$ are kept at temperature zero and until the temperature distribution becomes $u(x,0) = x$. Prove that the temperature $u(x,t)$ at a subsequent time is given by

$$u(x,t) = \frac{2}{\pi} \sum_{n=1}^{\infty} \frac{(-1)^n}{n} \, e^{-n^2 \pi^2 t} \sin \, n\pi x \; .$$

[The following result may be used for inversion of Laplace transformation :

$$L^{-1}\left[\frac{sh\ x\sqrt{p}}{p\ sh\sqrt{p}}\ ;\ p \to t\right] = x + \frac{2}{\pi}\sum_{n=1}^{\infty}\frac{(-1)^n}{n}\ e^{-n^2\pi^2 t}\ \sin(n\pi x)\ ] \ .$$

(52) An infinite string having one end at $x = 0$ is initially at rest on the x-axis. The end $x = 0$ undergoes a periodic transverse displacement $A_0 \sin nt$. Find the displacement of any point on the string at $t > 0$. Assume that the displacement $u(x, t)$ satisfies the wave equation $\frac{\partial^2 u}{\partial t^2} = c^2\frac{\partial^2 u}{\partial x^2}$.

[Ans. $u(x, t) = A_0 \sin n\left(t - \frac{x}{c}\right) H\left(t - \frac{x}{c}\right)$.]

(53) Solve the BVP $\frac{\partial^2 u}{\partial t^2} = a^2\frac{\partial^2 u}{\partial x^2} - g$, $x > 0$, $t > 0$ under conditions $u(x, 0) = u_t(x, 0) = 0$, $u(0, t) = 0$, $\lim_{x\to\infty} u_x(x, t) = 0$, $t \geqslant 0$ to show that $u(x, t) = \frac{1}{2}\ g\left(t - \frac{x}{a}\right)^2 H\left(t - \frac{x}{a}\right) - \frac{1}{2}\ g\ t^2$.

(54) Prove that, the temperature $u(x, t)$ in the semi-infinite medium $x > 0$, when the end $x = 0$ is maintained at zero temperature and the initial distribution of temperature is $f(x)$, is given by

$$u(x, t) = \frac{2}{\pi}\int_0^{\infty}\bar{f}(p)\ e^{-c^2 p^2 t}\ \sin\ xp\ dp \ .$$

where $\bar{f}(p)$ is Laplace transform of $f(x)$.

(55) If the initial temperature of an infinite bar is given by

$$\theta(x, 0) = \begin{cases} \theta_0 & , & |x| < a \\ 0 & , & |x| > a \end{cases} \ , \quad \text{determine the temperature at}$$

any point x and at any instant t to prove that

$$\theta(x, t) = \frac{\theta_0}{2}\left[erf\left(\frac{a + x}{2c\sqrt{t}}\right) + erf\left(\frac{a - x}{2c\sqrt{t}}\right)\right]$$

(56) Prove that $y(x, t) = f(t - \frac{x}{c})$ is the displacement of an infinitely long string having one end at $x = 0$ and is initially at rest along x-axis, when the end $x = 0$ is given a transverse displacement $f(t)$, $t > 0$.

(57) Show that $u(x,y) = L_2^{-1}\left[\frac{\bar{f}(p,q)}{p+q} ; (p,q) \rightarrow (x,y)\right]$ is the solution of the BVP defined by

$$\frac{\partial u}{\partial x} + \frac{\partial u}{\partial y} = f(x,y) \text{ in } x \geqslant 0, \ y \geqslant 0 \text{ and which vanishes on the}$$

co-ordinate axes.

(58) Use the method of double Laplace transform to find the solution in the positive quadrant $(x \geqslant 0, \ y \geqslant 0)$ of the PDE

$$\frac{\partial^2 u(x,y)}{\partial x \partial y} + u(x,y) = 0$$

if $u(0,y) = u(x,0) = 1$.

$$\left[\text{Ans. } u(x,y) = L_2^{-1}\left[\frac{2}{p+q} ; (p,q) \rightarrow (x,y)\right]\right]$$

(59) Use eqn. (4.26) of article 4.4 to evaluate the double Laplace transform of (i) $\sin(ax+by)$ and of (ii) $\cos(ax+by)$.

(60) Solve the following difference equations with prescribed initial conditions :

(a) $u_{n+1} - 2\,u_n = 0$, $u_0 = 1$ [Ans. $u_n = 2^n$]

(b) $y_{k+2} - 5\,y_{k+1} + 6\,y_k = 2, y_0 = 0, y_1 = 2$ [Ans. $y_k = 1 - 2^k + 3^k$]

(c) $2\,y_{k+1} - y_k - 4 = 0$, $y_0 = 3$ $\left[\text{Ans. } y_k = 4 - \left(\frac{1}{2}\right)^k\right]$

(d) $u'(t) - \alpha\,u(t-1) = \beta$, $u(0) = 0$

$$\left[\text{Ans. } u(t) = \beta\left\{t + \frac{\alpha(t-1)^2}{\Gamma(3)} + \alpha^2\frac{(t-2)^3}{\Gamma(4)} + \cdots + \frac{\alpha^n(t-n)^{n+1}}{\Gamma(n+2)}\right\}\right]$$

(e) $\Delta^2 u_n + 3u_n = 0$, $u_0 = 0$, $u_1 = 1$ [Ans. $u_n = n\,2^n$.]

(61) Prove that the general solution of the difference equation

$$u_{n+2} - 4\,u_{n+1} + 3\,u_n = 0$$

is $u_n = A \cdot 3^n + B \cdot 2^n$, where $A = u_1 - 2u_0$ and $B = 3u_0 - u_1$. If $u_0 = 2$ and $u_1 = -2$, find the general solution u_n.

(62) Solve the following integral equations :

(a) $f(t) = \sin 2t + \int_0^t f(t - \tau) \sin \tau \, d\tau$

(b) $f(t) = \dfrac{t}{2} \sin t + \int_0^t f''(\tau) \sin(t - \tau) d\tau$, $f(0) = f'(0) = 0$

(c) $\int_0^t f(\tau) \, J_0\{a(t - \tau)\} \, d\tau = \sin at$

(d) $f(t) = \sin t + \int_0^t f(\tau) \sin(2t - 2\tau) d\tau$

(e) $f(t) = t^2 + \int_0^t f'(t - \tau)e^{-a\tau} d\tau$, $f(0) = 0$

[Ans. (a) $\dfrac{1}{2}\left(t + \dfrac{3}{2}\sin 2t\right)$ (b) $(t - \cos t)$ (c) $a \, J_0(at)$

(d) $3\sin t - \sqrt{2}\sin(\sqrt{2}t)$ (e) $t^2 + \dfrac{2t}{a}$]

(63) Using Laplace transform evaluate the following :

(a) $\displaystyle\int_0^\infty \dfrac{\sin tx}{x(x^2 + a^2)} dx, (a, t > 0)$ (b) $\displaystyle\int_{-\infty}^{+\infty} \dfrac{\cos tx \, dx}{x^2 + a^2}$, $(a, t > 0)$

(c) $\displaystyle\int_{-\infty}^{+\infty} \dfrac{x \sin xt}{x^2 + a^2} dx$, $(a, t > 0)$ (d) $\displaystyle\int_0^\infty \exp(-tx^2)dx$, $(t > 0)$

$\left[\text{Ans. (a) } \dfrac{\pi}{2a^2}(1 - e^{-at}) \text{ (b) } \dfrac{\pi}{a}e^{-at} \text{ (c) } 2\pi \, e^{-at} \text{ (d) } \sqrt{\dfrac{\pi}{4t}}\right]$

(64) Solve the following difference equations using Laplace transform :

(a) $\Delta u_n - 2u_n = 0$, $u_0 = 1$

(b) $\Delta^2 u_n - 2u_{n+1} + 3u_n = 0$, $u_0 = 0$, $u_1 = 1$

(c) $u_{n+2} - 4u_{n+1} + 4u_n = 0$, $u_0 = 1$, $u_1 = 4$

(d) $u_{n+2} - 5u_{n+1} + 6u_n = 0$, $u_0 = 1$, $u_1 = 4$

(e) $\Delta^2 u_n + 3u_n = 0$, $u_0 = 0$, $u_1 = 0$

[Ans. (a) $u_n = 3^n$ (b) $u_n = n \, 2^{n-1}$ (c) $u_n = (n + 1)2^n$
(d)$u_n = 2(3^n - 2^{n-1})$ (e) $u_n = n2^n$]

Chapter 5

Hilbert and Stieltjes Transforms

5.1 Introduction.

Both Hilbert and Stieltjes transforms appear in many problems of Applied Mathematics. Specially, Hilbert transform plays important role in solving problems of fluid mechanics, electronics and signal processing etc. Also while solving problems of fracture mechanics through integral equation technique use of Stieltjes transform may be found useful due to its simple inversion formula. At first this chapter deals with Hilbert transform.

5.2 Definition of Hilbert Transform

Let $f(t)$ be defined in the real line $-\infty < t < \infty$. The Hilbert transform of $f(t)$ denoted by $\bar{f}_H(x)$, is defined by

$$\bar{f}_H(x) = H[f(t) \; ; \; t \to x] = \frac{1}{\pi} \int_{-\infty}^{+\infty} \frac{f(t)dt}{t-x} \tag{5.1}$$

The integral on the right hand side of eqn. (5.1) is defined in the Cauchy principal value sense.

The inverse Hilbert transform is derived formally by use of Fourier transform in the following steps.

Let us rewrite the formula in eqn. (5.1) as an integral equation of the convolution type for the determination of the function $f(t)$ given by

$$\bar{f}_H(x) = \frac{1}{\pi} \int_{-\infty}^{+\infty} \frac{f(t)dt}{t-x} \equiv \frac{1}{\sqrt{2\pi}} \int_{-\infty}^{+\infty} f(t) \; g(x-t) \; dt \tag{5.2}$$

where $g(x) = \sqrt{\frac{2}{\pi}} \left(-\frac{1}{x}\right)$ and on the assumption that $\bar{f}_H(x)$ is a known function of $x \in (-\infty \, , \, \infty)$.

$$\text{Since} \quad g(x) = - \sqrt{\frac{2}{\pi}} \; x^{-1}$$

its Fourier transform $\bar{g}(\xi)$ is given by

$$\bar{g}\,(\xi) = -\,i \; sgn\,(\xi)$$

Now, taking Fourier transform of eqn.(5.2), it is found that

$$\bar{\bar{f}}_H\,(\xi) = -\,i \; sgn\,(\xi) \cdot \bar{f}\,(\xi)\,, \tag{5.3}$$

where $\bar{\bar{f}}_H(\xi)$ and $\bar{f}(\xi)$ are the Fourier transforms of $\bar{f}_H(x)$ and $f(x)$ respectively. Now, from eqn. (5.3) one gets

$$\bar{f}(\xi) = \frac{-1}{i\; sgn(\xi)} \; \bar{\bar{f}}_H(\xi) = i\; sgn(\xi) \; \bar{\bar{f}}_H(\xi)$$

Taking Inverse Fourier transform we finally get

$$H^{-1}\left[\bar{f}_H(x)\right] \equiv f(x) = -\frac{1}{\pi} \int_{-\infty}^{+\infty} \frac{\bar{f}_H(t)dt}{t - x}\,, \tag{5.4}$$

as the required inversion formula for Hilbert transform after reversing all the arguments used above.

For a rigorous proof of this formula one may be referred to the text books of Titchmarsh (Oxford Univ. Press , 1948) or Integral equations by Tricomi (Q.J.Maths, Oxford, 2, 199, 1951).

5.3 Some Important properties of Hilbert Transforms.

From the definition of Hilbert transform in eqn. (5.1) and its inversion formula in eqn. (5.4) it can be derived in operator forms that

$$H^{-1} = -H \tag{5.5}$$

$$\text{Also,} \quad H[\,f(t+a)\,;\,t \to x\,] = \frac{1}{\pi} \int_{-\infty}^{+\infty} \frac{f(t+a)dt}{t - x}$$

$$= \frac{1}{\pi} \int_{-\infty}^{+\infty} \frac{f(y)dy}{y - a - x} = \frac{1}{\pi} \int_{-\infty}^{\infty} \frac{f(y)dy}{y - (a + x)}$$

$$= H[\,f(t)\,;\,t \to x + a\,] = \bar{f}_H\,(x + a) \tag{5.6}$$

$$H[\,f(at)\,;\,t \to x\,] = \frac{1}{\pi} \int_{-\infty}^{+\infty} \frac{f(at)dt}{t-x}$$

$$= \frac{1}{\pi} \int_{-\infty}^{+\infty} \frac{f(y)dy}{a[\frac{y}{a}-x]} = \frac{1}{\pi} \int_{-\infty}^{+\infty} \frac{f(y)dy}{y-(ax)} , \quad (a>0)$$

$$= H[\,f(t)\,;\,t \to ax\,] \equiv \bar{f}_H(ax) , \quad (a>0) \qquad (5.7)$$

$$\text{Also,} \quad H[\,f(-at)\,;\,t \to x\,] = \frac{1}{\pi} \int_{-\infty}^{+\infty} \frac{f(-at)dt}{t-x}$$

$$= \frac{1}{\pi} \int_{-\infty}^{+\infty} \frac{f(y)dy}{a\left[-\frac{y}{a}-x\right]} , \qquad (a>0)$$

$$= \frac{-1}{\pi} \int_{-\infty}^{+\infty} \frac{f(y)dy}{y-(-ax)} = -H[\,f(t)\,;\,t \to -ax\,] , \quad (a>0)$$

$$\equiv -\bar{f}_H(-ax) \qquad (5.8)$$

$$H[\,f'(t)\,;\,t \to x\,] = \frac{1}{\pi} \int_{-\infty}^{+\infty} \frac{f'(t)dt}{t-x}$$

$$= \frac{1}{\pi} \left[\left\{ \frac{1}{t-x} f(t) \right\}_{t \to -\infty}^{t \to +\infty} + \int_{-\infty}^{+\infty} \frac{f(t)dt}{(t-x)^2} \right] , \text{integrating by parts}$$

$$= \frac{1}{\pi} \int_{-\infty}^{+\infty} \frac{f(t)dt}{(t-x)^2} \equiv \frac{d}{dx} \bar{f}_H(x), \qquad (5.9)$$

$$\text{since} \quad \frac{d}{dx}[\,\bar{f}_H(x)\,] = \frac{d}{dx} \left\{ \frac{1}{\pi} \int_{-\infty}^{+\infty} \frac{f(t)dt}{t-x} \right\}$$

$$= \frac{1}{\pi} \int_{-\infty}^{+\infty} \frac{f(t)}{(t-x)^2} \, dt$$

and assuming the existence of all the integrals involved here in the Cauchy principal value sense.

$$\text{Finally,} \quad H[tf(t);t \to x] = \frac{1}{\pi} \int_{-\infty}^{+\infty} \frac{tf(t)}{t-x} dt = \frac{1}{\pi} \int_{-\infty}^{+\infty} \frac{t-x+x}{t-x} f(t)dt$$

$$= \frac{1}{\pi} \int_{-\infty}^{+\infty} f(t)dt + \frac{x}{\pi} \int_{-\infty}^{+\infty} \frac{f(t)dt}{t-x} = x\,\bar{f}_H(x) + (\pi)^{-1} \int_{-\infty}^{+\infty} f(t)dt$$

$$\qquad (5.10)$$

Theorem 5.1. If $f(t)$ is an even function of t, then

$$\bar{f}_H(x) = \frac{x}{\pi} \int_{-\infty}^{+\infty} \frac{f(t) - f(x)}{t^2 - x^2} \, dt \qquad (5.11)$$

Proof. Since by Cauchy principal value sense

$$\int_{-\infty}^{+\infty} \frac{dt}{t - x} = 0,$$

we have $\quad \bar{f}_H(x) = \frac{1}{\pi} \int_{-\infty}^{+\infty} \frac{f(t)dt}{t - x}$

$$= \frac{1}{\pi} \int_{-\infty}^{+\infty} \frac{f(t) - f(x)}{t - x} \, dt$$

$$= \frac{1}{\pi} \int_{-\infty}^{+\infty} \frac{(t + x)[\, f(t) - f(x)\,]}{t^2 - x^2} \, dt$$

$$= \frac{1}{\pi} \int_{-\infty}^{+\infty} \frac{x[\, f(t) - f(x)\,]}{t^2 - x^2} \, dt + \frac{1}{\pi} \int_{-\infty}^{+\infty} \frac{t[\, f(t) - f(x)\,]}{t^2 - x^2} \, dt$$

So, $\quad \bar{f}_H(x) = \frac{x}{\pi} \int_{-\infty}^{+\infty} \frac{f(t) - f(x)}{t^2 - x^2} \, dt$

as because the second integral of the right-hand side is zero, the integrand is being an odd function of t.

Example 5.1.

Find the Hilbert transform of

$$f(t) = \begin{cases} 1, & \text{for} \quad |t| < a \\ 0, & \text{for} \quad |t| > a \end{cases}$$

Solution. We have by definition

$$\bar{f}_H(x) = \frac{1}{\pi} \int_{-a}^{+a} \frac{dt}{t - x}$$

If $|t| < a$, the integrand has a singularity at $t = x$ and hence

$$\bar{f}_H(x) = \frac{1}{\pi} \lim_{\epsilon \to 0} \left[\int_{-a}^{x-\epsilon} \frac{dt}{t - x} + \int_{x+\epsilon}^{a} \frac{dt}{t - x} \right]$$

$$= \frac{1}{\pi} \lim_{\in \to 0} [\, \log |\in| - \log |(a+x)| + \log |(a-x)| - \log |\in|\,], \text{ for } |x| < a$$

$$= \frac{1}{\pi} \log \left| \frac{a-x}{a+x} \right| , \text{ for } |x| < a$$

If $|t| > a$, the integrand has no singularity in $-a < t < a$ and therefore,

$$\bar{f}_H(x) = \frac{1}{\pi} \int_{-a}^{a} \frac{dt}{t-x} = \frac{1}{\pi} [\, \log |t-x| \,]_{-a}^{a}$$

$$= \frac{1}{\pi} \log \left| \frac{a-x}{a+x} \right| , \text{ for } |x| > a$$

Thus for all x, in this case

$$\bar{f}_H(x) = \frac{1}{\pi} \log \left| \frac{a-x}{a+x} \right|$$

Example 5.2. Find the Hilbert transform of

$$f(t) = \frac{t}{t^2 + a^2} , \quad a > 0$$

Solution. We have by definition

$$\bar{f}_H(x) = \frac{1}{\pi} \int_{-\infty}^{+\infty} \frac{t\, dt}{(t^2 + a^2)(t-x)}$$

$$= \frac{1}{\pi} \int_{-\infty}^{+\infty} \frac{1}{a^2 + x^2} \left[\frac{a^2}{t^2 + a^2} + \frac{x}{t-x} - \frac{xt}{t^2 + a^2} \right] dt$$

$$= \frac{1}{\pi(a^2 + x^2)} \left[a^2 \int_{-\infty}^{+\infty} \frac{dt}{t^2 + a^2} + x \int_{-\infty}^{+\infty} \frac{dt}{t-x} - x \int_{-\infty}^{+\infty} \frac{t\, dt}{t^2 + a^2} \right]$$

$$= \frac{1}{\pi(a^2 + x^2)} \cdot (a\pi) + 0 + 0 , \text{ by Cauchy principal value of the}$$
$$\text{second and the third integrals}$$

$$= \frac{a}{x^2 + a^2} .$$

Example 5.3. Find the Hilbert transforms of

(i) $f(t) = \cos wt$ and

(ii) $f(t) = \sin wt$.

Solution.

(i) From the definition of Hilbert transform, we get

$$
\begin{aligned}
\bar{f}_H\,(x) &= \frac{1}{\pi} \int_{-\infty}^{+\infty} \frac{\cos wt}{t-x}\,dt \\
&= \frac{1}{\pi} \int_{-\infty}^{+\infty} \frac{\cos\{w(t-x)+wx\}}{t-x}\,dt \\
&= \frac{\cos wx}{\pi} \int_{-\infty}^{+\infty} \frac{\cos w(t-x)}{t-x}\,dt - \frac{\sin wx}{\pi} \int_{-\infty}^{+\infty} \frac{\sin w(t-x)}{t-x}\,dt \\
&= \frac{\cos wx}{\pi} \int_{-\infty}^{+\infty} \frac{\cos w\alpha}{\alpha}\,d\alpha - \frac{\sin wx}{\pi} \cdot (\pi) \\
&= 0 - \sin wx = -\sin wx
\end{aligned}
$$

(ii) It can similarly be shown that

$$
\bar{f}_H\,(x) = \cos wx.
$$

5.4 Relation between Hilbert Transform and Fourier Transform.

If we write

$$
a(t) \;=\; \frac{1}{\pi} \int_{-\infty}^{+\infty} f(\alpha)\,\cos\,(\alpha t)\,d\alpha
$$

$$
b(t) \;=\; \frac{1}{\pi} \int_{-\infty}^{+\infty} f(\alpha)\,\sin\,(\alpha t)\,d\alpha \tag{5.12}
$$

Fourier integral can be expressed as

$$
\begin{aligned}
f(x) \;&=\; \frac{1}{2} \int_{-\infty}^{+\infty} [\,a(t) \,+\, i\,b(t)\,]\,e^{-ixt}\,dt \\
&=\; \int_{0}^{\infty} [\,a(t)\,\cos\,(xt) \,+\, b(t)\,\sin\,(xt)\,]\,dt
\end{aligned}
$$

Let us now define a function of the complex variable z as

$$
\begin{aligned}
\phi\,(z) \;&=\; \int_{0}^{\infty} [\,a(t) - i\,b(t)\,]\,e^{izt}\,dt\ ,\quad \text{where } z = x + iy \\
&\equiv\; U(x,y) \,+\, i\,V(x,y)\ ,\ \text{say,}
\end{aligned}
$$

We have then

$$\lim_{y \to 0} U(x, y) = f(x)$$

and also $-\lim_{y \to 0} V(x, y) = \int_0^\infty [\, b(t) \cos (xt) - a(t) \sin (xt) \,]\, dt$

Substituting the values of $a(t)$ and $b(t)$ from eqn. (5.12), we find that

$$
\begin{aligned}
-\lim_{y \to 0} V(x, y) &= \frac{1}{\pi} \int_0^\infty dt \left[\int_{-\infty}^{+\infty} f(\alpha) \sin(\alpha - x)t \, d\alpha \right] \\
&= \lim_{\lambda \to \infty} \frac{1}{\pi} \int_{-\infty}^{+\infty} \frac{1 - \cos \lambda(\alpha - x)}{\alpha - x} f(\alpha) \, d\alpha \\
&= \lim_{\lambda \to \infty} \frac{1}{\pi} \int_0^\infty \frac{1 - \cos \lambda t}{t} \{\, f(x + t) - f(x - t) \} \, dt
\end{aligned}
$$

If $f(t)$ is sufficiently smooth, we have

$$\lim_{\lambda \to \infty} \int_0^\infty \cos \lambda t \, \frac{f(x + t) - f(x - t)}{t} \, dt = 0$$

by the Riemann-Lebesgue lemma. Therefore,

$$
\begin{aligned}
-\lim_{y \to 0} V(x, y) &= \frac{1}{\pi} \int_0^\infty \frac{f(x + t) - f(x - t)}{t} \, dt \qquad (5.13) \\
&\equiv H[\, f(t) \,;\, t \to x \,] = \bar{f}_H(x) \, .
\end{aligned}
$$

This last step is obtained by changing the variable of integration in the second integral in eqn. (5.13) and combining it with the first integral.

5.5   Finite Hilbert Transform.

The finite Hilbert transform is defined by Tricomi (1951) as

$$H[\, f(t) \,,\, a, b\,] = \bar{f}_H(x, a, b) = \frac{1}{\pi} \int_a^b \frac{f(t)}{t - x} \, dt \qquad (5.12)$$

when $x \in (a < x < b)$. In studying the airfoil theory such transform do arise in Aerodynamics. Also this transform is used in finding the solution of singular integral equation of the type

$$\frac{1}{\pi} \int_{-1}^1 \frac{f(t)}{t - x} \, dt = \bar{f}_H(x) \,, \qquad (5.13)$$

where $\bar{f}_H\,(x)$ and $f(t)$ are respectively known and unknown functions satisfying Hölder conditions on $(-1, 1)$ of the variables. Eqn. (5.13) arise in boundary value problems in solid and in fracture mechanics. In this respect works of the authors like Muskhelishvili (1963), Gakhov (1966), Peters (1972), Chakraborty & Williams (1980), Comninou (1977), Das, Patra and Debnath (2004) and others may be referred to.

Several authors including Okikiolu (1965) and Kober (1967) introduced the modified Hilbert transform defined by

$$H_\alpha\,[f(t)] = \bar{f}_{H\alpha}\,(x) = \frac{\mathrm{cosec}\left(\frac{\pi\alpha}{2}\right)}{2\,\Gamma(\alpha)}\int_{-\infty}^{+\infty}(t-x)^{\alpha-1}\,f(t)\,dt\;,$$

(5.14)

where $\quad 0 < \alpha < 1 \quad$ and $\quad x \in (-\infty\,,\,\infty)\,\cdot$

The Parseval's relation of this modified Hilbert transform is given in the following theorem.

Theorem 5.2. If $H_\alpha\,[f(t)] = \bar{f}_{H\alpha}\,(x)$, then

$$\int_{-\infty}^{+\infty}H_\alpha\,[\,f(t), x\,]\,h(x)\,dx = -\int_{-\infty}^{+\infty}H_\alpha\,[\,h(t), x\,]\,f(x)\,dx \quad (5.15)$$

Proof.

We do not persue with the proof of this theorem due to lack of space and scope of this treatise.

5.6 One-sided Hilbert Transform.

The two-sided Hilbert transform of $f(t)$ in $-\infty < t < x$, defined in eqn (5.1), can be written as

$$\begin{aligned}
\bar{f}_H\,(x) &= \frac{1}{\pi}\int_{-\infty}^{+\infty}\frac{f(t)dt}{t-x}\\
&= \frac{1}{\pi}\int_0^\infty\frac{f(t)}{t-x}\,dt - \frac{1}{\pi}\int_0^\infty\frac{f(-t)}{t+x}\,dt \quad (5.16)
\end{aligned}$$

when $x > 0$, where the second integral is actually the Stieltjes transform of $[-f(-t)]$ to be defined in this chapter. A similar expression for $\bar{f}_H\,(x)$ can also be written for $x < 0$. We now define one-sided Hilbert transform of $f(t)$ as

$$H^+ [f(t)] = \bar{f}_H^+ (x) = \frac{1}{\pi} \int_0^\infty \frac{f(t)}{t-x} \, dt \qquad (5.17)$$

Further, Mellin transform of $H^+\{f(t)\}$, to be defined in the next chapter, is given by

$$
\begin{aligned}
M\left[H^+\{f(t)\} \right] &= \int_0^\infty x^{p-1} \left[\frac{1}{\pi} \int_0^\infty \frac{f(t)dt}{t-x} \right] dx \\
&= \frac{1}{\pi} \int_0^\infty f(t) \left[\int_0^\infty \frac{x^{p-1} \, dx}{t-x} \right] dt \\
&= \cot(\pi p) \int_0^\infty x^{p-1} \, f(t) \, dt \\
&= \cot(\pi p) \, M\{f(t)\}
\end{aligned}
$$

Taking inverse Mellin transform, we obtain

$$H^+ [\, f(t) \,] = \bar{f}^+ (x) = \frac{1}{2\pi i} \int_{c-i\infty}^{c+i\infty} x^{-p} \cot(\pi p) \, \bar{f}(p) \, dp \qquad (5.18)$$

where $f(x)$ is defined on $0 < x < \infty$ and p may be a complex number.

5.7   Asymptotic Expansions of one-sided Hilbert Transform.

The one-sided Hilbert transform of a function $f(t)$ that is analytic for $0 < t < \infty$ and having the asymptotic expansion with known a_n, A_n, B_n and $w > 0$ in the form

$$f(t) = \sum_{n=1}^\infty \frac{a_n}{t^n} + \cos wt \sum_{n=1}^\infty \frac{A_n}{t^n} + \sin wt \sum_{n=1}^\infty \frac{B_n}{t^n} \ , \ \text{ as } t \to \infty, \quad (5.19)$$

is defined as

$$H^+ [\, f(t) \,;\, t \to x \,] = \bar{f}_H^+ (x) = \int_0^\infty \frac{f(t)}{t-x} \, dt \qquad (5.20)$$

Therefore, to find the asymptotic expansion of $\bar{f}_H^+ (x)$ for such f given in (5.19), we need to find the expansions of

$$P \int_0^\infty \frac{\cos wt}{t-x} \, dt \quad \text{and} \quad P \int_0^\infty \frac{\sin wt}{t-x} \, dt$$

Now, since $\quad P \int_0^\infty \dfrac{e^{iwt}}{t-x} \, dt = \pi i \, (\exp iwx) + \int_0^\infty \dfrac{e^{iwt}}{t-x} \, dt \ ,$

$$(5.21)$$

where the contour in the right hand side integral is being indented above at $t = x$. Deforming this contour into the positive imaginary axis by setting $t = iu$, $u > 0$ we get

$$
\begin{aligned}
\int_0^\infty \frac{e^{iwt}}{t-x} \, dt &= -i \int_0^\infty \frac{e^{-wu}}{x - iu} \, du \\
&= -\sum_{n=0}^\infty \frac{i^{n+1}}{x^{n+1}} \int_0^\infty u^n \, e^{-wu} \, du \ , \quad \text{by Watson's lemma} \\
&= -\sum_{n=0}^\infty \frac{n!}{(wx)^{n+1}} \exp\left\{ (n+1) \, i \, \frac{\pi}{2} \right\}
\end{aligned}
$$
$$(5.22)$$

From (5.17) and (5.18), separating the real and imaginary parts we get

$$
P \int_0^\infty \frac{\cos wt}{t-x} \, dt \sim -\pi \sin wx - \sum_{n=0}^\infty \frac{n!}{(wx)^{n+1}} \cos\left\{ (n+1)\frac{\pi}{2} \right\} \quad \text{as } x \to \infty
$$

and

$$
P \int_0^\infty \frac{\sin wt}{t-x} \, dt \sim -\pi \cos wx - \sum_{n=0}^\infty \frac{n!}{(wx)^{n+1}} \sin\left\{ (n+1)\frac{\pi}{2} \right\} \quad \text{as } x \to \infty
$$
$$(5.23)$$

Then the asymptotic expansion of one-sided Hilbert transform of $f(t)$ satisfying (5.19) is given by

$$
\begin{aligned}
\bar{f}_H^+ (x) = \int_0^\infty \frac{f(t)dt}{t-x} \sim{}& -\sum_{n=1}^\infty \frac{d_n}{x^n} - \log x \sum_{n=1}^\infty \frac{a_n}{x^n} \\
& -(\pi \sin wx) \sum_{n=1}^\infty \frac{A_n}{x^n} + (\pi \cos wx) \sum_{n=1}^\infty \frac{D_n}{x^n} \quad \text{as } x \to \infty
\end{aligned}
$$
$$(5.24)$$

where $\quad d_n = M\{f(t), n\} = \displaystyle\int_0^\infty t^{n-1} f(t) \, dt \qquad\qquad (5.25)$

The detailed of the derivation of (5.24) and (5.25) is not taken up here. For interested readers, the paper entilled "Integrals with a large parameter : Hilbert transforms (1983), Math Proc. Camb, Phil. Soc., 93, 141-149" may be referred to.

5.8   The Stieltjes Transform.

From the definition of Laplace Transform we have

$$\bar{f}(p) = L[\ f(t)\ ;\ t \to p\ ] = \int_0^\infty e^{-pt} f(t)\ dt \qquad (5.26)$$

and so Laplace Transform $\bar{f}(p)$ in variable p to y is given by

$$
\begin{aligned}
L[\ \bar{f}(p)\ ;\ p \to y\ ] &= \int_0^\infty e^{-py}\ \bar{f}(p)\ dp \\
&= \int_0^\infty e^{-py}\ dp \int_0^\infty f(t)\ e^{-pt}\ dt\ , \qquad (5.27)
\end{aligned}
$$

provided that the right hand side integral exists.

Now since, $\qquad \int_0^\infty e^{-pt-py}\ dp = \dfrac{1}{t+y}$

we see that

$$L\ [\ L\{f(t)\ ;\ t \to p\}\ ;\ p \to y\ ] = \int_0^\infty \frac{f(t)}{t+y}\ dt \qquad (5.28)$$

If the right hand side integral of eqn. (5.28) is convergent for complex values of y belonging to some region Ω of y-plane then we represent eqn. (5.28) as

$$
\begin{aligned}
\bar{f}_2\ (y) &= \int_0^\infty \frac{f(t)}{t+y}\ dt\ , \qquad \text{for} \quad y \in \Omega \\[2mm]
&\equiv S[\ f(t)\ ;\ t \to y\ ] \qquad (5.29)
\end{aligned}
$$

This is called Stieltjes transform of $f(t)$ for $y \in \Omega$. In particular, $\bar{f}_2\ (y)$ is analytic in the complex y-plane with a branch cut along the negative y-axis.

5.9   Some Deductions.

Since, $\dfrac{t}{t+y} = 1 - \dfrac{y}{t+y}$,   we have formally

$$S\,[\,t\,f(t)\ ;\ t \to y\,] = \int_0^\infty f(t)\,dt - y\,\bar{f}_2\,(y) \qquad (5.30)$$

Also since, $\dfrac{1}{(t+a)(t+y)} = \dfrac{1}{a-y}\left[\dfrac{1}{t+y} - \dfrac{1}{t+a}\right]$, we have

$$S\left[\dfrac{f(t)}{t+a}\ ;\ t \to y\right] = \dfrac{1}{a-y}\,[\,\bar{f}_2\,(y) - \bar{f}_2\,(a)\,] \qquad (5.31)$$

Again, from the change of scale we can easily deduce that

$$S\,[\,f(ax)\ ;\ x \to y\,] = \bar{f}_2\,(y)\ , \quad \text{for}\ \ a > 0 \qquad (5.32)$$

Also, putting $x = \frac{1}{u}$ we can show that

$$\int_0^\infty \dfrac{\frac{1}{x}\,f\left(\frac{1}{x}\right)}{x+y}\,dx = \int_0^\infty \dfrac{f(u)du}{1+uy}$$

This result implies that

$$S\left[\,x^{-1}\,f\left(x^{-1}\right)\ ;\ x \to y\,\right] = y^{-1}\,\bar{f}_2\,\left(y^{-1}\right) \qquad (5.33)$$

and using (5.32) in (5.33) it is obtained that

$$S\left[\,x^{-1}\,f\left(\dfrac{a}{x}\right)\ ;\ x \to y\,\right] = \bar{y}\,\bar{f}_2\left(\dfrac{a}{y}\right) \qquad (5.34)$$

Substituting $t = u^2$ it is deduced that

$$S\left[f\left(t^{\frac{1}{2}}\right)\ ;\ t \to y\right] = 2\int_0^\infty \dfrac{u f(u)}{u^2 + y}\,du$$

Again since, $\dfrac{2u}{u^2 + y} = \dfrac{1}{u + i\sqrt{y}} + \dfrac{1}{u - i\sqrt{y}}$,   we obtain

$$S\left[f\left(t^{\frac{1}{2}}\right)\ ;\ t \to y\right] = \bar{f}_2(i\sqrt{y}) + \bar{f}_2(-i\sqrt{y}) \qquad (5.35)$$

Stieltjes transform of the derivative of $f(t)$ is given by

$$S\,[f'(t)\ ;\ t \to y]\ =\ \int_0^\infty \dfrac{f'(t)dt}{t+y}$$

$$=\ \left[\dfrac{f(t)}{t+y}\right]_0^\infty + \int_0^\infty \dfrac{f(t)\,dt}{(t+y)^2}$$

$$=\ -\dfrac{1}{y}\,f(0) - \dfrac{d}{dy}\,\bar{f}_2(y) \qquad (5.36)$$

If $|Arg\ a| < \pi$ and $|Arg\ y| < \pi$, we have from definition

$$S\left[\frac{1}{t+a}\ ;\ t \to y\right] = \frac{1}{a-y}\ \log\left|\frac{a}{y}\right| \tag{5.37}$$

If $S[\ f(t)\ ;\ t \to y\ ] = \displaystyle\int_0^\infty \frac{f(t)\ dt}{t+y}$, then

$$\frac{d^n}{dy^n}[\ S\{f(t)\ ;\ t \to y\}\ ] = (-1)^n\ n!\ \int_0^\infty \frac{f(t)\ dt}{(t+y)^{n+1}}\ ,\ n = 1, 2, 3, \cdots$$

As a matter of fact Stieltjes transform can be looked upon as a repeated Laplace transform and the calculation of Stieltjes transform of functions can be found easily. For example, we know that

$$L\left[\frac{1}{\sqrt{t}}\ e^{-at}\ ;\ t \to p\right] = \sqrt{\pi}\ (p+a)^{-\frac{1}{2}}$$

$$L\left[\sqrt{\pi}\ (p+a)^{-\frac{1}{2}}\ ;\ p \to y\right] = \frac{\pi}{\sqrt{y}}\ e^{ay}\ Erfc\ (\sqrt{ay})$$

These results imply that

$$S\left[\frac{1}{\sqrt{t}}\ e^{-at}\ ;\ t \to y\right] = \frac{\pi}{\sqrt{y}}\ e^{ay}\ Erfc\ (\sqrt{ay}) \tag{5.38}$$

5.10   The Inverse Stieltjes Transform.

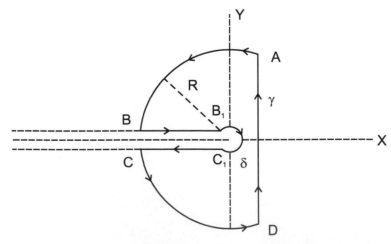

Fig. 1 z - plane : contour C

We describe below an informal proof of inverse Stieltjes transform. A rigorous proof of it may be found in the book of Widder (The Laplace Transform, Princeton University Press, 1941).

Let us consider the contour integral of $g(z)\ e^{pz}$ over an anticlockwise contour C in the z-plane, consisting of a straight line γ from $c - iR$ to $c + iR$, the circular arc AB of large radius R, a branch cut at $z = 0$ along the negative z-axis from B to B_1, with a small circle $\gamma_1 : |z| = \delta$ and then from C_1 to C and the circular arc CD of large radius R as shown in the above figure 1 where the function $g(z)$ is assumed to be analytic except for its branch singularities on the negative real z-axis. The function is also assumed to be such that

$$\lim_{|z| \to \infty} g(z) e^{pz} = 0 \ , \ \mathrm{Re}\ z \leqslant c \ , \ (\text{ c is arbitrary }) \qquad (5.39)$$

$$\lim_{|z| \to 0} z\ g(z) = 0 \qquad (5.40)$$

Then clearly,

$$\int_C g(z)\ e^{pz}\ dz = 0$$

This result implies that

$$\int_{c-iR}^{c+iR} g(z)\ e^{pz}\ dz + \int_{\mathrm{arc}AB} g(z)\ e^{pz}\ dz + \int_R^\delta g(x\ e^{i\pi})\ e^{-px}(-dx)$$

$$+ \oint_{\gamma_1} g(z)\ e^{pz}\ dz + \int_\delta^R g(x\ e^{-i\pi})\ e^{-px}(-dx) + \int_{\mathrm{arc}\ CD} g(z)\ e^{pz}\ dz = 0$$

$$(5.41)$$

Making $R \to \infty$ and $\delta \to 0$ when on $\gamma_1 : z = \delta\ e^{i\theta}$, we get from above that

$$\lim_{\delta \to 0} \int_{\gamma_1} g(z)\ e^{pz}\ dz = 0 \ , \ \text{by eqn. (5.40) above}$$

$$\lim_{R \to \infty} \int_{\mathrm{arc}AB} g(z)\ e^{pz}\ dz = 0 \ \text{and} \ \lim_{R \to \infty} \int_{\mathrm{arc}\ CD} g(z)\ e^{pz}\ dz = 0, \ \text{by (5.39)}$$

above

Therefore, eqn. (5.41) reduces to

$$\int_{c-i\infty}^{c+i\infty} g(z)\ e^{pz}\ dz \ = \ -\int_0^\infty g\left(x\ e^{i\pi}\right)\ e^{-px} dx + \int_0^\infty g\left(x\ e^{-i\pi}\right)\ e^{-px} dx$$

$$= \ \int_0^\infty \left[\ g\left(x\ e^{-i\pi}\right) - g\left(x\ e^{i\pi}\right)\ \right]\ e^{-px}\ dx$$

The last equation can be rewritten as

$$2\pi i \, L^{-1} \, [g(z) \; ; \; z \to p]$$

$$= L \, [\{ \, g \left(x \, e^{-i\pi} \right) - g \left(x \, e^{i\pi} \right) \} \; ; \; x \to p]$$

Therefore,

$$g(z) = L \left[\left[\, L \, (2\pi i)^{-1} \{ g \left(x \, e^{-i\pi} \right) - g \left(x \, e^{i\pi} \right) \} \; ; \; x \to p \right] \; ; \; p \to z \right]$$

$$(5.42)$$

Again, it is known from double Laplace transform that

$$L \, [\, L \, \{ \, f(x) \; ; \; x \to p \, \} \; ; \; p \to y \,] = S \, [\, f(x) \; ; \; x \to y \,] \qquad (5.43)$$

Therefore, comparing eqns. (5.42) and (5.43), we can write

$$S^{-1} \, [\, g(y) \; ; \; y \to x \,]$$

$$= \frac{1}{2\pi i} \{ g \left(x \, e^{-\pi i} \right) - g \left(x \, e^{i\pi} \right) \} \qquad (5.44)$$

as the formula for inverse stieltjes transform

5.11　Relation between Hilbert Transform and Stieltjes Transform.

Since Hilbert transform and Stieltjes transform are closely related we shall find a relation between them as below.

We know that

$$\bar{f}_H \, (x) = H \, [\, f(t) \; ; \; t \to x \,] = \frac{1}{\pi} \int_{-\infty}^{+\infty} \frac{f(t) \, dt}{t - x} \, ,$$

where x is real and the integral is in the sense of Cauchy principal value.

If $x > 0$, we may write

$$\bar{f}_H \, (x) = \frac{1}{\pi} \int_0^\infty \frac{f(t) \, dt}{t - x} + \frac{1}{\pi} \int_{-\infty}^0 \frac{f(t) \, dt}{t - x}$$

$$= \frac{1}{2\pi} \int_0^\infty \frac{f(t) \, dt}{t + x \, e^{i\pi}} + \frac{1}{2\pi} \int_0^\infty \frac{f(t) \, dt}{t + x \, e^{-i\pi}} - \frac{1}{\pi} \int_0^\infty \frac{f(-t)}{(t + x)} \, dt$$

$$\therefore \ \bar{f}_H\,(x) \ = \ (2\pi)^{-1}\ S\ \big[\ f(t)\ ;\ t \to x\ e^{i\pi}\ \big]$$

$$+\ (2\pi)^{-1}\ S\ \big[\ f(t)\ ;\ t \to x\ e^{-i\pi}\ \big] - \frac{1}{\pi}\ S\ \big[\ f(-t)\ ;\ t \to x\ \big]$$

$$\text{for } x > 0 \qquad (5.45)$$

Similarly, when $x < 0$, we have $x = -|x|$ and hence we have

$$\bar{f}_H\,(x) \ = \ \frac{1}{\pi}\int_0^\infty \frac{f(t)\ dt}{t + x} + \frac{1}{\pi}\int_{-\infty}^0 \frac{f(t)\ dt}{t - x}$$

$$= \ (\pi)^{-1}\ S\ \big[\ f(t)\ ;\ t \to |x|\ \big] - \frac{1}{\pi}\int_0^\infty \frac{f(-t)}{t + x}\ dt$$

$$= \ (\pi)^{-1}\ S\ \big[\ f(t)\ ;\ t \to -x\ \big] - \frac{1}{2\pi}\int_0^\infty \frac{f(-t)\ dt}{t + (-x\ e^{i\pi})}$$

$$-\ \frac{1}{2\pi}\int_0^\infty \frac{f(-t)\ dt}{t + (-x\ e^{-i\pi})}\quad \text{for } x < 0 \qquad (5.46)$$

Excercises

(1) Find Hilbert transform of the following functions :

$$\text{(a)}\quad f(t) \ = \ \frac{1}{(a^2 + t^2)}\ ,\quad Re\ a > 0$$

$$\text{(b)}\quad f(t) \ = \ \frac{t^\alpha}{(t + a)}\ ,\quad |\ Re\ \alpha\ | < 1$$

$$\text{(c)}\quad f(t) \ = \ exp\ (-at)$$

$$\text{(d)}\quad f(t) \ = \ t^{-\alpha}\ exp\ (-at)\ ,\quad Re\ a > 0\ ,\ Re\ \alpha < 1$$

$$\text{(e)}\quad f(t) \ = \ \frac{1}{t}\ \sin\ (a\sqrt{t})\ ,\quad a > 0$$

$$\text{(f)}\quad f(t) \ = \ \frac{\sin t}{t}$$

[**Ans.** (a) $\quad (a^2 + x^2)^{-1}\left[\dfrac{\pi x}{2a} - \log\left(\dfrac{x}{a}\right)\right]$

(b) $\quad (a - x)^{-1}(a^\alpha - x^\alpha)\ \pi\ cosec\ \pi\alpha$

(c) $\quad -\ exp\ (ax)\ Ei\ (-ax)$

(d) $\quad \Gamma(1 - \alpha)\ x^{-a}\ exp\ (ax)\ \Gamma\ (\alpha\ ,\ ax)$

(e) $\quad \dfrac{\pi}{x}[1 - exp(-a\sqrt{x})]$

(f) $\quad \dfrac{(\cos x - 1)}{x}$]

(2) Find Stieltjes transform of the following functions.

$$\text{(a)} \ f(t) \ = \ \frac{t^{\alpha-1}}{(t+a)}$$

$$\text{(b)} \ f(t) \ = \ \frac{1}{(t^2+a^2)}$$

$$\text{(c)} \ f(t) \ = \ \frac{t}{(t^2+a^2)}$$

[**Ans.** (a)$(a-x)^{-1}\left(x^{\alpha-1}-a^{\alpha-1}\right)\pi\ cosec\ (\alpha\pi)$

$$\text{(b)} \quad (a^2+x^2)^{-1}\left[\frac{\pi x}{2a}-\log\left(\frac{x}{a}\right)\right]$$

$$\text{(c)} \quad (a^2+x^2)^{-1}\left[\frac{\pi a}{2}+x\ \log\left(\frac{x}{a}\right)\right]\ ]$$

(3) Show that

$$\text{(a)} \ S\left[\frac{1}{\sqrt{t}}\ \cos\ a\sqrt{t}\ ;\ t\to y\right]=\frac{\pi}{\sqrt{y}}\ exp\left(-a\sqrt{x}\right),\ x>0$$

$$\text{(b)} \ S\left[\sin\left(k\sqrt{t}\right)\ ;\ t\to y\right]=\pi\ exp\left(-k\sqrt{y}\right),\ k>0$$

(4) Prove that

$$\lim_{\epsilon\to 0+}\frac{1}{\pi}\int_{\epsilon}^{\infty}\left[\frac{f(x+u)-f(x-u)}{u}\right]du=\frac{1}{\pi}\ P\int_{-\infty}^{\infty}\frac{f(t)}{t-x}\ dt$$

(5) Prove that if $f(t)=H(t)$, then

$$\text{(a)} \ [H\{f(t-a)-f(t-b)\}\ ;\ t\to x]=\frac{1}{\pi}\ \log\left|\frac{b-x}{a-x}\right|,\ b>a>0$$

$$\text{(b)} \ H\left[\frac{f(t-a)}{t}\ ;\ t\to x\right]=\frac{1}{\pi x}\ \log\frac{a}{|a-x|},\ a>0$$

$$\text{(c)} \ H\left[\frac{f(t)}{at+b}\ ;\ t\to x\right]=\frac{1}{\pi(ax+b)}\ \log\frac{b}{a|x|},\left(a,b>0\ ,\ x\neq\frac{b}{a}\right)$$

$$\text{(d)} \ H\left[\frac{\alpha t+\beta a}{t^2+a^2}\ ;\ t\to x\right]=\frac{\alpha a-\beta x}{x^2+a^2},\ a>0$$

(6) Prove that

(a) $S\left[\dfrac{1}{(a+x)(b+x)}\; ;\; x \to y\right] = \begin{cases} \dfrac{(a-b)\log(\frac{a}{y})-(a-y)\log(\frac{a}{b})}{(a-y)(b-y)(a-b)} &,\ b \neq a \\[12pt] \dfrac{a\log(\frac{y}{a})-(a-y)}{(a-y)^2 a} &,\ b = a \end{cases}$

(b) $S\left[x^{\nu-1}\; ;\; x \to y\right] = \pi \,\dfrac{y^{\nu-1}}{\sin\,\nu\,\pi}\ ,\quad 0 < \nu < 1$

(c) $S\left[(a+x)^{-1}\; ;\; x \to y\right] = (a-y)^{-1}\,\log\left(\dfrac{a}{y}\right)$

(7) Show that $f(t) = A\left[\dfrac{\Gamma'(\frac{1}{2})}{\sqrt{\pi t}} - \dfrac{2\log t}{\sqrt{t}}\right]$ is the solution of the integral equation $f(s) = -\dfrac{1}{\sqrt{\pi}}\int_0^\infty e^{-st}\, f(t)\, dt$.

Chapter 6

Hankel Transforms

6.1 Introduction.

Hankel transform arises in the discussion of axisymmetric problems in the cylindrical coordinates involving Bessel functions. This chapter deals with the definitions, elementary properties, the inversion theorem and parseval relations of this transform and its relation with other transforms. Applications of this transform to various axisymmetric problems in applied Mathematics are also discussed in this chapter.

6.2 The Hankel Transform.

The (infinite) Hankel transform of order ν of $f(r)$ is formally defined and denoted by

$$\left.\begin{aligned} \int_0^\infty r\ f(r)\ J_\nu(\xi r)\ dr &\equiv H_\nu[\ f(r);\ \xi\ ]\ , \\ &= \bar{f}_\nu(\xi), \end{aligned}\right\} \tag{6.1}$$

if $f(r)$ is integrable over $(0, \infty)$

Theoretically, ν may be zero, integer or a non-integral number. But in physical application ν is usually either zero or a positive integer. Accordingly, in discussions of elementary properties of Hankel transform ν is taken as $\nu = n$, $n = 0$ or a positive integer.

6.3 Elementary properties

Theorem 6.1. If $f_1(r)$ and $f_2(r)$ be two integrable functions in $(0, \infty)$ and c_1, c_2 are constants, then

$$H_n\ [\ c_1\ f_1(r) + c_2\ f_2(r)\ ] = c_1\ H_n\ [\ f_1(r)\ ] + c_2\ H_n\ [\ f_2(r)\ ] \tag{6.2}$$

This property is called the linearity property of Hankel transform.

Proof. We have by definition,

$$H_n\left[\,c_1\,f_1(r)+c_2\,f_2(r)\,\right] = \int_0^\infty r\left[\,c_1\,f_1(r)+c_2\,f_2(r)\,\right]J_n\,(\xi r)\,dr$$

$$= c_1\int_0^\infty f_1(r)\,J_n\,(\xi r)\,dr + c_2\int_0^\infty f_2(r)\,J_n\,(\xi r)\,dr$$

$$= c_1\,H_n\left[\,f_1(r)\,;\,\xi\,\right] + c_2\,H_n\left[\,f_2(r)\,;\,\xi\,\right] \tag{6.2}$$

Theorem 6.2. If a is a constant and $H\left[\,f(r)\,;\,\xi\,\right]$ is Hilbert transform of $f(r)$ of order ν, then

$$H_\nu\left[\,f(ar)\,;\,\xi\,\right] = \frac{1}{a^2}\,H_\nu\left[f(r)\,;\,\frac{\xi}{a}\right] = \frac{1}{a^2}\,\bar{f}_\nu\left(\frac{\xi}{a}\right)$$

Proof. By definition of Hankel transform in (6.1), we have

$$\begin{aligned}
H_\nu\left[\,f(ar)\,;\,\xi\,\right] &= \int_0^\infty f(ar)\cdot r\,J_\nu(\xi r)\,dr\\
&= \int_0^\infty f(x)\frac{x}{a}J_\nu\left(\frac{\xi}{a}\,x\right)\frac{1}{a}\,dx\ ,\ \text{taking } ar=x\\
&= \frac{1}{a^2}\int_0^\infty x\,f(x)\,J_\nu\left(\frac{\xi}{a}\,x\right)\,dx\\
&= \frac{1}{a^2}\,H_\nu\left[f(x)\,;\,\frac{\xi}{a}\right] = \frac{1}{a^2}\,\bar{f}_\nu\left(\frac{\xi}{a}\right) \tag{6.3}
\end{aligned}$$

This property of Hankel transform is known as the change of scale property of the transform.

Theorem 6.3. If $\bar{f}_n(\xi)$ be Hankel transform of $f(r)$ of order n, then Hankel transform of $f'(r)$ of order n is given by

$$H_n\left[\,f'(r)\,;\,\xi\,\right] = \frac{\xi}{2n}\left[\,(n-1)\bar{f}_{n+1}(\xi)-(n+1)\bar{f}_{n-1}(\xi)\,\right],n\geqslant 1 \tag{6.4}$$

$$H_1\left[\,f'(r)\,;\,\xi\,\right] = -\xi\,\bar{f}_0\,(\xi) \tag{6.5}$$

provided $[rf(r)]$ vanishes as $r\to\infty$ and as $r\to 0$.

Proof. We have

$$\begin{aligned}
H_n\left[\,f'(r)\,;\,\xi\,\right] &= \int_0^\infty r J_n(\xi r)f'(r)dr\\
&= [\,r J_n(\xi r)f(r)\,]_0^\infty - \int_0^\infty \frac{d}{dr}\{\,r J_n(\xi r)\,\}\,f(r)\,dr
\end{aligned}$$

Using the properties of Bessel function we have

$$\frac{d}{dr}\left[\, rJ_n(\xi r)\,\right] = (1-n)J_n(\xi r) + \xi r J_{n-1}(\xi r)$$

Therefore, if $\{\, rJ_n(\xi r)f(r)\,\}$ tends to zero as $r \to \infty$ and also as $r \to 0$, then

$$
\begin{aligned}
H_n\left[\, f'(r)\,;\,\xi\,\right] &= -\int_0^\infty \{\,(1-n)J_n(\xi r) + \xi r J_{n-1}(\xi r)\,\}\, f(r) dr \\
&= (n-1)\int_0^\infty f(r)J_n(\xi r)dr - \xi \int_0^\infty rf(r)J_{n-1}(\xi r)dr
\end{aligned}
$$

Again, using the recurrence relation

$$J_n(\xi r) = \frac{\xi r}{2n}\left[J_{n-1}(\xi r) + J_{n+1}(\xi r)\right]$$

of Bessel function, we get

$$
\begin{aligned}
H_n\left[\, f'(r)\,;\,\xi\,\right] &= -\xi \bar{f}_{n-1}(\xi) + \frac{\xi(n-1)}{2n}\left[\int_0^\infty rf(r)J_{n-1}(\xi r)\ dr \right.\\
&\qquad\qquad\qquad\qquad \left. +\int_0^\infty rf(r)J_{n+1}(\xi r)dr\right] \\
&= -\xi \bar{f}_{n-1}(\xi) + \xi \frac{(n-1)}{2n}\bar{f}_{n-1}(\xi) + \xi\frac{(n-1)}{2n}\bar{f}_{n+1}(\xi) \\
&= -\frac{(n+1)}{2n}\xi \bar{f}_{n-1}(\xi) + \frac{(n-1)}{2n}\,\xi\,\bar{f}_{n+1}(\xi) \\
H_n\left[\, f'(r)\,;\,\xi\,\right] &= \frac{\xi}{2n}\left[\,(n-1)\bar{f}_{n+1}(\xi) - (n+1)\bar{f}_{n-1}(\xi)\,\right]
\end{aligned}
$$

When $n = 1$, it follows that

$$H_1\left[\, f'(r)\,;\,\xi\,\right] = -\xi\,\bar{f}_0\,(\xi)$$

Thus, the eqns. (6.4) and (6.5) are proved.

Repeated application of (6.4) gives

$$
\begin{aligned}
H_n\left[\, f''(r)\,;\,\xi\,\right] &= \frac{\xi}{2n}\left[\,(n-1)\bar{f}'_{n+1}(\xi) - (n+1)\bar{f}'_{n-1}(\xi)\,\right] \\
&= \frac{\xi^2}{4}\left[\frac{n+1}{n-1}\bar{f}_{n-2}(\xi) - 2\frac{n^2-3}{n^2-1}\bar{f}_n\,(\xi) + \frac{n-1}{n+1}\bar{f}_{n+2}\,(\xi)\right]
\end{aligned}
$$

$$(6.6)$$

Theorem 6.4. If $\bar{f}_n(\xi)$ be Hankel transform of order n of $f(r)$, then Hankel transform of order n of $f''(r) + \frac{1}{r}f'(r) - \frac{n^2}{r^2}f(r)$ is $-\xi^2 \bar{f}_n(\xi)$, provided that $rf'(r) \to 0$ and $rf(r) \to 0$ when $r \to 0$ and when $r \to \infty$

Proof. From the definition of Hankel transform we have

$$H_n \left[f''(r) + \frac{1}{r}f'(r) - \frac{n^2}{r^2}f(r) \; ; \; \xi \right]$$

$$= \int_0^\infty \left\{ f''(r) + \frac{1}{r}f'(r) - \frac{n^2}{r^2}f(r) \right\} r \, J_n \, (\xi r) \, dr$$

Now, integrating by parts the first integral gives

$$\int_0^\infty r \frac{d^2}{dr^2} f(r) J_n \, (\xi r) \, dr = -\int_0^\infty \frac{df(r)}{dr} J_n \, (\xi r) \, dr$$
$$- \int_0^\infty r \frac{df(r)}{dr} J_n' \, (\xi r) \, dr$$

and therefore,

$$\int_0^\infty \left\{ \frac{d^2 f}{dr^2} + \frac{1}{r} \frac{df}{dr} \right\} r \, J_n \, (\xi r) \, dr = -\int_0^\infty \frac{df}{dr} r \, J_n' \, (\xi r) \, dr$$
$$= \int_0^\infty f(r) \frac{d}{dr} \left\{ r \, J_n'(\xi r) \right\} dr, \quad \text{on integrating by parts}$$

Since, $J_n(r)$ is Bessel function of order n, we have

$$\frac{d}{dr} \left[r \frac{d}{dr} J_n(r) \right] + \left(1 - \frac{n^2}{r^2} \right) r J_n(r) = 0$$

Here replacing r by ξr we get,

$$\frac{d}{dr} \left[r J_n'(\xi r) \right] = - \left(\xi^2 - \frac{n^2}{r^2} \right) r J_n(\xi r)$$

Using this result we get

$$\int_0^\infty \left\{ f''(r) + \frac{1}{r}f'(r) - \frac{n^2}{r^2}f(r) \right\} r J_n(\xi r) \, dr = -\xi^2 \int_0^\infty r f(r) J_n(\xi r) dr$$

Or $\quad \int_0^\infty \left\{ f''(r) + \frac{1}{r}f'(r) - \frac{n^2}{r^2}f(r) \right\} r J_n(\xi r) \, dr = -\xi^2 \bar{f}_n(\xi) \qquad (6.7)$

The last equation proves the required result under the given condition.

6.4   Inversion formula for Hankel Transform.

The valid proof for Hankel inversion theorem for general order ν is complicated. In application of Hankel transform to the solution of boundary value problems in mathematical physics it is usually only the transforms of order zero and unity are involved. So we present here a proof of Hankel inversion theorem of non-negative order n.

Theorem 6.5. If $\bar{f}_n(\xi)$ is Hankel transform of the function $f(r)$

implying $$\bar{f}_n(\xi) = H[\, f(r) \; ; \; n\,] = \int_0^\infty r\, f(r) J_n(\xi r)\, dr$$

then $$f(r) = \int_0^\infty \xi \bar{f}_n(\xi) J_n(\xi r)\, d\xi \tag{6.8}$$

is called Hankel inversion formula of $\bar{f}_n(\xi)$ and we express it as

$$f(r) = H^{-1}\left[\, \bar{f}_n(\xi) \; ; \; r\,\right] \tag{6.9}$$

Proof. Extending the results of general Fourier transform and its inversion formula of a function of one variable for function of two variables, we get

$$\bar{f}(s,t) = \int_{-\infty}^{+\infty} \int_{-\infty}^{+\infty} f(x,y)\, e^{i(sx+ty)} dx\, dy \tag{6.10}$$

$$f(x,y) = \frac{1}{4\pi^2} \int_{-\infty}^{+\infty} \int_{-\infty}^{+\infty} \bar{f}(s,t)\, e^{-i(sx+ty)}\, ds\, dt \tag{6.11}$$

Taking the following change of variables

$$x = r\cos\theta, \; y = r\sin\theta \; ; \; s = \xi\cos\alpha, \; t = \xi\sin\alpha$$

in eqns. (6.10) and (6.11) respectively, we get the transform pair as

$$\bar{f}(\xi,\alpha) = \int_0^\infty r\, dr \int_0^{2\pi} f(r,\theta)\, e^{i\xi r\,\cos(\theta-\alpha)}\, d\theta \tag{6.12}$$

and $$f(r,\theta) = \frac{1}{4\pi^2} \int_0^\infty \xi\, d\xi \int_0^{2\pi} \bar{f}(\xi,\alpha)\, e^{-i\xi r\cos(\theta-\alpha)}\, d\alpha \tag{6.13}$$

Let us now choose

$$f(r,\; \theta) = f(r) e^{-in\theta} \tag{6.14}$$

Then $$\bar{f}(\xi,\alpha) = \int_0^\infty f(r) r\, dr \int_0^{2\pi} e^{i\{-n\theta+\xi r\cos(\theta-\alpha)\}}\, d\theta \tag{6.15}$$

Now putting $\phi = \alpha - \theta - \dfrac{\pi}{2}$, we get

$$\int_0^{2\pi} e^{i\{-n\theta+\xi r\cos(\theta-\alpha)\}}\,d\theta = e^{in\left(\frac{\pi}{2}-\alpha\right)} \int_0^{2\pi} e^{i(n\phi-\xi r\sin\phi)}\,d\phi$$

$$= e^{in\left(\frac{\pi}{2}-\alpha\right)} \cdot 2\pi\, J_n(\xi r), \text{ since } J_n(\xi r)$$

$$= \frac{1}{2\pi} \int_0^{2\pi} e^{i(n\phi-\xi r\sin\phi)}\,d\phi$$

Therefore, eqn. (6.15) reduces to

$$\bar{f}(\xi,\alpha) = \int_0^\infty r\, f(r)\, e^{in\left(\frac{\pi}{2}-\alpha\right)} \cdot 2\pi\, J_n(\xi r)\, dr \qquad (6.16)$$

$$= 2\pi\, e^{in\left(\frac{\pi}{2}-\alpha\right)} \cdot \int_0^\infty r\, f(r)\, J_n(\xi r)\, dr$$

$$= 2\pi\, e^{in\left(\frac{\pi}{2}-\alpha\right)} \cdot \bar{f}(\xi) \qquad (6.17)$$

Thus using eqns, (6.14) and (6.17) in eqn. (6.13), we get

$$f(r)\, e^{-in\theta} = \frac{1}{2\pi} \int_0^\infty \left[\xi\, \bar{f}(\xi)\, e^{-in\theta} \cdot 2\pi\, J_n(\xi r)\right] d\xi$$

$$= e^{-in\theta} \int_0^\infty \xi\bar{f}(\xi) J_n(\xi r)\, d\xi$$

Or $$f(r) = \int_0^\infty \xi\, \bar{f}(\xi)\, J_n(\xi r)\, dr \qquad (6.18)$$

This result is the required inversion formula for the nth order Hankel transformed function $\bar{f}_n(\xi)$.

Note 6.1. For the case of general order $\nu > -\frac{1}{2}$ if $\sqrt{r}f(r)$ is piecewise continuous and absolutely integrable on the positive real line, Hankel inversion theorem is given by

$$\int_0^\infty \xi\, \bar{f}_\nu(\xi)\, J_\nu(\xi r)d\xi = \frac{1}{2}[f(r+0) + f(r-0)], \qquad (6.19)$$

where $$\bar{f}\nu(\xi) = H_\nu[f(r)\,;\,\xi]\,.$$

6.5   The Parseval Relation for Hankel Transforms.

Theorem 6.6. The Parseval theorem.

Let $\bar{f}(\xi)$ and $\bar{g}(\xi)$ be Hankel transforms of the functions $f(r)$ and $g(r)$ respectively. Then

$$\int_0^\infty r\, f(r)\, g(r)\, dr = \int_0^\infty \xi\, \bar{f}(\xi)\, \bar{g}(\xi)\, d\xi$$

Proof. It is known from eqn. (6.19) that

$$f(r) = \int_0^\infty \bar{f}(\xi)\, \xi\, J_n(\xi r)\, d\xi$$

and also that $\bar{g}(\xi) = \int_0^\infty r\, g(r)\, J_n(\xi r)\, dr$ from the definition of Hankel transform.

Therefore,

$$\int_0^\infty \xi\, \bar{f}(\xi)\, \bar{g}(\xi) d\xi$$

$$= \int_0^\infty \xi \bar{f}(\xi) \left[\int_0^\infty r g(r) J_n(\xi r) dr \right] d\xi$$

$$= \int_0^\infty r g(r) \left[\int_0^\infty \xi \bar{f}(\xi) J_n(\xi r) d\xi \right] dr,$$

by interchanging the order of integration.

$$= \int_0^\infty r g(r) f(r) dr \tag{6.20}$$

Thus the Parseval relation is formally established for the order n of Hankel transform.

Note 6.2. For the case of general order ν of Hankel transform, the corresponding parseval relation is given by

$$\int_0^\infty \xi\, \bar{f}(\xi)\, \bar{g}(\xi) d\xi = \int_0^\infty r f(r) g(r) dr \tag{6.21}$$

where $\quad \bar{f}(\xi) = \int_0^\infty r f(r) J_\nu(\xi r) dr$

and $\quad \bar{g}(\xi) = \int_0^\infty r g(r) J_\nu(\xi r) dr$

6.6 Illustrative Examples.

Example 6.1.

Find Hankel transform of order zero of $f(x) = \begin{cases} 1, & 0 < x < a \\ 0, & x > a \end{cases}$

Solution. We know from definition of Hankel transform that

$$
\begin{aligned}
H_0\left[f(x)\right] &= \int_0^\infty x f(x) J_0\left(\xi x\right) dx \\
&= \int_0^a x J_0\left(\xi x\right) dx \\
&= \left[\frac{x}{\xi} J_1(\xi x) \right]_0^a
\end{aligned}
$$

by the recurrence relation $\quad \dfrac{d}{dx}\left[x^n J_n(x)\right] = x J_{n-1}(x)$

$$
= \frac{a}{\xi} J_1(\xi a)
$$

Example 6.2. Evaluate $(a) H_1\left[\dfrac{e^{-ax}}{x} \; ; \; \xi \right]$

$$
(b) H_0\left[\frac{1}{x} \; ; \; \xi \right]
$$

Solution.

(a) We have, $\quad H_1\left[\dfrac{e^{-ax}}{x} \; ; \; \xi \right] = \int_0^\infty x \cdot \dfrac{e^{-ax}}{x} J_1\left(\xi x\right) dx$

$$
\begin{aligned}
&= \int_0^\infty e^{-ax} J_1\left(\xi x\right) dx \\
&= \frac{(a^2 + \xi^2)^{\frac{1}{2}} - a}{\xi(a^2 + \xi^2)^{\frac{1}{2}}}
\end{aligned}
$$

(b) $\quad H_0\left[\dfrac{1}{x} \; ; \; \xi \right] = \int_0^\infty J_0\left(\xi x\right) dx = \int_0^\infty e^{-0x} J_0\left(\xi x\right) dx$

$$
= (\xi^2)^{-\frac{1}{2}} = \frac{1}{\xi}, \text{ since } \int_0^\infty e^{-ax} J_0(bx) dx = \frac{1}{\sqrt{a^2 + \xi^2}}
$$

Example 6.3. Find Hankel transform of order zero of $\dfrac{e^{-iax}}{x}$. and hence evaluate $H_0\left[\dfrac{\cos\,ax}{x} \; , \; \dfrac{\sin\,(ax)}{x} \right]$

Solution. We have from definition

$$H_0\left[\frac{e^{-iax}}{x} ; \xi\right] = \int_0^\infty (\cos ax - i \sin ax) J_0(\xi x) \, dx$$

$$= \int_0^\infty \cos ax J_0(\xi x) dx - i \int_0^\infty \sin ax \, J_0(\xi x) \, dx$$

$$= Re(-a^2 + \xi^2)^{-\frac{1}{2}} - Im(\xi^2 - a^2)^{-\frac{1}{2}}$$

$$\therefore \quad H_0\left[\frac{\cos ax}{x} ; \xi\right] = \begin{cases} (\xi^2 - a^2)^{-\frac{1}{2}} & , \quad \xi > a \\ 0 & , \quad 0 < \xi < a \end{cases}$$

$$\text{and} \quad H_0\left[\frac{\sin ax}{x} ; \xi\right] = \begin{cases} 0 & , \quad \xi > a \\ (a^2 - \xi^2)^{-\frac{1}{2}} & , \quad \xi < a \end{cases}$$

Example 6.4. If

$$f(x) = \begin{cases} x^n, & 0 < x < a, \\ 0, & x > a \end{cases}$$

find Hankel transform of order n of $f(x)$.

Solution. We have, by definition

$$H_n[f(x)] = \int_0^\infty x J_n(\xi x) f(x) \, dx$$

$$= \int_0^a x^n . x \, J_n(\xi x) \, dx$$

By recurrence relation of Bessel function we have

$$\frac{d}{dx}\left[x^{n+1} J_{n+1}(x)\right] = x^{n+1} J_n(x)$$

$$\Rightarrow \frac{1}{\xi}\frac{d}{dx}\left[x^{n+1} J_{n+1}(\xi x)\right] = x^{n+1} J_n(\xi x)$$

Therefore,

$$H_n[f(x)] = \int_0^a \frac{1}{\xi}\frac{d}{dx}\left[x^{n+1} J_{n+1}(\xi x) \right] \, dx$$

$$= \frac{a^{n+1}}{\xi} J_{n+1}(\xi a).$$

Example. 6.5. Find Hankel transform of order zero of

$$\left[\frac{d^2}{dx^2} + \frac{1}{x}\frac{d}{dx} \right] \left(\frac{e^{-ax}}{x} \right).$$

Solution. We know from (6.7) for $n = 0$ that

$$\int_0^\infty \left[f''(r) + \frac{1}{r}f'(r) \right] r J_0(\xi r) dr = -\xi^2 \bar{f}_0(\xi)$$

Therefore,
$$\int_0^\infty \left[\frac{d^2}{dx^2} + \frac{1}{x}\frac{d}{dx} \right] \left(\frac{e^{-ax}}{x} \right) . x J_0(\xi x) \, dx$$

$$= -\xi^2 \int_0^\infty e^{-ax} J_0(\xi x) \, dx$$

$$= -\frac{\xi^2}{(a^2 + \xi^2)^{\frac{1}{2}}} .$$

Example. 6.6. Deduce that

$$H_\nu[x^\nu(a^2 - x^2)^{\mu-\nu-1} H(a - x) ; \xi] = 2^{\mu-\nu-1} \Gamma(\mu - \nu) a^\mu \xi^{\nu-\mu} J_\mu(\xi a)$$

and hence prove that

$$H_\nu\left[x^{\nu-\mu} J_\mu(ax); \xi \right] = \frac{\xi^\nu(a^2 - \xi^2)^{\mu-\nu-1}}{2^{\mu-\nu-1} \Gamma(\mu - \nu) a^\mu} H(a - \xi).$$

Solution. From definition of Hankel transform we have

$$H_\nu\left[x^\nu (a^2 - x^2)^{\mu-\nu-1} H(a - x) ; \xi \right]$$

$$= \int_0^a x^{\nu+1}(a^2 - x^2)^{\mu-\nu-1} J_\nu(x\xi) \, dx$$

$$= \sum_{r=0}^\infty \frac{(-1)^r (\frac{\xi}{2})^{\nu+2r}}{r! \, \Gamma(r + \nu + 1)} \int_0^a x^{2\nu+2r+1}(a^2 - x^2)^{\mu-\nu-1} \, dx,$$

$$\text{for } \mu > \nu \geqslant 0 \, , \, a > 0 \, . \quad (6.22)$$

The integral on the right hand side of this last equation has the value

$$\frac{1}{2} a^{2\nu+2r} . \frac{\Gamma(r + \nu + 1) \Gamma(\mu - \nu)}{\Gamma (r + \mu + 1)}$$

and therefore,

$$H_\nu\left[x^\nu(a^2 - x^2)^{\mu-\nu-1} H(a - x) ; \xi \right]$$

$$= 2^{\mu-\nu-1}\Gamma(\mu-\nu)a^{\mu}\xi^{\nu-\mu}\sum_{r=0}^{\infty}\frac{(-1)^r(\frac{1}{2}\xi a)^{\mu+2r}}{r!\,\Gamma\,(r+\mu+1)}$$

$$= 2^{\mu-\nu-1}\Gamma(\mu-\nu)a^{\mu}\xi^{\nu-\mu}J_{\mu}(\xi a)$$

$$(6.23)$$

Now, using Hankel inversion theorem we have if $\mu > \nu \geqslant 0$

$$2^{\mu-\nu-1}\,\Gamma(\mu-\nu)a^{\mu}\,H_{\nu}\left[\,\xi^{\nu-\mu}J_{\mu}(\xi a)\right] = x^{\nu}(a^2-x^2)^{\mu-\nu-1}\,H(a-x)$$

This result in an alternative form can be expressed as

$$H_{\nu}\left[x^{\nu-\mu}J_{\mu}(ax)\,;\,\xi\right] = \frac{\xi^{\nu}(a^2-\xi^2)^{\mu-\nu-1}}{2^{\mu-\nu-1}\,\Gamma\,(\mu-\nu)\,a^{\mu}}\,H(a-\xi) \left.\begin{array}{r}\\[2mm]\end{array}\right\}$$

$$\Rightarrow \int_0^{\infty} x^{1-\mu+\nu}J_{\mu}(ax)J_{\nu}\,(\xi x)\,dx = \frac{\xi^{\nu}(a^2-\xi^2)^{\mu-\nu-1}}{2^{\mu-\nu-1}\,\Gamma\,(\mu-\nu)\,a^{\mu}}\,H(a-\xi) \left.\begin{array}{r}\\[2mm]\end{array}\right\} \quad (6.24)$$

Example 6.7. Using eqns.(6.23) and (6.24) in example 6.6, deduce the corresponding results for the following particular cases

$$\text{(a) } \mu = \nu+1\,, \quad \text{(b) } \mu = \nu+\frac{1}{2}\,, \quad \text{(c) } \nu = 0\,, \quad \text{(d) } \nu = \frac{1}{2}$$

Solution. (a) Puttintg $\mu = \nu+1$ in eqns. (6.23) and (6.24) we obtain

$$H_{\nu}\left[\,x^{\nu}H\,(a-x)\,;\xi\,\right] = \frac{a^{\nu+1}}{\xi}J_{\nu+1}(\xi a) \qquad ,\,a>0$$

$$H_{\nu}\left[\,x^{-1}J_{\nu+1}(ax)\,;\,\xi\,\right] = \frac{\xi^{\nu}}{a^{\nu+1}}\,H\,(a-\xi) \qquad ,\,a>0$$

(b) Putting $\mu = \nu+\frac{1}{2}$ in the pair of eqns. (6.23) and (6.24) we get,

$$H_{\nu}\left[x^{\nu}(a^2-x^2)^{-\frac{1}{2}}\,H\,(a-x)\,;\,\xi\right] = \sqrt{\frac{\pi}{2\xi}}\,a^{\nu+\frac{1}{2}}\,J_{\nu+\frac{1}{2}}\,(\xi a)$$

$$H_{\nu}\left[\,x^{-\frac{1}{2}}J_{\nu+\frac{1}{2}}\,(ax)\,;\xi\,\right] = \sqrt{\frac{2}{\pi}}\,\frac{\xi^{\nu}\,H\,(a-\xi)}{a^{\nu+\frac{1}{2}}\sqrt{a^2-\xi^2}} \qquad ,\,a>0,\nu\geqslant 0$$

$$(6.25)$$

(c) Taking $\nu = 0$, the above two equations become

$$H_0\left[\,\frac{J_1(ax)}{x}\,;\,\xi\,\right] = \frac{H\,(a-\xi)}{a}\,, \qquad\qquad a>0 \qquad (6.26)$$

$$H_0\left[\,\frac{\sin(ax)}{x}\,;\,\xi\,\right] = \frac{H\,(a-\xi)}{\sqrt{a^2-\xi^2}}\,, \qquad\qquad a>0,\,\xi<a.$$

(d) Similarly taking $\nu = \frac{1}{2}$, the above pair of equations give

$$\int_0^\infty \sqrt{x}\, J_1\,(ax)\, J_{\frac{1}{2}}\,(x\xi)\, dx = \sqrt{\frac{2}{\pi}}\,\sqrt{\frac{\xi}{a^2 - \xi^2}}\,\frac{H(a-\xi)}{a}$$

Interchanging a and ξ, we have

$$H_1\left[\frac{\sin(ax)}{x}\,;\,\xi\right] = \frac{aH\,(\xi - a)}{\xi\sqrt{\xi^2 - a^2}}$$

Example 6.8. Prove that $H_0\left[\dfrac{1 - J_0(ax)}{x^2}\,;\,\xi\right] = \log\dfrac{a}{\xi}\cdot H\,(a-\xi)$

Solution. From eqn. (6.26) we can write

$$\int_0^\infty J_1\,(ax)\, J_0\,(\xi x)\, dx = \frac{H\,(a-\xi)}{a} \tag{6.27}$$

Here, using the results $\displaystyle\int_0^a J_1\,(\,ax\,)\, da = \frac{1 - J_0\,(ax)}{x}$

and $\displaystyle\int_0^a \frac{H\,(a-\xi)}{a}\, da = H\,(a-\xi)\,\log\left(\frac{a}{\xi}\right),$

we find on integrating (6.27) with respect to a from 0 to a that

$$\int_0^\infty \frac{1 - J_1\,(ax\,)}{x}\, J_0\,(\xi x)\, dx = H(a-\xi)\,\log\left(\frac{a}{\xi}\right)$$

$$\Rightarrow \qquad H_0\left[\,x^{-2}\,\{\,1 - J_0\,(ax)\,\}\,;\,\xi\right] = H(a-\xi)\,\log\left(\frac{a}{\xi}\right).$$

Example 6.9. Using the result

$$L\left[\,x^\nu J_\nu\,(xa)\,;\,x \to p\,\right] = \frac{2^{\,\nu}a^\nu\,\Gamma\,(\nu + \frac{1}{2})}{\sqrt{\pi}(p^2 + a^2)^{\nu + \frac{1}{2}}}\,,$$

prove that $\qquad H_\nu\left[\,x^{\nu-1}\,e^{-px}\,;\,\xi\right] = \dfrac{2^{\,\nu}\xi^\nu\,\Gamma\,(\nu + \frac{1}{2})}{\sqrt{\pi}(p^2 + \xi^2)^{\nu + \frac{1}{2}}}$

Solution. From the definition of Hankel transform we know that

$$
\begin{aligned}
H_\nu\left[\,x^{\nu-1}\,e^{-px}\,;\,\xi\right] &= \int_0^\infty x\cdot x^{\nu-1}\,e^{-px}\, J_\nu\,(\xi x)\, dx \\
&= \int_0^\infty \{\,x^\nu\cdot J_\nu\,(\xi x)\}\, e^{-px}\, dx \\
&= L\left[\,x^\nu J_\nu(\xi x)\,;\,x \to p\,\right] \\
&= \frac{2^\nu \xi^\nu\,\Gamma\,(\,\nu + \frac{1}{2})}{\sqrt{\pi}\,(p^2 + \xi^2)^{\nu + \frac{1}{2}}}\,,\ \text{by using the given condition.}
\end{aligned}
$$

Example 6.10. Evaluate $H_\nu \left[x^{-2} J_\nu(ax) ; \xi \right]$ when $\nu > -\frac{1}{2}$

Solution. Consider the following two functions

$$f(x) = x^\nu H(a-x) \, , \quad g(x) = Hx^\nu \, H(b-x) \, , \quad \nu = -\frac{1}{2}$$

Then they give, $\bar{f}_\nu(\xi) = \frac{a^{\nu+1}}{\xi} J_{\nu+1}(\xi a), \qquad \bar{g}_\nu(\xi) = \frac{b^{\nu+1}}{\xi} J_{\nu+1}(\xi b),$

by using the result of Example 6.7(a). Therefore, using Parseval relation of $f(x)$ and $g(x)$ we have

$$(ab)^{\nu+1} \int_0^\infty \xi^{-1} \, J_{\nu+1}(\xi a) J_{\nu+1}(\xi b) \, d\xi = \int_0^{\min \, (a,b)} x^{2\nu+1} \, dx$$

Now, suppose $0 < a < b$. Then the last equation becomes

$$\int_0^\infty \xi^{-1} \, J_{\nu+1} \, (\xi a) \, J_{\nu+1} \, (\xi b) \, d\,\xi = \frac{1}{2(\nu+1)} \left(\frac{a}{b} \right)^{\nu+1} \, , \quad \nu > -\frac{1}{2}$$

$$\Rightarrow H_\nu \left[x^{-2} \, J_\nu(ax) ; \xi \right] = \begin{cases} \frac{1}{2\nu} \left(\frac{\xi}{a}\right)^\nu \, , & 0 < \xi < a \\ \frac{1}{2\nu} \left(\frac{a}{\xi}\right)^\nu \, , & \xi > a \end{cases} \, , \quad \nu > -\frac{1}{2}$$

Example 6.11.   Find Hankel inversion of order 1 for the following functions

$$\text{(a) } \frac{1}{\xi} \, e^{-a\xi} \qquad\qquad \text{(b) } \frac{1}{\xi^2} \, e^{-a\xi}$$

Solution.

$\text{(a) } H^{-1} \left[\frac{1}{\xi} \, e^{-a\xi} ; \xi \to x \right] = \int_0^\infty \frac{e^{-a\xi}}{\xi} \cdot \xi \, J_1 \, (\xi x) \, d\xi$

$$= \int_0^\infty e^{-a\xi} \, J_1 \, (\xi x) \, d\xi$$

$$= \frac{1}{x} - \frac{a}{x \, (a^2 + x^2)^{\frac{1}{2}}}, \quad \text{by table of integrals.}$$

$\text{(b) } H^{-1} \left[\frac{1}{\xi^2} \, e^{-a\xi} ; \xi \to x \right] = \int_0^\infty \frac{e^{-a\xi}}{\xi} \, J_1 \, (a\xi) \, d\xi$

$$= \frac{(a^2 + x^2)^{\frac{1}{2}} - a}{x}, \quad \text{by example (a) above}$$

Example 6.12. Find Hankel transform of order zero of $\frac{1}{x}$. Then apply the inversion formula to get the original functions.

Solution.

$$H_0\left[\frac{1}{x} \; ; \; \xi\right] = \int_0^\infty \frac{1}{x} \cdot x \, J_0\,(\xi x)\, dx$$

$$= \int_0^\infty J_0\,(\xi x)\, dx = \frac{1}{\xi}, \quad \text{by table of integrals}$$

Thus, $H_0^{-1}\left[\frac{1}{\xi} \; ; \; x\right] = \int_0^\infty \frac{1}{\xi} \cdot \xi\, J_0\,(\xi x)\, d\xi$

$$= \int_0^\infty J_0\,(\xi x)\, dx = \frac{1}{x} \, .$$

Example 6.13. Obtain zero-order Hankel transforms of

$$\text{(a)} \; r^{-1}e^{-ar} \qquad \text{(b)} \; \frac{\delta(r)}{r} \qquad \text{(c)} \; H(a - r)$$

Solution.

(a) $H_0\left[\, r^{-1}e^{-ar} \; ; \; \xi \,\right] = \int_0^\infty e^{-ar}\, J_0\,(\xi r)\, dr = \dfrac{1}{\sqrt{\xi^2 + a^2}}$

(b) $H_0\left[\dfrac{\delta(r)}{r} \; ; \; r \to \xi\,\right] = \int_0^\infty \delta(r)\, J_0\,(\xi r)\, dr = 1$

(c) $H_0\left[H(a - r) \; ; \; \xi\,\right] = \int_0^\infty r H(a - r) J_0(\xi r)\, dr = \int_0^a r J_0\,(\xi r)\, dr$

$$= \frac{1}{\xi^2}\int_0^{ar} p\, J_0\,(p)\, dp = \frac{1}{\xi^2}\,[p\, J_1(p)]_0^{a\xi} = \frac{a}{\xi}\, J_1(a\xi).$$

Example 6.14. Find the nth order Hankel transforms for the functions

$$\text{(a)} \; r^n\, H(a - r) \qquad \text{(b)} \; r^n e^{-ar^2}$$

Solution.

(a) $H_n\left[\, r^n\, H(a - r) \; ; \; \xi\,\right] = \int_0^\infty r^{n+1}\, H(a - r)\, J_n\,(\xi r)\, dr$

$$= \int_0^a r^{n+1}\, J_n(\xi r)\, dr = \frac{a^{n+1}}{\xi}\, J_{n+1}\,(a\xi)$$

(b) $H_n\left[r^n\, e^{-ar^2} \; ; \; \xi\,\right] = \int_0^\infty r^{n+1}\, e^{-ar^2}\, J_n(\xi r)\, dr$

$$= \frac{\xi^n}{(2a)^{n+1}}\, \exp\left(-\frac{\xi^2}{2a}\right)\, , \quad \text{from the table of integrals}$$

Example 6.15. Solve the differential equation

$$\frac{\partial^2 u}{\partial r^2} + \frac{1}{r}\frac{\partial u}{\partial r} + \frac{\partial^2 u}{\partial z^2} = 0 \ , \ r \geqslant 0 \ , \ z \geqslant 0$$

satisfying the conditions (i) $u \to 0$ as $z \to \infty$ and as $r \to \infty$,

(ii) $u = f(r)$, on $z = 0$, $r \geqslant 0$.

Solution. Let $\bar{u}_0(\xi, z)$ be Hankel transform of zero order of $u(r, z)$. Then taking the zeroth order Hankel transform to

$$\frac{\partial^2 u}{\partial r^2} + \frac{1}{r}\frac{\partial u}{\partial r} + \frac{\partial^2 u}{\partial z^2} = 0$$

we get it as

$$-\xi^2 \, \bar{u}_0 \, (\xi, z) + \frac{d^2 \, \bar{u}_0}{\partial z^2} \, (\xi, z) = 0$$

Solving the last ODE we have

$$\bar{u}_0 \, (\xi, z) = A \, e^{\xi z} + B \, e^{-\xi z} \qquad (6.28)$$

where A, B are constants to be calculated. Now Hankel transform of order zero of given condition (i) becomes $\bar{u}_0 \, (\xi, z) = 0$, as $z \to \infty$ and hence (6.28) gives $A = 0$ implying

$$\bar{u}_0 \, (\xi, z) = B \, e^{-\xi z} \qquad (6.29)$$

Again, taking Hankel transform of order zero to the given condition (ii), it becomes

$$\bar{u}_0 \, (\xi, 0) = \bar{f}_0 \, (\xi) \equiv H_0 \, [\, f(r) \, ; \, r \to \xi \,] \qquad (6.30)$$

Then, eqn. (6.29.) by virtue of eqn. (6.30.) gives

$$B = \bar{f}_0 \, (\xi)$$

Thus, $\qquad \bar{u} \, (\xi, z) = \bar{f}_0 \, (\xi) \cdot e^{-\xi z}$

Applying Hankel inversion formula we have, therefore,

$$u(r, z) = \int_0^\infty \bar{f}_0 \, (\xi) \, e^{-\xi z} \, \xi \, J_0 \, (\xi r) \, d\xi$$

as the required solution of the B V P.

Example 6.16. The vibration $u(r,t)$ of a large membrane is given by

$$\frac{\partial^2 u}{\partial r^2} + \frac{1}{r}\frac{\partial u}{\partial r} = \frac{1}{c^2}\frac{\partial^2 u}{\partial t^2} \ , \ r \geqslant 0 \ , \ t \geqslant 0$$

with the conditions

$$u(r,0) = f(r) \ , \ \frac{\partial u(r,0)}{\partial t} = g(r).$$

Find $u(r,t)$ for $t > 0$.

Solution. Taking zeroth order Hankel transform, the given PDE governing the vibration becomes

$$-\xi^2 \ \bar{u}_0 \ (\xi,t) = \frac{1}{c^2}\frac{d^2 \ \bar{u}_0}{dt^2} \ (\xi,t)$$

Its general solution is

$$\bar{u}_0 \ (\xi,t) = A \ \cos \ (c \ \xi \ t) \ + \ B \ \sin \ (c \ \xi \ t)$$

Let $\bar{f}_0 \ (\xi)$ and $\bar{g}_0 \ (\xi)$ be Hankel transform of order zero of the given functions $f(r)$ and $g(r)$ respectively.

Then from the given conditions

$$\bar{u}_0 \ (\xi,0) \ = \bar{f}_0 \ (\xi) = \int_0^\infty f(r) \ r \ J_0 \ (\xi r) \ dr \tag{6.31}$$

and $\quad \dfrac{\partial}{\partial t}\bar{u}_0 \ (\xi,0) = \bar{g}_0 \ (\xi) = \displaystyle\int_0^\infty g \ (r) \ r J_0 \ (\xi r) \ dr \tag{6.32}$

From the general solution and (6.31) we get

$$\bar{f}_0(\xi) = A$$

and hence the general solution and (6.32) finally give

$$\bar{g}_0(\xi) = Bc\xi$$

Therefore, the solution $\bar{u}_0 \ (\xi,t)$ ultimately becomes

$$\bar{u}_0 \ (\xi,t) \ = \bar{f}_0 \ (\xi) \ \cos \ (c \ \xi t) + \frac{\bar{g}_0 \ (\xi)}{c \ \xi} \ \sin \ (c\xi t)$$

Now, applying the inversion formula of Hankel transform of order zero we get

$$u \ (r,t) \ = \int_0^\infty \left[\bar{f}_0 \ (\xi) \ \cos \ (c \ \xi t) + \frac{\bar{g}_0 \ (\xi)}{c \ \xi} \ \sin \ (c \ \xi \ t) \right] \xi \ J_0 \ (\ \xi \ r) \ dr$$

Example 6.17. Consider the axisymmetric Dirichlet problem for a thick plate $|z| \leqslant b$ determined by the function $u(\rho, z)$ which is harmonic and satisfies the boundary conditions.

$$u(\rho, b) = f(\rho) , \qquad u(\rho, -b) = g(\rho) \tag{6.33}$$

Find the solution $u(\rho, z)$. If $f(\rho) = g(\rho)$, evaluate $u(\rho, 0)$ and $\dfrac{\partial}{\partial z} u(\rho, 0)$

Solution. Since $u(\rho, z)$ satisfies the PDE

$$\left[\frac{\partial^2}{\partial \rho^2} + \frac{1}{\rho} \frac{\partial}{\partial \rho} + \frac{\partial^2}{\partial z^2} \right] [u(\rho, z)] = 0 \tag{6.34}$$

applying zeroth order Hankel transform to (6.34), we get

$$\frac{d^2}{dz^2} \bar{u}_o \, (\xi, z) - \xi^2 \, \bar{u}_0 \, (\xi, z) = 0$$

The appropriate solution of this equation is

$$\bar{u}_0 \, (\xi, z) = A \, (\xi) \, \sinh \, \{\xi \, (b + z)\} + B \, (\xi) \, \sinh \, \{ \, \xi \, (b - z)\} \tag{6.35}$$

Now, Hankel transformed boundary conditions in eqn. (6.33) are

$$\bar{u}_0(\xi, b) = \bar{f}(\xi) \text{ and } \bar{u}_0(\xi, -b) = \bar{g}_0 \, (\xi)$$

Using these conditions, the solution results in

$$A \, (\xi) \, \sinh \, (2 \, \xi \, b) = \bar{f}_0 \, (\xi)$$

and $\qquad B \, (\xi) \, \sinh \, (2\xi b) = \bar{g}_0 \, (\xi) \, .$

Thus the solution in (6.35) takes the form

$$\bar{u}_0 \, (\xi, z) = \text{cosech} \, (2 \, \xi \, b) \left[\bar{f}_0 \, (\xi) \, \sinh \, \{\xi(z + b)\} + \bar{g}_0 \, (\xi) \, \sinh \, \{ \, \xi \, (b - z)\} \right]$$

Applying Hankel inversion theorem we obtain the solution as

$$u(\rho \, , \, z) = H_0 \left[\bar{f}_0 \, (\xi) \frac{\sinh \, \xi(b + z)}{\sinh \, (2 \, \xi \, b)} + \bar{g}_0 \, (\xi) \frac{\sinh \, \xi(b - z)}{\sinh \, (2 \, \xi \, b)} \, ; \, \xi \to \rho \right] \tag{6.36}$$

If, in particular, $f(\rho) = g(\rho) \Rightarrow \bar{f}_0(\xi) = \bar{g}_0(\xi)$, then

$$\bar{u}(\xi \, , \, z) = \bar{f}_0(\xi) \, \frac{\cosh \, (\xi z)}{\cosh \, (\xi b)}$$

and the solution is

$$u(\rho , z) = H_0 \left[\bar{f}_0(\xi) \, \frac{\cosh(\xi z)}{\cosh(\xi b)} \, ; \, \xi \to \rho \right] \tag{6.37}$$

From (6.36) we can derive the general formulae

$$u(\rho, 0) = H_0 \left[\frac{1}{2} \{ \bar{f}_0 \, (\xi) + \bar{g}_0 \, (\xi) \} \, \text{sech} \, (\xi b) \, ; \, \xi \to \rho \right]$$

$$\frac{\partial u}{\partial z} \, (\rho, 0) = H_0 \left[\frac{1}{2} \xi \{ \bar{f}_0 \, (\xi) - \bar{g}_0 \, (\xi) \} \text{cosech} \, (\xi b) \, ; \, \xi \to \rho \right]$$

Also, from (6.37) we can derive the particular solution

$$u(\rho, 0) = H_0 \, [\bar{f}_0 \, (\xi) \, \text{sech} \, (\xi b) \, ; \, \xi \to \rho]$$

$$\frac{\partial u(\rho, 0)}{\partial z} = 0$$

Example 6.18. Consider the transverse displacement $z(\rho, t)$ of a large membrane which satisfies the non-homogeneous wave equation

$$\frac{\partial^2 z}{\partial \rho^2} + \frac{1}{\rho} \frac{\partial z}{\partial \rho} = \frac{1}{c^2} \frac{\partial^2 z}{\partial t^2} - \frac{p \, (\rho, t)}{T} \tag{6.38}$$

where T is the tension in the membrane, $p \, (\rho, t)$ is the symmetrical pressure applied normally to the membrane and $c^2 = T/\sigma$, where σ is the density per unit area of the membrane.

If the initial conditions are

$$z \, (\rho , \, 0) = f(\rho) , \quad \frac{\partial z(\rho, 0)}{\partial t} = g(\rho), \tag{6.39}$$

find the solution of the problem in the Hankel transformed domain.

Discuss the particular case of the problem if there is no applied pressure and if the initial displacement is $\in \, (1 + \rho^2/a^2)^{-1/2}$ to the membrane.

Solution. Introducing Hankel transform of order zero, the equation (6.38) becomes

$$\frac{d^2 \bar{z} \, (\xi, t)}{dt^2} + c^2 \, \xi^2 \, \bar{z} \, (\xi, t) = \frac{\bar{p}_0(\xi, t)}{\sigma} \tag{6.40}$$

where, $\bar{p}_0 \, (\xi, t) = H_0 \, [\, p \, (\rho, t) \, ; \, \rho \to \xi \,]$

Let Hankel transform of order zero of the initial conditions in (6.39) be given by

$$\bar{z}\,(\xi, 0) = \bar{f}_0\,(\xi)\ ,\ \frac{d}{dt}\bar{z}(\xi, 0) = \bar{g}_0\,(\xi) \tag{6.41}$$

Then, the solution of (6.40) under the initial conditions (6.41) is given by

$$\bar{z}\,(\xi, t) = \bar{f}_0\,(\xi)\cos\,(c\,t\,\xi) + (c\,\xi)^{-1}\,\bar{g}_0\,(\xi)\,\sin\,c\,t\,\xi$$
$$+(c\,\sigma\,\xi)^{-1}\int_0^t \bar{p}_0\,(\xi, \tau)\,\sin\,\{c\,\xi\,(t-\tau)\}\,d\tau \tag{6.42}$$

Particular case.

In this case $\bar{p}_0\,(\xi, t) = 0$ and $\bar{g}_0\,(\xi) = 0$. Also $\bar{f}_0\,(\xi) = \in\,a\,\xi^{-1}\,e^{-\xi a}$. Then the solution in eqn. (6.42) becomes

$$\bar{z}\,(\xi, t)\ =\ \in\,a\,\xi^{-1}\,e^{-\xi a}\,\cos\,(c\,t\,\xi)$$

Now, using the inversion theorem of Hankel transform of order zero to the above transformed solution, we get

$$\begin{aligned}
z\,(\rho, t)\ &=\ \in\,a\,H_0\left[\xi^{-1}\,e^{-\xi a}\,\cos\,(c\,t\,\xi)\ ;\ \xi \to \rho\right]\\
&=\ \in\,a\,Re\left[\int_0^\infty e^{-\xi(a+ict)}\,J_0\,(\xi\rho)\,d\xi\right]\\
&=\ \in\,a\,Re\left[\rho^2 + (a+ict)^2\right]^{-\frac{1}{2}}\\
&=\ \in\,a\,R^{-1}\,\cos\,\varphi
\end{aligned}$$

where R and φ are given by

$$R^4 = (a^2 + \rho^2 - c^2t^2)^2 + 4a^2c^2t^2$$
$$\text{and}\qquad \tan\,2\varphi = \frac{(2\,act)}{(a^2 + \rho^2 - c^2t^2)}\ .$$

Example 6.19. Find the potential $V(r, z)$ of a field due to a flat circular disc of unit radius with its center at the origin, axis along the z-axis and satisfying the differential equation

$$\frac{\partial^2 V}{\partial r^2} + \frac{1}{r}\,\frac{\partial V}{\partial r} + \frac{\partial^2 V}{\partial z^2} = 0\ ,\ 0 \leqslant r < \infty\ ,\ z \geqslant 0 \tag{6.43}$$

and the boundary conditions

$$V(r, z) = V_0,\quad \text{when}\ z = 0,\ 0 \leqslant r < 1 \tag{6.44}$$
$$\text{and}\qquad \frac{\partial V(r, z)}{\partial z} = 0,\quad \text{when}\ z = 0, r > 1 \tag{6.45}$$

Solution. Introducing Hankel transform of order zero, eqn. (6.43) becomes

$$\frac{d^2 \overline{V}_0}{dz^2} - \xi^2 \, \overline{V}_0 = 0$$

Its general solution is

$$\overline{V}_0 \, (\xi, z) = A(\xi) \, e^{\xi z} + B(\xi) \, e^{-\xi z}$$

where $A(\xi)$, $B(\xi)$ are arbitrary functions of ξ .

On physical grounds, $\overline{V}_0 \, (\xi, z) \to 0$ as $\xi \to \infty$, since $V(r, z) \to 0$ as $r \to \infty$. Therefore, $A(\xi) = 0$ implying that

$$\overline{V}_0 \, (\xi, z) = B(\xi) \, e^{-\xi z}$$

Applying Hankel inversion formula, we get

$$V(r, z) = \int_0^\infty B(\xi) \, \xi \, e^{-\xi z} \, J_0(\xi r) \, d\xi \qquad (6.46)$$

and $\quad \dfrac{\partial V(r, z)}{\partial z} = - \int_0^\infty B(\xi) \, \xi^2 \, e^{-\xi z} \, J_0(\xi r) \, d\xi \qquad (6.47)$

Now, the boundary conditions in eqns. (6.44) and (6.45) can be expressed as

$$\left. \begin{array}{l} \int_0^\infty \xi \, B(\xi) \, J_0 \, (\xi r) \, d\xi = V_0 \; , \; 0 \leqslant r < 1 \\[3mm] \text{and} \quad \int_0^\infty \xi^2 \, B(\xi) \, J_0 \, (\xi r) \, d\xi = 0 \; , \; r > 1 \end{array} \right\} \qquad (6.48)$$

after utilising eqns. (6.46) and (6.47).

But it is known that

$$\int_0^\infty J_0 \, (\xi r) \, \frac{\sin \xi}{\xi} \, d\xi \; = \; \frac{\pi}{2} \; , \; 0 \leqslant r < 1$$

and $\quad \displaystyle\int_0^\infty J_0 \, (\xi r) \, \sin \xi \, d\xi \; = \; 0 \; , \; r > 1$

Therefore, comparing the last two equations with the eqns. in (6.48), we get

$$B(\xi) = \frac{2V_0 \sin \xi}{\pi \xi^2}$$

So, $\quad V(r, z) = \dfrac{2V_0}{\pi} \displaystyle\int_0^\infty e^{-\xi z} \, \dfrac{\sin \xi}{\xi} \, J_0(\xi r) \, d\xi.$

Exercises.

(1) Find Hankel transform of order 1 of $x^{-2} e^{-x}$.

$$\left[\textbf{Ans.} \quad \frac{\left\{(1+\xi^2)^{\frac{1}{2}}-1\right\}}{\xi}\right]$$

(2) If $f(x) = \begin{cases} a^2 - x^2 & , \quad 0 \leqslant x < a, \\ 0 & , \quad x > a \end{cases}$

calculate Hankel transform of $f(x)$ of order zero.

$$\left[\textbf{Ans.} \quad \frac{4a}{\xi^3} J_1(a\xi) - \frac{2a^2}{\xi^2} J_0(\xi a)\right].$$

(3) (a) Prove that $H_0\left[\frac{df}{dx} ; x \to \xi\right] = -\xi\sqrt{a^2 + \xi^2}$, if $f(x) = \frac{1}{x} e^{-ax}$.

(b) Prove that $H_0\left[\frac{\delta(r)}{r} ; r \to \xi\right] = 1$.

(c) Prove that $H_1\left[e^{-ar} ; r \to \xi\right] = \frac{\xi}{(a^2+\xi^2)}$.

(d) Prove that $H_n\left[r^n e^{-ar^2} ; r \to \xi\right] = \frac{\xi^n}{(2a)^{n+1}} e^{-\frac{r^2}{4a}}$.

(4) Show that the solution of the boundary value problem defined by

$$\frac{\partial^2 u}{\partial r^2} + \frac{1}{r}\frac{\partial u}{\partial r} + \frac{\partial^2 u}{\partial z^2} = 0, \ 0 \leqslant r < \infty, \ z > 0,$$

$u(r, 0) = u_0$ for $0 \leqslant r < a$, where u_0 is a constant and
$u(r, z) \to 0$ as $z \to \infty$ is

$$u(r, z) = a u_0 \int_0^\infty J_1(a\xi) J_0(\xi r) e^{-z\xi} d\xi$$

(5) Solve the Neumann problem for the Laplace equation

$$u_{rr} + \frac{1}{r} u_r + u_{zz} = 0, \ 0 < r < \infty, \ 0 < z < \infty \text{ when}$$

$$u_z(r, 0) = -\frac{1}{\pi a^2} H(a - r), \ 0 < r < \infty, \ u(r, z) \to 0 \text{ as } z \to \infty$$

to prove that $u(r, z) = \frac{1}{\pi a} \int_0^\infty \frac{1}{\xi} J_1(a\xi) J_0(r\xi) e^{-\xi z} d\xi$

(6) Heat is supplied at a constant rate Q per unit area per unit time
over a circular area of radius a in the plane $z = 0$ to an infinite solid

of thermal conductivity k, the rest of the plane is being kept at zero temperature. Solve the steady state heat conduction problem to find the temperature field $u(r, z)$ satisfying the equation

$$u_{rr} + \tfrac{1}{r} u_r + u_{zz} = 0 \ , \ 0 < r < \infty \ , \ -\infty < z < \infty$$

with the boundary conditions $u \to 0$ as $r \to \infty$, $u \to 0$ as $|z| \to \infty$ and $-k\, u_z = \tfrac{1}{2}\, Q\, H(a - r)$ when $z = 0$. Finally, prove that

$$u(r, z) = \tfrac{Qa}{2k} \int_0^\infty \xi^{-1} \, \exp\left(-|z|\xi\right) J_1(a\xi)\, J_0(r\xi)\, d\xi \ .$$

Chapter 7

Finite Hankel Transforms

7.1 Introduction.

Finite Hankel transform arises in discussing the solution of certain special type of boundary value problems. Sneddon (Phil. Mag., 37, 1946) was the first author who introduced this transform. Later application of this transform was found in the works of several other authors in discussing solutions of axisymmetric physical problems in long circular cylinders and membranes.

7.2 Expansion of some functions in series involving cylinder functions : Fourier-Bessel Series.

Suppose $f(r)$ is a real piecewise continuous function in $(0, a)$ and is of bounded variation in every subinterval $[r_1, r_2]$ where $0 < r_1 < r_2 < a$. Then, if the integral $\int_0^a \sqrt{r}\, |f(r)| dr$ is finite, a Fourier series like series expansion to $f(r)$ represented by

$$f(r) = \sum_{m=1}^{\infty} c_m\, J_\nu \left(x_{\nu m}\, \frac{r}{a} \right) , \qquad \nu \geqslant -\frac{1}{2} \tag{7.1}$$

where $x_{\nu m}$ is given by $J_\nu\, (x_{\nu m}\,) = 0, m = 1, 2, 3, \cdots$ do exist at every point of continuity of $f(r)$ and to

$$\frac{1}{2}[\, f(r+0) + f(r-0)\,] \tag{7.2}$$

at every point of discontinuity of $f(r)$.

The co-efficient c_m of the expansion of $f(r)$ in (7.1) can be determined by using the orthogonality property of the system of functions

$$J_\nu \left(x_{\nu m}\, \frac{r}{a}\, , \right)\ m = 1, 2, \ldots . \tag{7.3}$$

To prove this assertion, let there exist two distinct non zero real numbers α, β such that $J_\nu \, (\alpha \, r)$, $J_\nu \, (\beta \, r)$ are the Bessel functions of first kind. Then we would have

$$\left[\frac{d^2}{dr^2} + \frac{1}{r} \frac{d}{dr} + \left(\alpha^2 - \frac{\nu^2}{r^2} \right) \right] J_\nu \, (\, \alpha \, r) = 0$$

$$\left[\frac{d^2}{dr^2} + \frac{1}{r} \frac{d}{dr} + \left(\beta^2 - \frac{\nu^2}{r^2} \right) \right] J_\nu \, (\, \beta \, r) = 0$$

Now, subtracting the second equation multiplied by $r \, J_\nu \, (\alpha \, r)$ from the first equation multiplied by $r \, J_\nu \, (\beta \, r)$, and integrating the result with respect to r from $r = 0$ to $r = a$, we get

$$\int_0^a r \, J_\nu \, (\alpha r) \, J_\nu \, (\beta r) \, dr$$
$$= \frac{[a\beta \, J_\nu \, (a\alpha) \, J_\nu' \, (\, a\beta) - a\alpha \, J_\nu \, (\beta a) \, J_\nu \, ' \, (\alpha a)]}{(\alpha^2 - \beta^2)} \qquad (7.4)$$

Setting, $a\alpha = x_{\nu m}$ and $a\beta = x_{\nu n}$ we get from equation (7.4) that

$$\int_0^a r \, J_\nu \, (\alpha r) \, J_\nu \, (\beta r) \, dr = 0, \quad \text{if} \quad m \neq n.$$

When $m = n$, taking the limit of eqn. (7.4) as $\beta \to \alpha$ we get after elimination of J_ν'' that

$$\int_0^a r \, J_\nu^2 \, (\alpha r) \, dr = \frac{a^2}{2} \left[J_\nu'^{\,2} \, (\alpha a) + \left(1 - \frac{\nu^2}{a^2 \alpha^2} \right) J_\nu^2 \, (\alpha a) \right] \qquad (7.5)$$

Or, $$\int_0^a r \, J_\nu^2 \, (x_m \frac{r}{a}) \, dr = \frac{a^2}{2} J_\nu'^{\,2} \, (x_{\nu n}) \qquad (7.6)$$

$$= \frac{a^2}{2} J_{\nu+1}^2 \, (x_{\nu n}) \qquad (7.7)$$

Thus assuming the possibility of an expansion of $f(r)$ of the form (7.1), multiplying it by $r \, J_\nu(x_{\nu n} \frac{r}{a})$ and integrating term by term from $r = 0$ to $r = a$, we obtain

$$c_m = \frac{2}{a^2 \, J_{\nu+1}^2 \, (x_{\nu m})} \int_0^a r f(r) \, J_\nu \left(x_{\nu m} \, \frac{r}{a} \right) dr, \; m = 1, 2, \cdots \quad (7.8)$$

The series (7.1) with co-efficients calculated from eqn.(7.8), is called the Fourier-Bessel series of $f(r)$.

We are now ready to define finite Hankel transform in the following section.

7.3   The Finite Hankel Transform.

Form I   If $f(r)$ satisfies the conditions laid down in article 7.1 over the finite interval $0 \leqslant r \leqslant a$, then finite Hankel transform of order n is denoted by $H_n[f(r)]$ and is defined by

$$H_n\left[\,f(r)\,\right] = \int_0^a r\, f(r)\, J_n\,(rp_i)\, dr \equiv \bar{f}_n\,(p_i) \qquad (7.9)$$

where p_i is a positive root of the equation

$$J_n\,(ap_i) = 0 \qquad (7.10)$$

The function $f(r)$ can be considered as inverse finite Hankel transform of $\bar{f}_n(p_i)$ and it is denoted by

$$f(r) \;=\; H_n^{-1}\left[\,\bar{f}_n\,(p_i)\,\right]$$

and is defined by

$$f(r) = \frac{2}{a^2} \sum_{i=1}^{\infty} \bar{f}_n\,(p_i)\, \frac{J_n\,(rp_i)}{J_{n+1}^2\,(ap_i)} \qquad (7.11)$$

where $p_i\;(0 < p_1 < p_2 < \cdots)$ are the roots of the equation $J_n(ap_i) = 0$. This means that

$$J_n'\,(ap_i) = J_{n-1}\,(ap_i) = -J_{n+1}\,(ap_i)$$

due to standard recurrence relation of Besel function. Here, the abbreviation ap_i is used instead of $x_{\nu i}$ as was used in article 7.2.

As a particular case, zeroth order finite Hankel transform of $f(r)$ and its inversion are then defined by

$$H_0\left[\,f(r)\,\right] \equiv \bar{f}_0\,(p_i) \;=\; \int_0^a r\, f(r)\, J_0\,(rp_i)\, dr \qquad (7.12)$$

$$H_0^{-1}\left[\,\bar{f}_0\,(p_i)\,\right] \equiv f(r) \;=\; \frac{2}{a^2} \sum_{i=1}^{\infty} \bar{f}_0\,(p_i)\, \frac{J_0\,(rp_i)}{J_1^2\,(ap_i)} \qquad (7.13)$$

where the summation is taken over all the positive roots of $J_0(ap) = 0$. Similarly, finite Hankel transforms of other different orders of $f(r)$ can be defined as shown above.

7.4 Illustrative Examples.

Example 7.1. Find finite Hankel transform of order n of $f(r) = r^n$, for $0 \leqslant r \leqslant a$.

Solution. We know that

$$\frac{d}{dx} [x^n \, J_n \, (x)] = x^n \, J_{n-1} \, (x)$$

Therefore, $\quad \dfrac{d}{dr} \left[r^{n+1} \, J_{n+1} \, (k_i \, r) \right] = k_i \, r^{n+1} \, J_n \, (k_i \, r)$

and hence $\quad H_n[r^n] = \displaystyle\int_0^a r^n \cdot r \, J_n \, (k_i \, r) \, dr$

$$= \frac{1}{k_i} \int_0^a \frac{d}{dr} \left[r^{n+1} \, J_{n+1} \, (k_i \, r) \right] \, dr$$

$$= \frac{a^{n+1}}{k_i} \, J_{n+1} \, (k_i \, a)$$

In particular if $n = 0$, then

$$H_n[\, 1 \,] = \frac{a}{k_i} \, J_1 \, (ak_i) \; .$$

Example 7.2. Find finite Hankel transform of $(a^2 - r^2)$ of order zero for $0 \leqslant r \leqslant a$.

Solution. We have

$$H_0 \left[\, (a^2 - r^2) \, \right] = \int_0^a r \, (a^2 - r^2) \, J_0 \, (r \, k_i) \, dr$$

$$= \int_0^a a^2 r \, J_0 \, (rk_i) \, dr - \int_0^a r^3 \, J_0 \, (rk_i) \, dr$$

$$= a^2 \int_0^a \frac{1}{k_i} \cdot \frac{d}{dr} \{ \, r J_1 \, (k_i r) \, \} \, dr$$

$$- \int_0^a r^2 \frac{1}{k_i} \cdot \frac{d}{dr} \{ \, r J_1 \, (k_i r) \, \} \, dr$$

$$= \frac{a^2}{k_i} [\, r J_1 \, (\, k_i r \,) \,]_0^a - \left[\frac{r^2}{k_i} \, r \, J_1 \, (k_i r) \, \right]_0^a + \int_0^a \frac{2r}{k_i} \, r J_1 \, (k_i \, r) \, dr$$

$$= \frac{a^3}{k_i} J_1 \, (k_i \, a) - \frac{a^3}{k_i} J_1 \, (k_i \, a) + \frac{2}{k_i^2} [\, r^2 \, J_2 \, (k_i \, r) \,]_0^a$$

$$= \frac{2a^2}{k_i^2} J_2 (k_i \, a)$$

$$= \frac{2a^2}{k_i^2} \left[\frac{2}{k_i \, a} J_1(k_i \, a) - J_0 \, (k_i \, a) \right], \text{by recurrence relation}$$

$$= \frac{4a}{k_i^3} \, J_1 \, (k_i \, a) - \frac{2a^2}{k_i^2} \, J_0 \, (k_i \, a) \ .$$

Example 7.3. Prove that $\int_0^1 \frac{J_n \, (\alpha \, x)}{J_n \, (\alpha)} \cdot x \, J_n \, (k_i \, x) \, dx = \frac{k_i}{\alpha^2 - k_i^2} \, J_n' \, (k_i)$, where k_i is the positive root of $J_n \, (k_i) \, = 0$.

Solution. The Bessel function $J_n(x)$ satisfies the ODE

$$\left[x^2 \frac{d^2}{dx^2} + x \frac{d}{dx} + (x^2 - n^2) \right] J_n(x) = 0 \qquad \text{(i)}$$

Replacing x by $k_i \, x$ and $\alpha \, x$ successively in (i), we get

$$x^2 \frac{d^2}{dx^2} \, J_n \, (k_i \, x) + x \frac{d}{dx} \, J_n \, (k_i \, x) + (k_i^2 x^2 - n^2) \, J_n \, (k_i \, x) = 0$$

and $\quad x^2 \dfrac{d^2}{dx^2} \, J_n \, (\alpha \, x) + x \dfrac{d}{dx} \, J_n \, (\alpha \, x) + (\alpha^2 x^2 - n^2) \, J_n \, (\alpha \, x) = 0$

Multiplying the first equation by $J_n(\alpha \, x)$ and the second by $J_n \, (k_i \, x)$ and then subtracting, we get

$$x \frac{d}{dx} \left[x \left\{ \, J_n \, (\alpha \, x) \frac{d}{dx} \, J_n \, (k_i \, x) - J_n \, (k_i x) \frac{d}{dx} J_n \, (\alpha \, x) \right\} \right]$$

$$- x^2 \, (\alpha^2 - k_i^2) \, J_n \, (k_i \, x) \, J_n \, (\alpha x) = 0.$$

Or $\qquad x J_n \, (k_i x) \, J_n \, (\alpha \, x)$

$$= \frac{1}{\alpha^2 - k_i^2} \frac{d}{dx}$$

$$\left[x \left\{ \, J_n \, (\alpha \, x) \frac{d}{dx} \, J_n \, (k_i \, x) - J_n \, (k_i \, x) \frac{d}{dx} \, J_n \, (\alpha x) \right\} \right]$$

$\therefore \qquad \displaystyle\int_0^1 \frac{J_n \, (\alpha \, x)}{J_n \, (\alpha)} \, x J_n \, (k_i \, x) \, dx$

$$= \frac{1}{\alpha^2 - k_i^2} \frac{1}{J_n \, (\alpha)}$$

$$\left[\, k_i \, J_n \, (\alpha) J_n' \, (k_i) - \alpha \, J_n \, (k_i) \, J_n' \, (\alpha) \, \right]$$

$$= \frac{k_i}{\alpha^2 - k_i^2} \, J_n' \, (k_i) \ , \text{ since } J_n \, (k_i) \, = 0$$

Example. 7.4. Prove that finite Hankel transform of order n of

$$f(x) = \frac{2^{1+n-m}}{\Gamma \, (m - n)} . x^n \, (1 - x^2)^{m-n-1} \ , \ 0 < x < 1,$$

is $\qquad p^{n-m} \, J_n \, (p).$

Solution. By definition

$$H_n \left[f \left(x \right) \right] = \int_0^1 \frac{2^{1+n-m}}{\Gamma \left(m - n \right)} \, x^n \left(1 - x^2 \right)^{m-n-1} x J_n \left(px \right) dx$$

$$= \sum_{r=0}^{\infty} \int_0^1 \frac{2^{1+n-m}}{\Gamma \left(m - n \right)} \, x^{n+1} \left(1 - x^2 \right)^{m-n-1} \frac{(-1)^r}{r! \Gamma \left(n + r + 1 \right)} \frac{\left(p \, x \right)^{n+2r}}{2^{n+2r}} \, dx$$

$$\therefore \; H_n \left[f(x) \right] = \frac{1}{\Gamma(m-n)} \sum_{r=0}^{\infty} \frac{(-1)^r \, p^{n+2r}}{\Gamma(n+r+1)2^{m+2r}} \int_0^1 y^{(n+r+1)-1}$$

$$(1-y)^{m-n-1} \, dy$$

$$= \frac{1}{\Gamma(m-n)} \sum_{r=0}^{\infty} \frac{(-1)^r \, p^{n+2r}}{\Gamma(n+r+1)2^{m+2r}} \cdot \frac{\Gamma \left(n+r+1 \right) \Gamma \left(m - n \right)}{\Gamma \left(n + r + 1 + m - n \right)}$$

$$= p^{n-m} \sum_{r=0}^{\infty} \frac{(-1)^r}{\Gamma(m+r+1)} \left(\frac{p}{2} \right)^{m+2r}$$

$$= p^{n-m} \, J_m \left(p \right)$$

Example. 7.5. Find $H^{-1} \left[\frac{c}{p} J_1 \left(ap \right) \right]$, where p is a positive root of $J_0 \left(ap \right) = 0$.

Solution. If $f(r)$ be the required function such that $H \left[f(r) \right] = \bar{f}_H \left(p \right) = \frac{c}{p} J_1 \left(ap \right)$, then from equation (7.13) we have

$$\left[\frac{2}{a^2} \sum_p \left\{ \frac{c}{p} J_1 \left(ap \right) \right\} \frac{J_0 \left(p \, r \right)}{J_1^2 \left(p \, a \right)} \right] = f(r)$$

Or $\quad \dfrac{2c}{a^2} \sum_p \dfrac{J_0 \left(pr \right)}{J_1(p \, a)} = f(r)$

7.5 Finite Hankel Transform of order n in $0 \leqslant x \leqslant 1$ of the derivatrive of a function.

For $0 \leqslant x \leqslant 1$. Hankel transform of first order derivative $\frac{df(x)}{dx}$ of order n is given by

$$H_n \left[\frac{df}{dx} \right] = \int_0^1 \frac{df}{dx} . x \, J_n \left(px \right) dx \; , \; \text{where } p \text{ is a root of } J_n(p) = 0$$

$$= \left[f(x) . x J_n \left(px \right) \right]_0^1 - \int_0^1 f(x) \frac{d}{dx} \left\{ x \, J_n \left(px \right) \right\} \, dx$$

$$= -\int_0^1 f(x) . \left[\frac{px}{2n} \{ (n+1) \, J_{n-1}(px) - (n-1) J_{n+1}(px) \} \right] dx$$

$$= \frac{p}{2n} \left[(n-1) \int_0^1 f(x) \, x \, J_{n+1}(px) \, dx \right.$$

$$\left. -(n+1) \int_0^1 f(x) \, x \, J_{n-1}(px) \, dx \right]$$

$$= \frac{p}{2n} \left[\, (n-1) \, H_{n+1}\{f(x)\} - (n+1) \, H_{n-1}\{f \, (x)\} \, \right] \quad (7.14)$$

Corollary. 7.1. Putting $n = 1$ in eqn. (7.14), we get

$$H_1 \left[\frac{df}{dx} \right] = - \, p \, H_o \, [f(x)] \qquad (7.15)$$

7.6 Finite Hankel Transform over $0 \leqslant x \leqslant 1$ of order n of $\frac{d^2f}{dx^2} + \frac{1}{x} \frac{df}{dx}$, when p is the root of $J_n(p) = 0$.

By definition of Hankel transform of order n in $0 \leqslant x \leqslant 1$, we have

$$H_n \left[\frac{d^2 f}{dx^2} + \frac{1}{r} \frac{df}{dx} \right] = \int_0^1 \left[\frac{d^2 f}{dx^2} + \frac{1}{x} \frac{df}{dx} \right] x \, J_n \, (px) \, dx$$

$$= \int_0^1 x \, f''(x) \, J_n \, (px) \, dx + \int_0^1 f'(x) \, J_n \, (px) \, dx$$

$$= \left[x \, J_n(px) \, f'(x) \right]_0^1 - \int_0^1 \frac{d}{dx} \{ x \, J_n(px) \} f'(x) \, dx + \int_0^1 f'(x) \, J_n'(px) dx$$

$$= - \int_0^1 px \, J_n'(px) \, f'(x) \, dx$$

$$= -\frac{p}{2} \int_0^1 x \, f'(x) [J_{n-1}(px) - J_{n+1}(px)] \, dx \; ,$$

$$\text{by } \frac{d}{dx} \, J_n(px) = \frac{p}{2} [J_{n-1}(px) - J_{n+1}(px)]$$

$$= \frac{p}{2} \, H_{n+1} \, [f'(x)] - \frac{p}{2} \, H_{n-1} \, [f'(x)] \; . \qquad (7.16)$$

7.7 Finite Hankel Transform of $f''(x) + \frac{1}{x}f'(x) - \frac{n^2}{x^2} \, f(x)$, where p is the root of $J_n(p) = 0$ in $0 \leqslant x \leqslant 1$.

We have, by the result of article 7.4 that

$$H_n \left[\, f''(x) + \frac{1}{x} f'(x) - \frac{n^2}{x^2} f(x) \, \right]$$

$$= \int_0^1 \left[f''(x) + \frac{1}{x} f'(x) \right] x \, J_n \, (px) \, dx - \int_0^1 \frac{n^2}{x^2} f(x). \; x J_n \, (px) \, dx$$

$$= -p \int_0^1 x \, J_n{}' \, (px) f' \, (x) \, dx - n^2 \int_0^1 \frac{1}{x} f(x) \, J_n \, (px) \, dx$$

$$= -p \left[\, \{ \, x J_n'(px) \; f(x) \, \}_0^1 - \int_0^1 \{ J_n'(px) + px \; J_n''(px) \} \; f(x) \, dx \right]$$

$$-n^2 \int_0^1 \frac{1}{x} f(x) J_n(px) dx$$

$$= -p \, J_n'(p) \; f(1) + p \left[\int_0^1 \left\{ J_n'(px) + px \left(\frac{n^2}{p^2 x^2} - 1 \right) J_n(px) - J_n'(px) \right\} \right.$$

$$\left. f(x) \, dx \right] - n^2 \int_0^1 \frac{1}{x} f(x) \, J_n \, (px) \, dx,$$

by $px \, J_n'' \, (px) + J_n'(px) = px \left(\dfrac{n^2}{p^2 x^2} - 1 \right) J_n \, (px).$

Therefore,

$$H_n \left[f''(x) + \frac{1}{x} f'(x) - \frac{n^2}{x^2} f(x) \right]$$

$$= -p \, J_n{}'(p) f(1) - p^2 \int_0^1 x \, J_n(x) \; f(x) \, dx, \quad \text{after simplification.}$$

$$(7.17)$$

Corollary 7.2 Putting $n = 0$ and using the result $J_0'(p) = -J_1(p)$ we get from the above result that

$$H_0 \left[f''(x) + \frac{1}{x} f'(x) \right] = p \, J_1 \, (p) \; f(1) \; - p^2 \, H_0 \, [f(x)] \qquad (7.18)$$

7.8 Other forms of finite Hankel Transforms.

For utility purpose in applications other two different forms of finite Hankel transforms of order n are introduced below.

Form II If $f(r)$ is a piecewise continuous function in $0 \leqslant r \leqslant 1$. then its finite Hankel transform of order n is denoted by $\bar{f}(p)$ and is defined by

$$H_n [\, f(r) \,] = \bar{f}(p) = \int_0^1 f(r). \, r \, J_n \, (pr) \, dr, \qquad (7.19)$$

where p is a positive root of

$$p \, J_n{}' \, (p) + h \, J_n \, (p) = 0 \qquad (7.20)$$

Correspondingly, inverse finite Hankel transform of $\bar{f}(p)$ is given by

$$H_n^{-1}\left[\bar{f}(p)\right] \equiv f(r) = 2\sum_p \frac{p^2\bar{f}(p)}{h^2 + p^2 - n^2} \cdot \frac{J_n\ (pr)}{\{J_n\ (p)\}^2} \qquad (7.21)$$

where the summation is being taken over all the positive roots of the equation (7.20).

Form III   If the field of variation of the variable r ranges finitely as $0 < a \leqslant r \leqslant b$, a third form of finite Hankel transform of order n of a function $f(r)$, which is piecewise continuous in the range, is defined by

$$H_n\left[f(r)\,\right] = \bar{f}(p) = \int_a^b r f(r)\ B_n\ (pr)\ dr, \qquad (7.22)$$

where $B_n\ (p\,r) = J_n(p\,r)\ Y_n(p\,a)\ -\ Y_n(pr)\ J_n(p\,a) \qquad (7.23)$

and $Y_n(p\,r)$ is the Bessel function of second kind of order n. Here p is a positive root of the equation

$$B_n\ (p\ b) = 0 \Rightarrow J_n(p\ b)\ Y_n(p\ a)\ =\ Y_n(p\ b)\ J_n(p\ a) \qquad (7.24)$$

Then inverse of $\bar{f}(p)$ in this case is defined and denoted by

$$H_n^{-1}\left[\,\bar{f}\ (p)\,\right] \equiv f(r) = \frac{\pi^2}{2}\sum_p \frac{p^2\{\ J_n(pb)\}^2\ \bar{f}\ (p)}{\{\ J_n(pa)\ \}^2 - \{\ J_n(pb)\ \}^2}\ B_n\ (pr),$$
$$\qquad (7.25)$$

where the summation is being taken over all the positive roots of the equation (7.25).

7.9   Illustrations.

(i)   As an illustration we evaluate finite Hankel transform of order zero in $(0,1)$ of $f''(x) + \frac{1}{x}f'(x)$, where, p is a positive root of $p\ J_0'(p) + h\ J_0(p) = 0$ below.

By definition, we have

$$H_0\left[f''(x) + \frac{1}{x}\ f'(x)\right]$$

$$= \int_0^1\left[f''(x) + \frac{1}{x}\ f'(x)\right]x J_0(px)dx$$

$$= [f'(x) \cdot x \, J_0(px) \,]_0^1 - \int_0^1 f'(x) \, [\, x \, p \, J_0'(px) + J_0(px) \,] \, dx$$

$$+ \int_0^1 f'(x) J_0(px) dx$$

$$= [f'(1) J_0(p)] - p \int_0^1 f'(x) \cdot x \, J_0'(px) dx$$

$$= f'(1) J_0(p) - p[\{f(x) \cdot x \, J_0'(px)\}_0^1 - \int_0^1 f(x) \, \frac{d}{dx}\{x \, J_0'(px)\} dx]$$

$$= f'(1) J_0(p) - p \, f(1) J_0'(p) + p \int_0^1 f(x)\{J_0'(px) + px J_0''(px)\} dx$$

$$= f'(1) J_0(p) - p \, f(1) J_0'(p) + p \int_0^1 (-px) \, J_0(px) \, f(x) dx$$

$$= f'(1) J_0(p) - p \, f(1) \, J_0'(p) - p^2 \, H_0[f(x)]$$

But since p is a root of $p \, J_0'(p) + h \, J_0(p) = 0$, we have

$$J_0'(p) = -h \, \frac{J_0(p)}{p}$$

and therefore,

$$H_0 \left[f''(x) + \frac{1}{x} \, f'(x) \right] = [f'(1) + h \, f(1)] J_0(p) - p^2 \, H_0[f(x)] \quad (7.26)$$

(ii) For illustration of the **form-III** of finite Hankel transform of order n in $0 < a \leqslant r \leqslant b$, of $f''(r) + \frac{1}{r} f'(r) - \frac{n^2 f(r)}{r^2}$, where p is a root of $J_n(pb) \, Y_n(pa) = Y_n(pb) \, J_n(pa)$, is stated without proof as

$$H_n \left[f''(r) + \frac{1}{r} f'(r) - \frac{n^2}{r^2} \, f(r) \right]$$

$$= \frac{2}{\pi} \left[f(b) \frac{J_n(pa)}{J_n(pb)} - f(a) \right] - p^2 \, H_n \, [f(r)] \quad (7.27)$$

7.10   Application of finite Hankel Transforms.

We discuss below the application of finite Hankel transform to solve some initial-boundary value problems.

Example 7.6   Solve the following initial-boundary value problem defined by

$$\frac{\partial^2 v}{\partial r^2} + \frac{1}{r} \frac{\partial v}{\partial r} = \frac{1}{\kappa} \frac{\partial v}{\partial t} \, , \quad 0 \leqslant r < 1 \, , \, t \geqslant 0$$

under the conditions $v(1,t) = v_0$, a costant, when $t > 0$ and
$v(r,0) = 0$, $0 \leqslant r < 1$, by using finite Hankel transform in $0 \leqslant r < 1$.

Solution. Let $\bar{v}_0 \, (p,t)$ be Hankel transform of $v(r,t)$ over $0 \leqslant r < 1$ of order zero. Then, taking Hankel transform of order zero, the given PDE under the given boundary condition $v(1.t) = v_0$ becomes

$$p \, v(1,t) \, J_1(p) - p^2 \, H_0[v] = \frac{1}{\kappa} \int_0^1 \frac{\partial v}{\partial t} \, r \, J_0(pr) \, dr$$

Hence, $\qquad p \, v_0 \, J_1(p) - p^2 \, \bar{v}_0 = \frac{1}{\kappa} \frac{d}{dt} \bar{v}_0$

Thus, $\bar{v}_0(p,t)$ is a solution of the linear ODE

$$\frac{d \, \bar{v}_0}{dt} + \kappa \, p^2 \, \bar{v}_0 = \kappa \, p \, v_0 \, J_1(p)$$

Its solutions is given by

$$\bar{v}_0(p,t) = \left(\frac{v_0}{p} \right) J_1(p) + c \, e^{-\kappa p^2 t}, \text{ where } c \text{ is a constant.}$$

Now, taking zero order Hankel transform of $v(r,0) = 0$ we get $\bar{v}_0(p,0) = 0$ and therefore, the above solution gives $c = - \left(\frac{v_0}{p} \right) J_1(p)$.
Hence,

$$\bar{v}_0(p,t) = \frac{v_0}{p} J_1(p) \left(1 - e^{-\kappa p^2 t} \right)$$

Applying inversion formula of Hankel transform we finally get the solution as

$$v(r,t) = 2 \sum_p \bar{v}_0 \, (p,t) \frac{J_0(pr)}{[J_0'(p)]^2}$$

$$= 2 \, v_0 \sum_p \left(1 - e^{-\kappa p^2 t} \right) \left\{ \frac{J_0(pr)}{p \, J_1(p)} \right\} \, , \text{ since } J_0'(p) = -J_1(p),$$

where summation is taken over all p satisfying $J_0(p) = 0$.

Example 7.7. Find the temperature distribution in a long circular cylinder defined by the axisymmetric heat conduction equation

$$\frac{1}{\kappa} \frac{\partial u}{\partial t} = \frac{\partial^2 u}{\partial r^2} + \frac{1}{r} \frac{\partial u}{\partial r} \, , \quad 0 \leqslant r \leqslant a \, , \, t > 0$$

under the boundary and initial conditions

$$u(r,t) = f(t), \quad \text{on } r = a, \ t > 0$$
$$u(r,0) = 0, \qquad 0 \leqslant r \leqslant a$$

Solution. Under finite Hankel transform of order zero, the given PDE becomes

$$\frac{1}{\kappa}\, \bar{u}_t(p,t) = -p^2 \bar{u}\,(p,t) + ap\, J_1(ap)\, f(t), \qquad \text{by (7.18)}$$

Also, $u(r,0) = 0$, $0 \leqslant r \leqslant a$ under the transformed domain results in

$$\bar{u}\,(p,\,0) = 0$$

The solution of the above first order *ODE* gives

$$\bar{u}\,(p,t) = \kappa\, ap\, J_1\,(ap) \int_0^t f(\tau)\, \exp\left\{-\kappa p^2(t-\tau)\right\} d\tau$$

Thus, the formal solution of the boundary value problem is given by

$$u(r,t) = \frac{2\kappa}{a} \sum_p \frac{p\, J_0(pr)}{J_1(pa)} \int_0^t f(\tau)\, \exp\left\{-\kappa p^2(t-\tau)\right\} d\tau,$$

where the summation is taken over all positive root p satisfying $J_0(ap) = 0$. In particular, if $f(t) = \text{constant} = T_0$, say, then

$$
\begin{aligned}
u(r,t) &= \frac{2\kappa\, T_0}{a} \sum_p \frac{p\, J_0(pr)}{J_1(pa)} \int_0^t \exp\left\{-\kappa p^2(t-\tau)\right\} d\tau \\
&= \frac{2T_0}{a} \sum_p \frac{J_0(pr)}{p\, J_1(pa)} \left[1 - \exp\left(-\kappa p^2 t\right)\right]
\end{aligned}
$$

Thus, the temperature distribution consists of a transient term which decays to zero as $t \to \infty$ together with a steady-state term.

Example 7.8. The axisymmetric unsteady motion of a viscous fluid in an infinitely long circular cylinder of radius a is governed by the PDE

$$\frac{\partial u}{\partial t} = \nu \left[\frac{\partial^2 u}{\partial r^2} + \frac{1}{r}\frac{\partial u}{\partial r} - \frac{u(r,t)}{r^2}\right], \quad 0 \leqslant r \leqslant a,\ t > 0,$$

where $u = u(r,t)$ is the tangential fluid velocity and ν is the constant kinematic viscosity of the fluid. The cylinder is initially at rest at $t = 0$, and it is then allowed to rotate with constant angular velocity Ω. Thus the respective boundary and initial conditions are

$$u(r,t) = a\,\Omega\ , \ t > 0$$

and $\qquad u(r,t) = 0\ , \quad t = 0$

Solution. Taking Laplace transform in t and finite Hankel transform of order one over $0 < r < a$ jointly defined by

$$\bar{\bar{u}}\,(p,s) = \int_0^\infty e^{-st}\,dt\,\left[\int_0^a r\,J_1(pr)\,u(r,t)\,dr\right]\,,$$

where p is the positive root of the equation $J_1(ap) = 0$ and using (7.17) the given PDE under the boundary and initial conditions becomes

$$s\bar{\bar{u}}\,(p,s) = -\nu\left[p^2\bar{\bar{u}}(p,s)\right] - \frac{\nu\,a^2\,p\,\Omega}{s}\,J_1'(ap)$$

Or $\qquad \bar{\bar{u}}\,(p,s) = -\dfrac{\nu\,a^2\,\Omega\,p\,J_1'(ap)}{s(s+\nu p^2)}$

The inverse Laplace transform gives

$$\bar{u}(p,t) = -\frac{a^2\,\Omega}{p}\,J_1'(ap)\left[1 - \exp\left(-\nu t p^2\right)\right]$$

Finally, using inverse Hankel transform with $J_1'(ap) = -J_2(ap)$ the above equation gives

$$u(r,t) = 2\Omega\sum_p \frac{J_1(rp)}{p\,J_2(ap)}[1 - \exp(-\nu t p^2)]$$

Since, $\qquad H_n\,[r^n] = \dfrac{a^{n+1}}{p}\,J_{n+1}(ap)\ ,$ for $n = 0,1,2,\cdots$

we have $\qquad r = H_1^{-1}\left[\dfrac{a^2}{p}\,J_2(ap)\right] = 2\sum_p \dfrac{J_1(rp)}{p\,J_2(ap)}$

Using this result we get the final solution of the initial-boundary value problem as

$$u(r,t) = r\Omega - 2\Omega\sum_p \frac{J_1(pr)}{p\,J_2(pa)}\,\exp\left(-\nu t p^2\right)$$

This solution consists of a steady part $r\Omega$ together with a transient part $-2\Omega\sum_p \frac{J_1(pr)}{p\,J_2(pa)}\,\exp\left(-\nu p^2 t\right)$ which tends to zero as $t \to \infty$.

Example 7.9. A viscous fluid of kinematic viscosity ν is contained between two infinitely long concentric circular cylinders of radi a and b. The inner cylinder is kept at rest and the outer cylinder suddenly started rotating with uniform angular velocity ω. Find the tangential velocity $u(r,t)$ of the fluid if the equation of motion is

$$\frac{\partial^2 u}{\partial r^2} + \frac{1}{r}\frac{\partial u}{\partial r} - \frac{u}{r^2} = \frac{1}{\nu}\frac{du}{dt} \, , \quad a < r < b \, , \ t > 0,$$

given that $u = b\,\omega$ when $r = b$, $u = 0$ when $r = a$ and also $u = 0$ when $t = 0$.

Solution. Given the boundary values and the initial condition of the initial boundary value problem as

$$u(b,t) = b\,\omega \, , \quad u(a,t) = 0 \tag{i}$$

$$\text{and} \quad u(r,0) = 0 \tag{ii}$$

Now, taking finite Hankel transform of order 1 over $a \leqslant r \leqslant b$ on r, let $\bar{u}(p,t)$ be given by

$$\bar{u}(p,t) = \int_a^b r\,B_1(pr)\,u(r,t)\,dr \tag{iii}$$

where p is a positive root of the equation

$$J_1(pb)\,Y_1(pa) = Y_1(pb)\,J_1(pa) \tag{iv}$$

Then, the given constitutive PDE becomes

$$H_1\left[\frac{\partial^2 u}{\partial r^2} + \frac{1}{r}\frac{\partial u}{\partial r} - \frac{u^2}{r^2}\right] = \frac{1}{\nu}\frac{d}{dt}\,\bar{u}\,(p\,,\,t)$$

Simplifying, we get

$$\frac{2}{\pi}\left[u(b,t)\frac{J_1(pa)}{J_1(pb)} - u(a,t)\right] - p^2\bar{u}(p,t) = \frac{1}{\nu}\frac{d\bar{u}}{dt} \, , \text{ by } (7.27)$$

$$\text{Or} \quad \frac{d\bar{u}}{dt} + p^2\nu\,\bar{u}(p,t) = \frac{2b\nu\omega}{\pi}\frac{J_1(pa)}{J_1(pb)}$$

This linear ODE has the solution given by

$$\bar{u}(p,t) = \frac{2b\omega}{\pi p^2}\frac{J_1(pa)}{J_1(pb)} + C\,e^{-\nu p^2 t} \tag{v}$$

But taking Hankel transform of order 1 on r, the initial condition in (ii) gives

$$\bar{u}(p\,,0) = 0 \qquad\qquad\qquad\qquad\qquad\text{(vi)}$$

Employing (vi) in (v), we get

$$C = -\frac{2b\omega}{\pi p^2}\,\frac{J_1(pa)}{J_1(pb)}$$

and thus (v) becomes

$$\bar{u}(p,t) = \frac{2b\omega}{\pi p^2}\,\frac{J_1(pa)}{J_1(pb)}\left[1 - e^{-\nu p^2 t}\right] \qquad\qquad\text{(vii)}$$

Now, inverting Hankel transform of order 1, the above equation leads to

$$u(r,t) = \pi\,\omega\,b\sum_p \frac{(1 - e^{-\nu p^2 t})\,J_1(pa)\,J_1(pb)}{[J_1(pa)]^2 - [J_1(pb)]^2}\,\cdot\,B_1(pr)$$

as the final solution of the initial-boundary value problem.

Example 7.10. The free symmetric vibration of a thin circular membrane of radius a is governed by the wave equation

$$\frac{\partial^2 u}{\partial r^2} + \frac{1}{r}\,\frac{\partial u}{\partial r} = \frac{1}{c^2}\,\frac{\partial^2 u}{\partial t^2}\ ,\ 0 < r < a\ ,\ t > 0$$

with boundary and initial condition

$$u(a,t) = 0 \qquad,\ t \geqslant 0$$
$$u(r,0) = f(r)\ ,\quad \frac{\partial u(r,0)}{\partial t} = g(r)\ ,\ 0 < r < a$$

Determine the solution of the problem

Solution. Applying zeroth order finite Hankel transform over r in $0 < r < a$ the given constitutive equation becomes

$$\frac{d^2\bar{u}}{dt^2} + c^2\,p^2\,\bar{u} = 0 \qquad\qquad\qquad\qquad\text{(i)}$$

and the given boundary conditions become

$$\bar{u}(p,0) = \bar{f}(p)\quad\text{ and }\quad \frac{d}{dt}\,[\,\bar{u}\,(p\,,0)\,] = \bar{g}(p) \qquad\text{(ii)}$$

The solution of this system of second order ODE in (i) under (ii) is then

$$\bar{u}(p,t) = \bar{f}(p) \, \cos \, (p \, c \, t) + \frac{\bar{g}(p)}{pc} \, \sin \, (p \, c \, t) \qquad \text{(iii)}$$

Finally the inverse Hankel transform yeilds the solution as

$$u(r,t) = \frac{2}{a^2} \sum_p \bar{f}(p) \, \cos \, (p \, c \, t) \cdot \frac{J_0(rp)}{J_1^2(ap)} + \frac{2}{ca^2} \sum_p \bar{g}(p) \, \sin \, (p \, c \, t) \, \frac{J_0(rp)}{p \, J_1^2(ap)},$$

where the summation is taken over all positive roots p of $J_0(pa) = 0$.

Exercises.

(1) Find the zeroth order Hankel transform of
   (a) $f(r) = r^2$ (b) $f(r) = J_0(\alpha \, r)$

$$\left[\textbf{Ans. (a) } \frac{a^2}{p^2} \left(ap - \frac{4}{ap} \right) J_1(ap) \quad \text{(b) } \frac{ap}{(\alpha^2 - p^2)} \, J_0(a\alpha) \, J_1(ap) \right]$$

(2) Prove that $H_0 \left[\frac{1}{r} \, f'(r) \right] = p \, H_1 \left\{ \frac{1}{r} \, f(r) \right\} - f(a)$, where p is a positive root of $J_0(ap) = 0$

(3) Find the solution of the forced symmetric vibrations of a thin elastic membrane that satisfy the initial boundary value problem

$$\frac{\partial^2 u}{\partial r^2} + \frac{1}{r} \frac{\partial u}{\partial r} - \frac{1}{c^2} \frac{\partial^2 u}{\partial t^2} = -\frac{p \, (r,t)}{T_0} \, ,$$

where $p(r,t)$ is the applied pressure for producing the vibration and the membrane is stretched by a constant tension T_0. The membrane is set into motion from rest from its equilibrium position so that

$$u(r,0) = 0 = \left[\frac{\partial u}{\partial t} \right]_{t=0}$$

(4) Find finite Hankel transform of $\frac{(r^2 - b^2)}{r}$ with the kernel $J_1(pr) \, Y_1(pb) - J_1(pb) \, Y_1(pr)$.

[Hint: Start with evalvating $\int_a^b (r^2 - b^2)\{J_1(pr)Y_1(pb) - Y_1(pr)J_1(pb)\}dr$]

[**Ans.** $H_1\left\{\dfrac{(r^2 - b^2)}{r}\right\} = \dfrac{b^2 - a^2}{ap^2}\dfrac{J_1(pb)}{J_1(pa)}$, where p is the positive

root of $Y_1(pb) J_1(ap) - J_1(pb) Y_1(pa) = 0$

and $Y_1(pb) J_1'(pa) - J_1(pb) Y_1'(pa) = \dfrac{1}{pa}\dfrac{J_n(pb)}{J_n(pa)}$] .

Chapter 8

The Mellin Transform

8.1   Introduction.

This chapter deals with the origin, theory and applications of Mellin transforms. We derive the transform from its natural way in connection with the solution of a problem of potential theory in an infinite wedge. Riemann was the first author to discuss this transform in 1876 and many others too have shown its applications later.

Let us consider a problem of potential theory in an infinite wedge defined by $\rho > 0$, $|\varphi| < \alpha$ where ρ, φ are plane polar co-ordinates. We wish to find the solutions $u(\rho, \varphi)$ of the two-dimensional Laplace equation

$$\frac{\partial^2 u}{\partial \rho^2} + \frac{1}{\rho}\frac{\partial u}{\partial \rho} + \frac{1}{\rho^2}\frac{\partial^2 u}{\partial \varphi^2} = 0 \tag{8.1}$$

under the conditions that

$$u\left(\rho, \pm \alpha\right) = f(\rho), \text{ (a given function)} \tag{8.2}$$

and $\quad \rho^s\, u\left(\rho, \varphi\right), \; \rho^{s+1}\dfrac{\partial u}{\partial \rho}$ $\tag{8.3}$

tend to zero as $\rho \to 0$ and also as $\rho \to \infty$ for some specified complex number s. From (8.1), we can write

$$\frac{\partial}{\partial \rho}\left[\rho\,\frac{\partial u}{\partial \rho}\right] + \frac{1}{\rho}\frac{\partial^2 u}{\partial \varphi^2} = 0$$

Now multiplying the above equation by ρ^s and integrating the result with respect to ρ from $\rho = 0$ to $\rho \to \infty$ we get

$$\int_0^\infty \rho^s\left[\frac{\partial}{\partial \rho}\left\{\rho\,\frac{\partial u}{\partial \rho}\right\}\right]d\rho + \int_0^\infty \rho^{s-1}\frac{\partial^2}{\partial \varphi^2}\,u(\rho, \varphi)\,d\rho = 0$$

If we use the formula for integration by parts in the first term we get the above equation as

$$\left[\rho^{s+1} \frac{\partial u}{\partial \rho} - s\, \rho^s\, u(\rho, \varphi) \right]_0^\infty + s^2 \int_0^\infty \rho^{s-1}\, u(\rho, \varphi)\, d\rho$$

$$+ \frac{d^2}{d\phi^2} \int_0^\infty \rho^{s-1}\, u(\rho, \varphi)\, d\rho = 0$$

Now, using the conditions (8.3) we get

$$\left[\frac{d^2}{d\varphi^2} + s^2 \right] \int_0^\infty \rho^{s-1}\, u(\rho, \varphi)\, d\rho = 0 \tag{8.4}$$

Defining Mellin transform of the function $u(\rho, \varphi)$ over $0 < \rho < \infty$ as

$$u^*(s, \varphi) \equiv M\, [\, u(\rho, \varphi)\,;\, \rho \to s\,] = \int_0^\infty \rho^{s-1}\, u(\rho, \varphi)\, d\rho$$

the equation (8.4) can be expressed as

$$\frac{d^2 u^*}{d\varphi^2}\, (s, \varphi) + s^2\, u^*(s, \varphi) = 0 \tag{8.5}$$

which is an ordinary differential equation.

If the conditions on the faces of wedge are as specified in (8.2) be used, we have

$$u^*(s\,, \pm\, \alpha) = f^*(s)$$

after taking Mellin transforms. Then the solution of the boundary value problem under prescribed conditions is

$$u^*(s, \varphi) = f^*(s)\, \cdot\, \frac{\cos\, s\varphi}{\cos\, s\alpha} \tag{8.6}$$

Therefore, the solution $u^*(s, \varphi)$ of the boundary-value problem satisfying the prescribed condition is a function whose Mellin transform is given by

$$f^*(s)\, \frac{\cos\, s\varphi}{\cos\, s\alpha}\, .$$

8.2   Definition of Mellin Transform.

Let $f(x)$ be defined for $0 < x < \infty$. Mellin transform of this function is a new function of the complex variable s defined by

$$f^*(s) = M\, [\, f(x)\,;\, x \to s\,] = \int_0^\infty f(x)\, x^{s-1}\, dx \tag{8.7}$$

for those values of s for which the right hand side improper integral in eqn. (8.7) converges.

We shall now discuss some of the basic important properties of Mellin transform.

Property I. Linearity property.

Let c_1, c_2 be constants and let $f_1(x)$ and $f_2(x)$ be two given functions. Then the linearity property of Mellin transform states that

$$M[c_1 f_1(x) + c_2 f_2(x) \; ; x \to s] = c_1 M[f_1(x) \; ; \; x \to s] + c_2 \, M[f_2(x) \; ; \; x \to s]$$

Proof.

The proof of this result is very simple and can be seen in the following steps.

$$M \, [c_1 f_1(x) + c_2 f_2(x) \; ; \; x \to s]$$
$$= \int_0^\infty x^{s-1} \, [c_1 f_1(x) + c_2 f_2(x)] \, dx$$
$$= c_1 \int_0^\infty f_1(x) \, x^{s-1} \, dx + c_2 \int_0^\infty f_2(x) \, x^{s-1} \, dx$$
$$= c_1 f_1^*(s) + c_2 f_2^*(s) \, , \qquad (8.8)$$

for those s for which both the improper integrals do exist.

Property II. Change of scale property.

If $\qquad M \, [f(x) \; ; \; x \to s] = f^*(s) \, ,$ then
$$M \, [f(ax) \; ; \; x \to s] = a^{-s} \, f^*(s)$$

Proof.

The proof of this result is deduced by a transformation of variable in the integral form of the transform.

$$M \, [f(ax) \; ; \; x \to s] \;\; = \;\; \int_0^\infty f(ax) x^{s-1} dx$$
$$= \int_0^\infty f(t) \cdot \frac{t^{s-1}}{a^s} \, dt \, , \quad \text{putting} \quad ax = t$$
$$= \;\; a^{-s} \, f^*(s) \qquad (8.9)$$

Property III.

If $\qquad M \, [f(x) \; ; \; x \to s] = f^*(s) \, ,$ then
$$M \, [x^a f(x) \; ; \; x \to s] = f^*(s + a)$$

Proof.

To prove the above result, we get

$$
\begin{aligned}
M\left[x^a f(x) \; ; \; x \to s\right] &= \int_0^\infty x^a f(x) \cdot x^{s-1} \, dx = \int_0^\infty f(x) \cdot x^{(s+a)-1} \, dx \\
&= f^*(s+a) \; , \; \text{by use of (8.7)} \tag{8.10}
\end{aligned}
$$

Property IV.

If $\qquad M\left[f(x) \; ; \; x \to s\right] = f^*(s)$, then

$$
M\left[\frac{1}{x} f\left(\frac{1}{x}\right) \; ; \; x \to s\right] = f^*(1-s) \; .
$$

Proof.

To prove the above result, we have by definition of Mellin transform

$$
\begin{aligned}
&M\left[\frac{1}{x} f\left(\frac{1}{x}\right) \; ; \; x \to s\right] \\
&= \int_0^\infty \frac{1}{x} f\left(\frac{1}{x}\right) x^{s-1} \, dx \\
&= \int_0^\infty t^{-s} f(t) \, dt \qquad , \text{ putting } \quad x = \frac{1}{t} \\
&= \int_0^\infty t^{(1-s)-1} f(t) \, dt \\
&= f^*(1-s) \; , \; \text{by (8.7)} \tag{8.11}
\end{aligned}
$$

Property V.

If $\qquad M\left[f(x) \; ; \; x \to s\right] = f^*(s)$, then

$$
M\left[f(x^a) \; ; \; x \to s\right] = \frac{1}{a} \, f^*\left(\frac{s}{a}\right) \; , \; a > 0
$$

Proof. To prove the above result, we have

$$
\begin{aligned}
M\left[f(x^a) \; ; \; x \to s\right] &= \int_0^\infty x^{s-1} f(x^a) \, dx \\
&= \frac{1}{a} \int_0^\infty t^{\left(\frac{s}{a}-1\right)} f(t) \, dt \; , \; \text{where } \quad x^a = t \\
&= \frac{1}{a} \, f^*\left(\frac{s}{a}\right) \; , \; \text{by (8.7)} \tag{8.12}
\end{aligned}
$$

Property VI. Derivative of Mellin Transform.

If $M[f(x) ; x \to s] = f^*(s)$, then $M[\log x.f(x) ; x \to s]$ is equal to $\frac{d}{ds} f^*(s)$.

Proof. We have, by definition of Mellin transform

$$M[\log x.f(x) ; x \to s] = \int_0^\infty x^{s-1} \log x.f(x) \, dx$$

But, from the given condition $\frac{d}{ds} [f^*(s)] = \int_0^\infty \frac{d}{ds} [x^{s-1} f(x)] \, dx$

$$= \int_0^\infty f(x) . \frac{1}{x} . x^s \log x \, dx = \int_0^\infty x^{s-1} . f(x) \log x.dx$$

Thus, $M[\log x.f(x) ; x \to s] = \frac{d}{ds} f^*(s)$ (8.13)

8.3   Mellin Transform of derivative of a function.

The following two sets of results are found to be true in these cases.

Results I   (a) $M[f'(x) ; x \to s] = -(s-1)f^*(s-1)$

(b) $M[f''(x) ; x \to s] = (s-1)(s-2)f^*(s-2)$

(c) $M[f^n(x); x \to s] = (-1)^n \frac{\Gamma(s)}{\Gamma(s-n)} f^*(s-n), n = 1, 2, 3, \cdots$

where $M[f(x) ; x \to s] = f^*(s)$ and $f^n(x) = \frac{d^n}{dx^n} f(x)$

Proof.

(a) By definition of Mellin transform

$$M[f'(x) ; x \to s] = \int_0^\infty x^{s-1} f'(x) \, dx$$

$$= [x^{s-1} f(x)]_0^\infty - (s-1) \int_0^\infty x^{s-2} f(x) \, dx$$

$$= -(s-1) \int_0^\infty x^{(s-1)-1} f(x) \, dx ,$$

for the existence of real numbers λ_1 and λ_2 such that

$$\lim_{s \to 0} x^{s-1} f(x) = 0 , \quad \lim_{s \to \infty} x^{s-1} f(x) = 0$$

for $\lambda_1 < Re \, s < \lambda_2$ and $f^*(s-1)$ exists.

Thus, $M[f'(x) ; x \to s] = -(s-1) f^*(s-1)$

(8.14)

(b) Again application of the result in (8.14) will give

$$M\left[f''(x) \; ; \; x \to s\right] = M\left[\varphi'(x) \; ; \; x \to s\right] , \text{ where } \varphi(x) = f'(x)$$
$$= -(s-1)\,\varphi^*(s-1) = -(s-1)\,M\left[f'(x) \; ; \; x \to s\right]$$
$$= -\{(s-1)\}\{-(s-2)\}\,M\left[f(x) \; ; \; x \to s\right]$$
$$= (s-1)(s-2)\,f^*(x) \tag{8.15}$$

(c) For the last case (c), if n is a positive integer, by using the Mathematical induction n times, we get

$$M\left[f^{(n)}(x) \; ; \; x \to s\right] = (-1)^n\,(s-1)(s-2)\cdots(s-n+1)f^*(s-n)$$
$$= (-1)^n\,\frac{\Gamma(s)}{\Gamma(s-n)}\,f^*(s-n) \tag{8.16}$$

Results II　　If $M\left[f(x) \; ; \; x \to s\right] = f^*(s)$, then

(a)　　$M\left[\left(x\dfrac{d}{dx}\right)f(x) \; ; \; x \to s\right] = -s\,f^*(s)$

(b)　　$M\left[\left(x\dfrac{d}{dx}\right)^2 f(x) \; ; \; x \to s\right] = s^2\,f^*(s)$

(c)　　$M\left[\left(x\dfrac{d}{dx}\right)^n f(x) \; ; \; x \to s\right] = (-1)^n\,s^n\,f^*(s)$, $n = 2, 3, \cdots$

Proof.

(a)　By (8.10) we know that

$$M\left[\,x\,\phi\,(x) \; ; \; x \to s\,\right] = \varphi^*\,(s+1)$$

Replacing $\varphi(x) \equiv f'(x)$, we get

$$M\left[\,x\,f'(x) \; ; \; x \to s\,\right] = M\left[\,f'(x) \; ; \; x \to s+1\,\right] \tag{8.17}$$

Also, since $M\left[\,f'(x) \; ; \; x \to s\,\right] = -(s-1)\,f^*(s-1)$, replacing s by $s+1$ here we get

$$M\left[\,f'(x) \; ; \; x \to s+1\,\right] = -s\,f^*(s) \tag{8.18}$$

Therefore, from eqn. (8.17) and (8.18) we get

$$M\left[\,x\,f'(x) \; ; \; x \to s\,\right] = -s\,f^*(s) , \text{ which is the result in (a).}$$
$$\tag{8.19}$$

(b) Let us assume here that

$$\left(x\frac{d}{dx}\right)f(x) = g(x)$$

Then,

$$
\begin{aligned}
M\left[x\,g'(x)\;;\;x \to s\right] &= M\left[g'(x)\;;\;x \to s+1\right],\; \text{by (8.17)} \\
&= -s\,g^*(s) \\
&= -s\left\{M\left[x\frac{d}{dx}\,f(x)\right]\right\} \\
&= -s\left\{-s\,f^*(s)\right\},\qquad\quad \text{by (8.18)} \\
&= s^2\,f^*(s) \qquad\qquad\qquad\qquad (8.20)
\end{aligned}
$$

(c) To prove this result, we use the method of Mathematical induction to obtain

$$M\left[\left(x\frac{d}{dx}\right)^n f(x)\;;\;x \to s\right] = (-1)^n\,s^n\,f^*(s) \qquad (8.21)$$

after the nth stage.

8.4 Mellin Transform of Integral of a function.

Result I.   If $M\left[f(x)\;;\;x \to s\right] = f^*(s)$, then

(a) $M\left[\displaystyle\int_0^x f(t)dt\;;\;x \to s\right] = -\dfrac{1}{s}\,f^*(s+1)$

(b) $M\left[\displaystyle\int_0^x dy\left\{\displaystyle\int_0^y f(t)dt\right\}\;;\;x \to s\right] = \dfrac{1}{s(s+1)}\,f^*(s+2)$

(c) $M\left[I_n\,f(x)\;;\;x \to s\right] = (-1)^n\,\dfrac{\Gamma(s)}{\Gamma(s+n)}\,f^*(s+n)$, where by

$I_n\,f(r)$ we mean $I_n\,f(x) - \displaystyle\int_0^x I_{n-1}\,f(t)\;dt$

Proof. The proof of these results are given below :

(a) Let $\int_0^x f(t)\,dt = G(x) \Rightarrow G'(x) = f(x)$

Now, since $M\left[G'(x)\;;\;x \to s\right] = -(s-1)G^*(s-1)$

$$
\begin{aligned}
&= -(s-1)M\left[G(x)\;;\;x \to s-1\right] \\
&= -(s-1)M\left[\int_0^x f(t)dt\;;\;x \to s-1\right]
\end{aligned}
$$

we get $M\left[f(x) \; ; \; x \to s+1\right] = -s\,M\left[\int_0^x f(t)dt \; ; \; x \to s\right]$

$\therefore \quad M\left[\int_0^x f(t)dt \; ; \; x \to s\right] = -\frac{1}{s}\,M\left[f(x) \; ; \; x \to s+1\right]$

$$= -\frac{1}{s}\,f^*(s+1) \qquad (8.22)$$

(b) Repeating the result in (a) we get

$M\left[\int_0^x dy \left\{\int_0^y f(t)dt\right\} \; ; x \to s\right] = -\frac{1}{s}M\left[\int_0^x f(t)dt; x \to s+1\right]$

$$= -\frac{1}{s}\left\{-\frac{1}{s+1}f^*(s+2)\right\}$$

$$= \frac{1}{s(s+1)}\,f^*(s+2) \qquad (8.23)$$

(c) For proving the result in this part the method of Mathematical induction will be applied on the index n.

Result II. If $M\left[f(x) \; ; \; x \to s\right] = f^*(s)$, then

(a) $\quad M\left[\int_x^\infty f(t)dt \; ; \; x \to s\right] = \frac{1}{s}\,f^*(s+1)$

(b) $\quad M\left[\int_x^\infty dy \left\{\int_y^\infty f(t)dt\right\} ; x \to s\right] = \frac{1}{s(s+1)}\,f^*(s+2)$

(c) $\quad M\left[I_n^\infty f(x) \; ; \; x \to s\right] = \frac{\Gamma(s)}{\Gamma(s+n)}\,f^*(s+n)$,

where $\quad I_n^\infty\, f(x) = \int_x^\infty I_{n-1}^\infty\, f(t)dt$.

Proof. As in result I, the result of these parts is shown below.

(a) Let $\quad \int_x^\infty f(t)dt = g(x) \Rightarrow -f(x) = g'(x)$

Therefore, $\quad M\left[-f(x) \; ; \; x \to s\right] = M\left[g'(x) \; ; \; x \to s\right]$

$$= -(s-1)M\left[\int_x^\infty f(t)dt \; ; \; x \to s-1\right]$$

$\Rightarrow \quad M\left[f(x) \; ; \; x \to s\right] = (s-1)\,M\left[\int_x^\infty f(t)dt \; ; \; x \to s-1\right]$

$\Rightarrow \quad M\left[f(x) \; ; \; x \to s+1\right] = s\,M\left[\int_x^\infty f(t)dt \; ; \; x \to s\right],$

$$(8.24)$$

which is the proof of the result (a).

(b)   Part (b) can be proved by using the result of part (a) once again as in Result I.

(c) To prove this part once again the method of Mathematical induction on index n will be used.

8.5   Mellin Inversion theorem.

Fourier Transform of $g(t)$, $-\infty < t < \infty$ is defined as

$$G(\xi) = \frac{1}{\sqrt{2\pi}} \int_{-\infty}^{+\infty} g(t) e^{i\xi t} \, dt$$

Substituting $x = e^t$, $s = c + i\xi$ in the above we get

$$G\,(ic - is) = \frac{1}{\sqrt{2\pi}} \int_0^\infty x^{-c} \, g(\log x) \, . \, x^{s-1} \, dx \qquad (8.25)$$

Again in the corresponding inversion formula of Fourier transform

$$g(t) = \frac{1}{\sqrt{2\pi}} \int_{-\infty}^{+\infty} G(\xi) \, e^{-i\xi t} \, d\xi \,,$$

if the above transform be applied we get

$$g(\log x) = \frac{1}{\sqrt{2\pi}} \, i \int_{c-i\infty}^{c+i\infty} G(ic - is) \, x^{c-s} \, ds \qquad (8.26)$$

Now, writing

$$f(x) = (2\pi)^{-\frac{1}{2}} \, x^{-c} \, g(\log x) \,, \quad f^*(s) = G(ic - is)$$

we have from eqns (8.25) and (8.26)

$$f^*(s) \;\; = \;\; \int_0^\infty x^{s-1} \, f(x) \, dx \,, \qquad (8.27)$$

$$f(x) \;\; = \;\; \frac{1}{2\pi i} \int_{c-i\infty}^{c+i\infty} f^*(s) \, x^{-s} \, ds \qquad (8.28)$$

as the definition of Mellin transform of $f(x)$ and its corresponding inversion formula respectively.

It may be noted here that the left hand side of eqn. (8.28) can also be expressed as

$$f(x) = M^{-1} \, [f^*(s) \,; \, s \to x] = \frac{1}{2\pi i} \int_{c-i\infty}^{c+i\infty} f^*(s) \, x^{-s} \, ds \,.$$

8.6 Convolution theorem of Mellin Transform.

Theorem 8.1. If $f^*(s)$ and $g^*(s)$ be Mellin Transforms of $f(x)$ and $g(x)$ respectively, then

$$M\left[f(x)\,g(x)\;;\;x \to s\right] = \frac{1}{2\pi i}\int_{c-i\infty}^{c+i\infty} f^*(z)\cdot g^*(s-z)\,dz\;.$$

Proof.

$$M\left[f(x)\,g(x)\;;\;x \to s\right]$$

$$= \int_0^\infty f(x)\,g(x)\cdot x^{s-1}\,dx$$

$$= \int_0^\infty f(x)\,x^{s-1}\left[\frac{1}{2\pi i}\int_{c-i\infty}^{c+i\infty} g^*(z)\,x^{-z}\,dz\right]dx$$

$$= \frac{1}{2\pi i}\int_{c-i\infty}^{c+i\infty} g^*(z)\left[\int_0^\infty f(x)\,x^{s-z-1}\,dx\right]dz$$

$$= \frac{1}{2\pi i}\int_{c-i\infty}^{c+i\infty} g^*(z)\,f^*(s-z)\,dz$$

Similarly, we would get

$$M\left[f(x)\,g(x)\;;\;x \to s\right] = \int_0^\infty f(x)\,g(x)\,x^{s-1}\,dx$$

$$= \frac{1}{2\pi i}\int_{c-i\infty}^{c+i\infty} f^*(z)\,g^*(s-z)\,dz$$

Thus we get

$$M\left[f(x)\,g(x)\;;\;x \to s\right]$$

$$\left.\begin{aligned}
&= \tfrac{1}{2\pi i}\int_{c-i\infty}^{c+i\infty} f^*(z)\,g^*(s-z)\,dz\\[2mm]
&= \tfrac{1}{2\pi i}\int_{c-i\infty}^{c+i\infty} g^*(z)\,f^*(s-z)\,dz
\end{aligned}\right\} \tag{8.29}$$

as the result of the convolution theorem.

Corollary 8.1. Another form of convolution theorem of Mellin transform is given below :

Theorem 8.2. If $f^*(s)$ and $g^*(s)$ are Mellin transforms of the functions $f(x)$ and $g(x)$ respectively, then

$$M^{-1}\left[f^*(s)\,g^*(s)\;;\;s \to x\right] = \int_0^\infty f\left(\frac{x}{t}\right)\,g(t)\,\frac{dt}{t}$$

Proof. We know that

$$M^{-1}\left[\,f^*(s)\,g^*(s)\;;\;s \to x\,\right]$$

$$= \frac{1}{2\pi i}\int_{c-i\infty}^{c+i\infty} x^{-s}\,f^*(s)\,g^*(s)\,ds\;,\;\text{ by (8.28)}$$

$$= \frac{1}{2\pi i}\int_{c-i\infty}^{c+i\infty} x^{-s}\,f^*(s)\left\{\int_0^\infty t^{s-1}\cdot g(t)dt\right\}ds$$

$$= \int_0^\infty \frac{g(t)}{t}\left\{\frac{1}{2\pi i}\int_{c-i\infty}^{c+i\infty} f^*(s)\left(\frac{x}{t}\right)^{-s}ds\right\}dt$$

$$= \int_0^\infty \frac{g(t)}{t}\,f\left(\frac{x}{t}\right)dt\;, \tag{8.30}$$

since $f\left(\dfrac{x}{t}\right) = \dfrac{1}{2\pi i}\displaystyle\int_{c-i\infty}^{c+i\infty}\left(\dfrac{x}{t}\right)^{-s} f^*(s)\,ds$ \hfill (8.31)

8.7 Illustrative solved Examples.

Example 8.1. Evaluate Mellin transform of

(a) $f(x) = e^{-nx}$

(b) $f(x) = (1+x)^{-a}$

(c) $f(x) = (1+x^a)^{-b}$, where $0 < \operatorname{Re}\,s < \operatorname{Re}\,(ab)$

(d) $f(x) = \cos kx$ and $f(x) = \sin kx$

(e) $f(x) = \dfrac{2}{(e^{2x}-1)}$

Solutions. By definition of Mellin transform, we have

$$\text{(a)}\quad M\left[e^{-nx}\;;\;x \to s\right] = \int_0^\infty x^{s-1}\,e^{-nx}\,dx$$

$$= \frac{1}{n^s}\int_0^\infty t^{s-1}\,e^{-t}\,dt,\quad\text{where}\quad nx = t$$

$$= \frac{\Gamma(p)}{n^s}$$

(b) $M\left[(1+x)^{-a}\ ;\ x\to s\right]\ =\ \displaystyle\int_0^\infty \frac{x^{s-1}}{(1+x)^{s+(a-s)}}\ dx$

$$= B\left(s\ ,\ a-s\right),\quad \text{for } Re\ s>0\ ,\ Re\ (a-s)>0$$
$$= \frac{\Gamma(s)\ \Gamma(a-s)}{\Gamma(a)}\ ,\quad \text{for } Re\ a> Re\ s>0$$

(c) $M\left[\dfrac{1}{(1+x^a)^b}\ ;\ x\to s\right]\ =\ \displaystyle\int_0^\infty \frac{x^{s-1}}{(1+x^a)^b}\ dx$

$$= \int_0^\infty \frac{(y)^{\frac{s}{a}-1}dy}{(1+y)^b}\ ,\quad \text{where } x^a=y,\ \frac{a}{x}\ dx=\frac{dy}{y}$$

$$= \frac{1}{a}\int_0^\infty \frac{y^{\left(\frac{s}{a}-1\right)}}{(1+y)^{\left(b-\frac{s}{a}\right)+\frac{s}{a}}}\ dy$$

$$= \frac{1}{a}\ B\left(b-\frac{s}{a}\ ,\ \frac{s}{a}\right)$$

$$= \frac{\Gamma\left(b-\frac{s}{a}\right)\Gamma\left(\frac{s}{a}\right)}{a\ \Gamma(b)}\ ,\quad \text{where } 0< Re\ s< Re\ (ab)$$

(d) $M\left[\cos\ kx\ ;\ x\to s\right]=\displaystyle\int_0^\infty x^{s-1}\cdot\frac{e^{ikx}+e^{-ikx}}{2}\ dx$

$$= \frac{1}{2}\int_0^\infty e^{-ikx}\cdot x^{s-1}\ dx+\frac{1}{2}\int_0^\infty e^{ikx}\ x^{s-1}\ dx$$

We know from (a) above that

$$\int_0^\infty e^{-nx}\cdot x^{s-1}\ dx=\frac{\Gamma(s)}{n^s}$$

Hence, $\displaystyle\int_0^\infty e^{-ikx}\cdot x^{s-1}\ dx=\frac{\Gamma(s)}{(ik)^s}\quad$ and

$$\int_0^\infty e^{ikx}\cdot x^{s-1}\ dx=\frac{\Gamma(s)}{(-ik)^s}$$

Therefore, $\displaystyle\int_0^\infty \cos\ kx\cdot x^{s-1}\ dx=\frac{\Gamma(s)}{k^s}\ \cos\ \frac{\pi s}{2}$

and similarly $\displaystyle\int_0^\infty x^{s-1}\sin\ kx\ dx=\frac{\Gamma(s)}{k^s}\ \sin\ \frac{\pi s}{2}$

(e)   we have $M\left[\dfrac{2}{e^{2x}-1} \; ; \; x \to s\right]$

$$= 2 \int_0^\infty x^{s-1} \frac{dx}{e^{2x}-1} \, dx$$

$$= 2 \sum_{n=1}^\infty \int_0^\infty x^{s-1} \cdot e^{-2nx} \, dx$$

$$= 2 \sum_{n=1}^\infty \frac{\Gamma(s)}{(2n)^s} = 2^{1-s} \, \Gamma(s) \sum_{n=1}^\infty \frac{1}{n^s}$$

$$= 2^{1-s} \, \Gamma(s) \, \zeta(s) \, ,$$

where $\zeta(s) = \displaystyle\sum_{n=1}^\infty \frac{1}{n^s}$, $Re \ s > 1$ is Riemann zeta function.

Example 8.2. If $f^*(s)$ and $g^*(s)$ are Mellin transforms of $f(x)$ and $g(x)$ respectively, prove that

$$\int_0^\infty f(x) \, g(x) \, dx = \frac{1}{2\pi i} \int_{c-i\infty}^{c+i\infty} f^*(z) \, g^*(1-z) \, dz \ .$$

Solution. We know, by convolution theorem of Mellin transform that

$$\int_0^\infty x^{s-1} \, f(x) \, g(x) \, dx = \frac{1}{2\pi i} \int_{c-i\infty}^{c+i\infty} f^*(z) \, g^*(s-z) \, dz$$

Now, putting $s = 1$ in this relation we get

$$\int_0^\infty f(x) \, g(x) \, dx = \frac{1}{2\pi i} \int_{c-i\infty}^{c+i\infty} f^*(z) \, g^*(1-z) \, dz \ .$$

Example 8.3. Find Mellin inversion of $\Gamma(s)$.

Solution. Let $f(x)$ be the required Mellin inversion of $\Gamma(s)$. Then

$$
\begin{aligned}
f(x) \ &= \ \frac{1}{2\pi i} \int_{c-i\infty}^{c+i\infty} x^{-s} \, \Gamma(s) \, ds \\[2mm]
&= \ \frac{1}{2\pi i} \int_{c-i\infty}^{c+i\infty} x^{-s} \cdot \frac{\pi \, ds}{\Gamma(1-s)\sin s\pi} \\[2mm]
&= \ \frac{1}{2i} \int_{c-i\infty}^{c+i\infty} \frac{x^{-s} \, ds}{\Gamma(1-s)\sin s\pi} \\[2mm]
&= \ \frac{1}{2i} \int_C \frac{x^{-s} \, ds}{\Gamma(1-s)\sin s\pi} \ ,
\end{aligned}
$$

where C is an anticlockwise closed large semi-circular contour with centre at $s = 0$ lying to the left of the imaginary s-axis. Then the integrand has singularities at its poles at $s = 0, -1, -2, \cdots$. Thus, the value of the integral is $2\pi i$ times sum of the residues of the integrand at its poles. Hence,

$$
\begin{aligned}
f(x) &= \frac{1}{2i}\, 2\pi i \sum_{n=0}^{\infty} \frac{x^n}{\Gamma(1+n)\left[\frac{d}{ds}\left(\sin\, s\pi\right)\right]_{s=-n}} \\
&= \sum_{n=0}^{\infty} (-1)^n\, \frac{x^n}{n!} \\
&= e^{-x}\ .
\end{aligned}
$$

Example 8.4. Solve the boundary value problem defined by

$$
x^2\, u_{xx} + x\, u_x + u_{yy} = 0\ ,\ 0 \leqslant x < \infty\ ,\ 0 < y < 1
$$

subject to boundary conditions $u(x,0) = 0\ ,\ u(x,1) = \begin{cases} A, & 0 \leqslant x \leqslant 1 \\ 0, & x > 1 \end{cases}$

where A is a constant, by applying Mellin transform over x.

Solution. Applying Mellin transform over x we get the PDE as

$$
u_{yy}^* + s^2\, u^* = 0\ ,\qquad 0 < y < 1
$$

and the boundary conditions transform to

$$
u^*(s,0) = 0\ ,\ u^*(s,1) = A \int_0^1 x^{s-1}\, dx = \frac{A}{s}
$$

Then the solution of the transformed ODE is

$$
u^*(s,y) = \frac{A}{s}\, \frac{\sin\, sy}{\sin\, s}\ ,\ 0 < Re\ s < 1\ .
$$

Taking inverse Mellin transform we get

$$
u(x,y) = \frac{A}{2\pi i} \int_{c-i\infty}^{c+i\infty} \frac{x^{-s}}{s}\, \frac{\sin\, sy}{\sin\, s}\, ds\ .
$$

The function $u^*(s,y)$ is analytic in $0 < Re\ s < \pi$ and thus $0 < c < \pi$. The above integrand has simple poles at $s = n\pi\ ,\ n = 1,2,3,\cdots$ and all those lie inside a large semi-circular contour in the right-half of s-plane. Therefore,

$$u(x, y) = \frac{A}{\pi} \sum_{n=1}^{\infty} \frac{1}{n}(-1)^n \, x^{-n\pi} \, \sin(n\pi y) \, .$$

Example 8.5. Consider the problem of determining the potential $\varphi(r, \theta)$ that satisfies the Laplace equation

$$r^2 \, \varphi_{rr} + r \, \varphi_r + \varphi_{\theta\theta} = 0$$

in an infinite wedge $0 < r < \infty$, $-\alpha < \theta < \alpha$ and the boundary conditions

$$\varphi(r, \alpha) = f(r) \, , \quad \varphi(r, -\alpha) = g(r) \, , \quad 0 \leqslant r < \infty$$
$$\varphi(r, \theta) \to 0 \quad \text{as} \quad r \to \infty \quad \text{for all} \quad \theta \quad \text{in} \quad -\alpha < \theta < \alpha \, .$$

Solution. Applying Mellin transform over r, the Laplace equation becomes

$$s^2 \, \varphi^*(s, \theta) + \frac{d^2 \, \varphi^*(s, \theta)}{d\theta^2} = 0$$

and the boundary conditions transform to

$$\varphi^*(s, \alpha) = f^*(s) \, , \quad \varphi^*(s, -\alpha) = g^*(s) \quad \text{and} \quad \varphi^*(s, \theta) \to 0 \quad \text{as} \quad s \to \infty$$

The general solution of the transformed ODE is

$$\varphi^*(s, \theta) = A \, \cos \, s\theta + B \sin \, s\theta \, ,$$

where A and B are arbitrary constants to be determined under the transformed boundary conditions. These give

$$A \, \cos \, s \, \alpha + B \, \sin \, s \, \alpha = f^*(s)$$
$$A \, \cos \, s \, \alpha - B \, \sin \, s \, \alpha = g^*(s)$$

Solving for A and B, we get

$$A = \frac{f^*(s) + g^*(s)}{2 \cos \, s \, \alpha} \, , \quad B = \frac{f^*(s) - g^*(s)}{2 \sin \, s \, \alpha}$$

Thus the solution becomes

$$\varphi^*(s, \theta) = f^*(s) \, \frac{\sin \, s(\alpha + \theta)}{\sin \, (2s\alpha)} + g^*(s) \, \frac{\sin \, s(\alpha - \theta)}{\sin \, (2s\alpha)}$$

Inverting the Mellin transformed equation, the solution of the problem can be expressed as

$$\varphi(r,\theta) = M^{-1}\left[f^*(s)\ h^*(s,\alpha+\theta)\right] + M^{-1}\left[g^*(s)\ h^*(s,\alpha-\theta)\right]$$

where $h^*(s,\theta) = \dfrac{\sin(s\theta)}{\sin(2s\alpha)}$

Now, $h(r,\theta) = M^{-1}\left[h^*(s,\theta)\right] = M^{-1}\left[\dfrac{\sin s\theta}{\sin 2s\alpha}\right]$

$$= \frac{1}{2\alpha}\ \frac{r^n\ \sin(n\theta)}{1+2r^n\ \cos n\theta + r^{2n}}\ ,$$

after simplification where $n = \dfrac{\pi}{2\alpha}$

Therefore, the above form of solution of the problem is represented as

$$\varphi(r,\theta) = \frac{n\ r^n\ \cos\ n\theta}{\pi}\left[\int_0^\infty \frac{u^{n-1}\ f(u)\ du}{u^{2n} - 2r^n u^n\ \sin\ n\theta + r^{2n}}\right]$$

$$+ \frac{nr^n\ \cos\ n\theta}{\pi}\left[\int_0^\infty \frac{u^{n-1}\ g(u)\ du}{u^{2n} + 2r^n u^n\ \sin n\theta + r^{2n}}\right]\ , \ -\frac{\pi}{2n} < \theta < \frac{\pi}{2n}$$

8.8   Solution of Integral equations.

We begin with the solution of the following integral equation of general type

$$\int_0^\infty f(x)\ K(xt)\ dx = g(t)\ , \ t > 0 \tag{8.32}$$

Taking Mellin transform of both sides of eqn. (8.32) we get

$$f^*(1-s)\ K^*(s) = g^*(s)$$

by using eqn. (8.30).

Now replacing s by $1-s$ we find

$$f^*(s) = g^*(1-s)\ L^*(s)\ , \quad \text{where} \quad L^*(s) = \frac{1}{K^*(1-s)}$$

This solution can be written in the form

$$f(x) = \int_0^\infty g(t)\ L(xt)\ dt, \tag{8.33}$$

provided that the inverse Mellin transform in

$$L(x) = M^{-1}\left[\frac{1}{K^*(1-s)} \; ; \; s \to x\right]$$

exists.

In particular, the equation (8.32) will have the solution

$$f(x) = \int_0^\infty g(t) \; K(xt) \; dt \quad , \quad x > 0$$

if $K^*(s) \; K^*(1-s) = 1$

Example 8.6.

Solve the integral equation

$$\int_0^\infty f(\xi) \; g\left(\frac{x}{\xi}\right) \frac{d\xi}{\xi} = h(x) \; , \; x > 0$$

Solution. Applying Mellin transform the given integral equation reduces to

$$f^*(s) = h^*(s) \; K^*(s) \; , \quad \text{where} \quad K^*(s) = \frac{1}{g^*(s)}$$

Then inverting the Mellin transform we get the required solution of the given integral equation as

$$\begin{aligned}
f(x) &= M^{-1}\left[h^*(s) \; K^*(s)\right] \\
&= \int_0^\infty h(\xi) \; K\left(\frac{x}{\xi}\right) \frac{d\xi}{\xi} \; , \quad \text{by use of (8.30)}
\end{aligned}$$

8.9   Application to Summation of Series.

A useful method of summing slowly convergent series has been given by Macfarlane (Phil Mag., 40,188,(1950)).

Theorem 8.3. If $f^*(s)$ is Mellin transform of $f(x)$, then

$$\sum_{n=0}^\infty f(n+a) = \frac{1}{2\pi i} \int_{c-i\infty}^{c+i\infty} f^*(s) \; \zeta(s,a) \; ds$$

where $\zeta(s,a)$ is the generalised zeta function of Riemann, defined when $Re \; s > 1$, by the equation

$$\zeta(s,a) = \sum_{n=0}^\infty (s+a)^{-n} \; , \; 0 < a \leqslant 1 \; , \; Re \; s > 1$$

Proof. It follows from inverse Mellin transform that

$$f(n+a) = \frac{1}{2\pi i} \int_{c-i\infty}^{c+i\infty} f^*(s) \, (n+a)^{-s} \, ds \qquad (8.34)$$

Summing this result over all n from $n=0$ to $n \to \infty$, it gives

$$\sum_{n=0}^{\infty} f(n+a) = \frac{1}{2\pi i} \int_{c-i\infty}^{c+i\infty} f^*(s) \, \zeta(s,a) \, ds \qquad (8.35)$$

Similarly proceeding, we can have

$$f(nx) = M^{-1}\left[n^{-s} f^*(s) \right] = \frac{1}{2\pi i} \int_{c-i\infty}^{c+i\infty} x^{-s} \, n^{-s} \, f^*(s) \, ds$$

Thus, $\displaystyle\sum_{n=0}^{\infty} f(nx) = \frac{1}{2\pi i} \int_{c-i\infty}^{c+i\infty} x^{-s} \, f^*(s) \, \zeta(s) \, ds = M^{-1}\left[f^*(s) \, \zeta(s) \right]$

$$(8.36)$$

where $\displaystyle\zeta(s) = \sum_{n=1}^{\infty} n^{-s}$

When $x = 1$, eqn. (8.36) reduces to

$$\sum_{n=1}^{\infty} f(n) = \frac{1}{2\pi i} \int_{c-i\infty}^{c+i\infty} f^*(s) \, \zeta(s) \, ds \qquad (8.37)$$

Illustrative Examples.

Example 8.7. Show that

$$\text{(a)} \qquad \sum_{n=1}^{\infty} (-1)^{n-1} \, n^{-s} = (1 - 2^{1-s}) \, \zeta(s)$$

$$\text{(b)} \qquad \sum_{n=1}^{\infty} \frac{\sin \, an}{n} = \frac{1}{2} \, (\pi - a) \, .$$

Solutions.

(a) We know that

$$\sum_{n=1}^{\infty} (-1)^{n-1} \, n^{-s} \cdot t^n = \sum_{n=1}^{\infty} (-1)^{n-1} \, t^n \cdot \frac{1}{\Gamma(s)} \int_0^{\infty} x^{s-1} \, e^{-nx} \, dx$$

$$= \frac{1}{\Gamma(s)} \int_0^{\infty} x^{s-1} \, dx \sum_{n=1}^{\infty} (-1)^{n-1} \, t^n \, e^{-nx}$$

$$= \frac{1}{\Gamma(s)} \int_0^\infty x^{s-1} \frac{t\, e^{-x}}{1 + t\, e^{-x}}\, dx$$

$$= \frac{1}{\Gamma(s)} \int_0^\infty x^{s-1} \frac{t}{e^x + t}\, dx$$

Thus in the limit as $t \to 1$, we finally get

$$\sum_{n=1}^\infty (-1)^{n-1}\, n^{-s} = \frac{1}{\Gamma(s)}\, M\left[\frac{1}{1 + e^x}\; ;\; x \to s\right]$$

$$= (1 - 2^{1-s})\, \zeta(s)$$

(b) Mellin transform of $f(x) = \frac{\sin\, ax}{x}$ is given by

$$M\left[\frac{\sin\, ax}{x}\; ;\; x \to s\right] = \int_0^\infty x^{s-2} \sin\, ax\, dx$$

$$= -\frac{\Gamma(s-1)}{a^{s-1}}\, \cos\, \frac{\pi s}{2}\ , \quad \text{using Fourier sine transform}$$

Thus, $\quad \displaystyle\sum_{n=1}^\infty \frac{\sin\, an}{n} = \frac{1}{2\pi i} \int_{c-i\infty}^{c+i\infty} \frac{\Gamma(s-1)}{a^{s-1}}\, \zeta(s)\, \cos\, \frac{\pi s}{2}\, ds\ .$

After using the functional equation for the zeta function

$$(2\pi)^s\, \zeta(1 - s) = 2\, \Gamma(s)\, \zeta(s)\, \cos\left(\frac{\pi s}{2}\right)$$

the above equation gives

$$\sum_{n=1}^\infty \frac{\sin\, an}{n} = -\frac{a}{2} \cdot \frac{1}{2\pi i} \int_{c-i\infty}^{c+i\infty} \left(\frac{2\pi}{a}\right)^s \frac{\zeta(1 - s)}{(s - 1)}\, ds$$

$$= \frac{1}{2}\, (\pi - a)\ ,$$

since the integrand has two simple poles at $s = 0$ and at $s - 1$ inside the large semicircular contour closed on the left side of the s-plane.

8.10   The Generalised Mellin Transform.

In order to extend the applicability of Mellin transform D. Naylor (J. Math. Mech. 12, 265 - 274, 1963) has introduced the generalised

Mellin transform in $a \leqslant r < \infty$ when the unknown function in the constitutive equation is prescribed at $r = a$. It is defined as

$$M_- [f(r) ; r \to s] = \int_a^\infty \left[r^{s-1} - \frac{a^{2s}}{r^{s+1}} \right] f(r) \, dr \equiv f_-^*(s) \qquad (8.38)$$

together with its inverse transform

$$M_-^{-1} [f^*(s) ; s \to r] = f(r) \equiv \frac{1}{2\pi i} \int_{c-i\infty}^{c+i\infty} r^{-s} f_-^*(s) \, ds \, , \, r \geqslant a, \quad (8.39)$$

where $f_-^*(s)$ is analytic in $|Re \, s| < \gamma$ with $c < \gamma$.

Again, on the other hand if the derivative of the unknown function in the constitutive equation is prescribed, then the corresponding transfrom pair is given by

$$M_+ [f(r) ; r \to s] = f_+^*(s) \equiv \int_a^\infty \left[r^{s-1} + \frac{a^{2s}}{r^{s+1}} \right] f(r) \, dr, \quad (8.40)$$

and $$M_+^{-1} [f_+^*(s) ; s \to r] = f(r) \equiv \frac{1}{2\pi i} \int_{c-i\infty}^{c+i\infty} r^{-s} f_+^*(s) \, ds \, , \, r \geqslant a$$

$$(8.41)$$

where c is defined as in eqn. (8.39)

Under (8.38), the differential operator expression over $f(r)$ becomes

$$M_- \left[r^2 \frac{\partial^2 f}{\partial r^2} + r \frac{\partial f}{\partial r} \right] = s^2 \, f_-^*(s) + 2sa^s \, f(a) \, , \qquad (8.42)$$

provided $f(r)$ is appropriately behaved at infinity. More precisely

$$\lim_{r \to \infty} \left[(r^s - a^{2s} \, r^{-s}) \, r \, f_-^*(s) - s \, (r^s + a^{2s} \, r^{-s}) \, f(r) \right] = 0 \, .$$

Also under the eqn. (8.40) the same expression over $f(r)$ becomes

$$M_+ \left[r^2 \frac{\partial^2 f}{\partial r^2} + r \frac{\partial f}{\partial r} \right] = s^2 \, f_+^*(s) - 2a^{s+1} \, f'(a) \, , \qquad (8.43)$$

where $f'(r)$ exists at $r = a$ and $f(r)$ is appropriately behaved at infinity.

These types of generalised Mellin transforms have been applied in solving physical problems on truncated conical region by B. Patra [Bull. De L'Academic Pol. des. Sci., 24, 10, 1976] and many others.

8.11   Convolution of generalised Mellin Transform.

Theorem 8.4.   If $M_+[f(r) ; r \to s] = f_+^*(s)$

and $M_+[g(r) ; r \to s] = g_+^*(s)$, then

$$M_+ [g(r) \, f(r) ; r \to s] = \frac{1}{2\pi i} \int_L f_+^*(\xi) \, g_+^*(s - \xi) \, d\xi .$$

Proof. Assuming that $f_+^*(s)$ and $g_+^*(s)$ are analytic in $|Re\ s| < \gamma$, we have

$$M_+ [f(r) \, g(r) ; r \to s] = \int_a^\infty \left[r^{s-1} + \frac{a^{2s}}{r^{s+1}} \right] f(r) \, g(r) \, dr$$

$$= \frac{1}{2\pi i} \int_L f_+^*(\xi) \, d\xi \left(\int_a^\infty r^{s-\xi-1} \, g(r) \, dr \right)$$

$$+ \frac{1}{2\pi i} \int_a^\infty \frac{a^{2s}}{r^{s+1}} \, g(r) \, dr \int_L r^{-\xi} \, f_+^*(\xi) \, d\xi$$

Replacing ξ by $-\xi$ and using $f_+^* (\xi) = a^{2\xi} \, f_+^* (-\xi)$ we obtain

$$\int_L r^{-\xi} \, f_+^* (\xi) \, d\xi = \int_L r^{\xi} \, a^{-2\xi} \, f_+^* (\xi) \, d\xi$$

The path of integration $L : Re\ \xi = c$ becomes $Re\ \xi = -c$, but these paths can be reconciled if $f_+^*(\xi) \to 0$ as $[Im\ \xi] >> 1$. Then

$$\int_a^\infty \frac{a^{2s}}{r^{s+1}} \, f(r) \, g(r) \, dr = \frac{1}{2\pi i} \int_L f_+^*(\xi) \, d\xi \int_a^\infty \frac{a^{2s-2\xi}}{r^{s-\xi+1}} \, g(r) \, dr$$

Therefore, using these results we get

$$M_+ [f(r) \, g(r) ; r \to s] = \frac{1}{2\pi i} \int_L f_+^* (\xi) \, d\xi \int_a^\infty r^{s-\xi-1} \, g(r) \, dr$$

$$+ \frac{1}{2\pi i} \int_L f_+^*(\xi) \, d\xi \int_a^\infty \frac{a^{2s-2\xi}}{r^{s-\xi+1}} \, g(r) \, dr \quad (8.44)$$

$$= \frac{1}{2\pi i} \int_L f_+^*(\xi) \, g_+^*(s - \xi) \, d\xi .$$

Thus the proof is complete.

8.12   Finite Mellin Transform.

We conclude this chapter with an account of a finite Mellin transform due to D. Naylor (1963) as defined below :

The finite Mellin transform of first kind over the range $0 \leqslant r \leqslant a$ is defined as

$$M_1 [\, f(r); r \rightarrow s \,] = \int_0^a \left[\frac{a^{2s}}{r^{s+1}} - r^{s-1} \right] f(r) \; dr \equiv f_1^*(s) \qquad (8.45)$$

and its corresponding inversion formula is defined as

$$M_1^{-1} [\, f_1^*(s) \,; \; s \rightarrow r \,] = \frac{1}{2\pi i} \int_L r^{-s} f_+^*(s) \; ds \equiv f(r), \qquad (8.46)$$

where the contour $L : \; (c - i\infty \,, \; c + i\infty)$ in the s-plane in some strip $|Re \; s| \leqslant \gamma$ and $0 < c < \gamma$.

Also for the finite range of r in $0 \leqslant r \leqslant a$, Mellin transform of second kind is defined as

$$M_2 [\, f(r) \,; \; r \rightarrow s \,] = \int_0^a \left[\frac{a^{2s}}{r^{s+1}} + r^{s-1} \right] f(r) \; dr \equiv f_2^*(s) \qquad (8.47)$$

and its corresponding inversion formula is

$$M_2^{-1} [\, f_2^*(s) \,; \; s \rightarrow r \,] = \frac{1}{2\pi i} \int_L r^{-s} f_2^*(s) \; ds \equiv f(r) \qquad (8.48)$$

The transform pairs shown above in eqns. (8.45) - (8.48) are useful in solving boundary value problems in finite conical regions.

Other finite Mellin transform pair, the third kind, suitable for analysis of boundary value problems in spherical polar co ordinates are similarly defined.

For example, in $0 \leqslant r \leqslant a$, the third kind Mellin transform and its inversion are defined as

$$M_3 [\, f(r) \,; \; r \rightarrow s] = \int_0^a \left[\frac{a^{2s+1}}{r^{s+1}} - r^s \right] f(r) \; dr \equiv f_3^*(s) \qquad (8.49)$$

$$M_3^{-1} [\, f_3^*(s) \,; \; s \rightarrow r] = \frac{1}{2\pi i} \int_L r^{-s-1} \; f_3^*(s) \; ds \equiv f(r) \qquad (8.50)$$

The finite Mellin transform of fourth kind and its inversion formula are defined as

$$M_4 [f(r) \,; \; r \rightarrow s] = \int_0^a \left[\frac{a^{2s+2}}{r^{s+1}} + r^{s+1} \right] f(r) \; dr \equiv f_4^*(s) \qquad (8.51)$$

$$M_4^{-1} [f_4^*(s) \,; \; s \rightarrow r] = \frac{1}{2\pi i} \int_L r^{-s-1} \; f_4^*(s) \; ds \equiv f(r) \qquad (8.52)$$

Illustrations.

(a) It is easy to verify by application of "integration by parts" that

$$M_1 \left[r^2 f_{rr} + r \, f_r \; ; \; r \to s \right]$$

$$= \int_0^a \left[\frac{a^{2s}}{r^{s+1}} - r^{s-1} \right] \left[r^2 \, f_{rr} + r \, f_r \right] \, dr$$

$$= \; 2s \; a^s \; f(a) - s^2 \; f_1^*(s) \tag{8.53}$$

Clearly, this transform is useful if $f(a)$ is prescribed as one of the boundary conditions in the corresponding boundary value problems.

(b) Similarly it may be verified that

$$M_2 \left[r^2 \, f_{rr} + r \, f_r \; ; \; r \to s \right]$$

$$= \int_0^a \left[\frac{a^{2s}}{r^{s+1}} + r^{s-1} \right] \left[r^2 \, f_{rr} + r \, f_r \right] \, dr$$

$$= \; 2a^{s-1} \; f'(a) + s^2 \; f_2^*(s) \; . \tag{8.54}$$

This transform is useful when $f'(a)$ is prescribed as one of the boundary conditions in the corresponding boundary value problems.

(c) When there is an axisymmetric differential operator

$$L = \left[r^2 \, \frac{d^2}{dr^2} + 2r \, \frac{d}{dr} \right]$$

in the constitutive Laplace equation of the boundary value problem with $0 \leqslant r \leqslant a$, we can similarly, as in (a) and (b), verify that

$$M_3 \left[Lf(r) \; ; \; r \to s \right] = s(s+1) \; f_3^*(s) + 2a^{s+1} \; f(a) \tag{8.55}$$

$$\text{and} \quad M_4 \left[Lf(r) \; ; \; r \to s \right] = s(s+1) f_4^*(s) + 2a^{s+3} \; f'(a) \tag{8.56}$$

These formulae in eqns. (8.55) or (8.56) are useful in transforming the portion $Lf(r)$ of the associated axisymmetric Laplace equation of the physical problem when either $f(a)$ or $f'(a)$ are prescribed as one of the respective boundary conditions in the corresponding boundary value problems.

Exercises.

(1) Show that

(a) $M\left[\cos x ; x \to s\right] = \Gamma(s) \cos \frac{\pi s}{2}$, $0 < Re\ s < 1$

(b) $M\left[\sin x ; x \to s\right] = \Gamma(s) \sin \frac{\pi s}{2}$, $0 < Re\ s < 1$

(c) $M\left[(1+x^2)^{-1} ; x \to s\right] = \frac{\pi}{2} \operatorname{cosec} \left(\frac{\pi s}{2}\right)$, $0 < Re\ s < 2$

(d) $M\left[\frac{1}{2} \operatorname{sech} x ; x \to s\right] = \Gamma(s) L[s]$, where $L(s) = \sum_{n=1}^{\infty} \frac{1}{(2n-1)^s}$

(2) Show that

$$M\left[e^{-x \cos \varphi} \cos(x \sin \varphi) ; x \to s\right] = \Gamma(s) \cos s\ \varphi,\ Re\ s > 0\ ,\ |\varphi| < \frac{\pi}{2}$$

$$M\left[e^{-x \cos \varphi} \sin(x \sin \varphi) ; x \to s\right] = \Gamma(s) \sin s\ \varphi,\ Re\ s > -1\ ,\ |\varphi| < \frac{\pi}{2}$$

Deduce that if $|\varphi| < \frac{\pi}{2}$,

$$M^{-1}\left[\frac{\cos s\ \varphi}{\sin\ s\ \pi} ; s \to x\right] = \frac{1}{\pi} \cdot \frac{1 + x\ \cos\ \varphi}{1 + 2x \cos \varphi + x^2}$$

$$M^{-1}\left[\frac{\sin s\ \varphi}{\sin\ s\ \pi} ; s \to x\right] = \frac{1}{\pi} \cdot \frac{x\ \sin\ \varphi}{1 + 2x \cos \varphi + x^2}$$

and hence that if $|\varphi| < \frac{\pi}{2n}$, $n > 0$

$$M^{-1}\left[\frac{\cos s\ \varphi}{\sin \frac{s\pi}{n}} ; s \to x\right] = \frac{n}{\pi} \cdot \frac{1 + x^n \cos (n\varphi)}{1 + 2x^n \cos (n\varphi) + x^{2n}}$$

$$M^{-1}\left[\frac{\sin s\ \varphi}{\sin \frac{s\pi}{n}} ; s \to x\right] = \frac{n}{\pi} \cdot \frac{x^n\ \sin (n\varphi)}{1 + 2x^n \cos (n\varphi) + x^{2n}}$$

(3) Prove that

(a) $\quad M\left[\frac{2}{\sqrt{\pi}} \int_0^x \frac{f(t)\ dt}{\sqrt{x^2 - t^2}} ; x \to s\right] = \frac{\Gamma\left(\frac{-s+1}{2}\right)}{\Gamma\left(1 - \frac{s}{2}\right)} f^*(s)$

(b) $\quad M\left[\frac{2}{\sqrt{\pi}} \int_x^\infty \frac{f(t)\ dt}{\sqrt{t^2 - x^2}} ; x \to s\right] = \frac{\Gamma\left(\frac{s}{2}\right)}{\Gamma\left(\frac{s+1}{2}\right)} f^*(s)$

(4) Show that the integral equation

$$f(x) = h(x) + \int_0^\infty g(x\xi)\ f(\xi)\ d\xi$$

has the formal solution

$$f(x) = \frac{1}{2\pi i} \int_{c-i\infty}^{c+i\infty} \left[\frac{h^*(s) + g^*(s)\, h^*(1-s)}{1 - g^*(s)\, g^*(1-s)} \right] x^{-s}\, ds$$

(5) Show that

$$\sum_{n=1}^{\infty} \frac{\cos kn}{n^2} = \frac{k^2}{4} - \frac{\pi k}{2} + \frac{\pi^2}{6}$$

Hence deduce that

$$\sum_{n=1}^{\infty} \frac{1}{n^2} = \frac{\pi^2}{6}$$

(6) If $f(x) = \sum_{n=1}^{\infty} a_n\, e^{-nx}$, show that

$$M\left[f(x)\,;\; x \to s\right] = f^*(s) = \Gamma(s)\, g^*(s)$$

where $\quad g^*(s) = \sum_{n=1}^{\infty} a_n\, n^{-s}$.

If $a_n = 1$, for all n, derive

$$f^*(s) = \Gamma(s)\, \zeta(s)$$

Show also that $\quad M\left[\dfrac{\exp(-ax)}{1 - e^{-x}}\,;\; x \to s\right] = \Gamma(s)\, \zeta(s, a)$.

(7) Show that

$$\sum_{n=1}^{\infty} \frac{\cos an}{n^3} = \frac{1}{12}\left[a^3 - 3\pi a^2 + 2\pi^2 a\right]$$

(8) Prove that

(a) $\quad M\left[\displaystyle\int_0^{\infty} \xi^n\, f(x\xi)\, g(\xi)\, d\xi\,;\; x \to s\right] = f^*(s)\, g^*(1 + n - s)$

(b) $\quad M\left[\displaystyle\int_0^{\infty} \xi^n\, f\left(\frac{x}{\xi}\right) g(\xi)\, d\xi\,;\; x \to s\right] = f^*(s)\, g^*(s + n + 1)$

(9) Show that the solution of the boundary value problem defined by

$$r^2\, \varphi_{rr} + r\, \varphi_r + \varphi_{\theta\theta} = 0\,,\quad 0 < r < \infty\,,\quad 0 < \theta < \pi$$
$$\varphi(r, 0) = \varphi(r, \pi) = f(r)$$

is $\quad \varphi(r, \theta) = \dfrac{1}{2\pi i} \displaystyle\int_{c-i\infty}^{c+i\infty} r^{-s}\, \dfrac{f^*(s)\cos\left\{s\left(\theta - \frac{\pi}{2}\right)\right\} ds}{\cos\left(\frac{\pi s}{2}\right)}$

Chapter 9

Finite Laplace Transforms

9.1 Introduction.

Laplace transform of a piecewise continuous function $f(t)$ for $0 \leqslant t < \infty$, where $|f(t)| < M e^{\sigma\,t}, M, \sigma$ being real constants defined in section 3.3 by

$$F(s) = \int_0^\infty f(t)\, e^{-st}\, dt$$

exists when $Re\ s > \sigma$. But from the above definition it can not be concluded that Laplace transform of any general form of $f(t)$ do exist. For example, if $f(t) = e^{at^2}$, $a > 0$, its Laplace transform does not exist. Again from the physical point of view there are situations where determining the response of a linear system with disturbance $f(t)$ of any general form in $0 \leqslant t \leqslant T$, T is finite is also a pertinent question for discussion.

To study all such situations raised above the power and elegance of Finite Laplace transform defined below may be found very useful. In this direction the papers of H.S. Dunn (Proc. Camb. Phil. Soc. 63, 1967) and L. Debnath and J. Thomas (ZAMM, 56, 1976) may be consulted as basic and pioneering works.

9.2 Definition of Finite Laplace Transform.

Finite Laplace transform of a continuous function or of a piecewise continuous function $f(t)$ in $0 \leqslant t \leqslant T$ satisfying the Lipschitz condition is denoted by

$$L_T\left[\, f\ (t)\ ;\ t \to s\ \right] = \bar{f}(s, T)$$

and is defined by

$$L_T[f(t)\ ;\ t \to s] = \bar{f}(s, T) = \int_0^T f(t)\, e^{-st}\, dt, \tag{9.1}$$

where s is any real or complex number and T is a finite real number.

The inverse finite Laplace transform is defined by the complex integral

$$f(t) = L_T^{-1} \left[\bar{f}(s,T) \; ; \; s \to t \right] = \frac{1}{2\pi i} \int_{c-i\infty}^{c+i\infty} \bar{f}(s,T) \, e^{st} \, ds, \qquad (9.2)$$

where the contour of integration denoted by Γ is a straight line joining points $c - iR$ and $c + iR$ as $R \to \infty$ and when the function $f(t)$ is continuous in $0 \leqslant t \leqslant T$. If $f(t)$ is a piecewise continuous function with a finite number of finite discontinuities in $0 \leqslant t \leqslant T$, then the corresponding inverse finite transform of $\bar{f}(s,T)$ at t is given by

$$\frac{1}{2\pi i} \int_{\Gamma} \bar{f}(s,T) \, e^{-st} \, ds = \frac{1}{2}[\, f(t-0) + f(t+0)] \qquad (9.3)$$

as $R \to \infty$. This formula (9.3) is obtained due to the fact that $\bar{f}(s,T)$ is an entire function of the complex variable s.

Note. 9.1. Let $\int f(t) \, e^{-st} dt = -F(s,t) \, e^{-st}.$

Then $\bar{f}(s,t) = \displaystyle\int_0^T f(t) \, e^{-st} \, dt = F(s,0) - F(s,T) \, e^{-st}, \qquad (9.4)$

where $\quad F(s,0) = \bar{f}(s) = \displaystyle\int_0^\infty f(t). \, e^{-st} \, dt \qquad (9.5)$

Therefore, from (9.2) and (9.4) we can write

$$f(t) = \frac{1}{2\pi i} \int_{\Gamma} F(s,0) \, e^{st} \, ds - \frac{1}{2\pi i} \int_{\Gamma} F(s,T) \, e^{s(t-T)} \, dt, \qquad (9.6)$$

where Γ is a line contour from $c - iR$ to $c + iR$ as $R \to \infty$.

The first integral in (9.6) may be closed on the left half plane. For $t < T$, the contour of the second integral is closed in the right half plane. So, selectintg Γ such that all the poles of $F(s,0)$ lie to the left of Γ and hence the first integral is the solution of the initial value problem, the second integral being zero in this case. When $t > T$, the second integral is closed in the left half of the complex s-plane so that $f(t) = 0$. Thus, there is no need to consider the second integral and this is identical with usual Laplace transform.

Note. 9.2. Following the definitions of usual Laplace transform and Finite Laplace transform it may be noted that if the usual Laplace transform of a function $f(t)$ exists, then its Finite Laplace transform must exist. This is so because

$$\bar{f}(s) = L[f(t); t \to s] = \int_0^\infty f(t)e^{-st}dt = \int_0^T e^{-st}f(t)dt + \int_T^\infty e^{-st}f(t)dt$$

and the existence of the left hand side integral implies the existence of the second integral of the right hand side of the above equation. But the converse of this statement is not true. Since, for example

$L[e^{at^2}; t \to s] = \int_0^\infty e^{at^2}e^{-st} dt$, for $a > 0$ does not exist though $\int_0^T e^{at^2} e^{-st}dt$ does exist for $a > 0$.

9.3   Finite Laplace Transform of elementary functions.

Determine finite Laplace transforms of $f(t)$ in the following examples supplied in 9.1-9.10

Example 9.1. $f(t) = 1$

Solution.

$$L_T[1; t \to s] = \int_0^T 1e^{-st} dt = \frac{1}{s}[1 - e^{-sT}] \tag{9.7}$$

Example 9.2. $f(t) = e^{at}$

Solution.

$$L_T[e^{at}; t \to s] = \int_0^T e^{-(s-a)t} dt = \frac{1 - e^{-(s-a)T}}{s - a} \tag{9.8}$$

Example 9.3. $f(t) = \sin at$

Solution.

$$L_T[\sin at; t \to s] = \int_0^T \sin at.\, e^{-st} dt$$

$$= \frac{a}{s^2 + a^2} - \frac{e^{-sT}}{s^2 + a^2}(s \sin aT + a \cos aT) \tag{9.9}$$

Example 9.4. $f(t) = \cos at$

Solution.

$$L_T[\cos at; t \to s] = \int_0^T \cos at.\, e^{-st} dt$$

$$= \frac{s}{s^2 + a^2} + \frac{e^{-sT}}{s^2 + a^2}[a \sin aT - s \cos aT] \tag{9.10}$$

Example 9.5. $f(t) = t$

Solution.

$$L_T[\, t;\ t \to s\,] \ = \ \int_0^T t\, e^{-st}\, dt$$

$$= \ \frac{1}{s^2} - \frac{e^{-sT}}{s} \ \ [\frac{1}{s} + T\,] \tag{9.11}$$

Example 9.6. $f(t) = t^n$, $n =$ positive integer.

Solution.

$$L_T[\, t^n;\ t \to s\,] = \int_0^T t^n\, e^{-st}\, dt$$

$$= \frac{n!}{s^{n+1}} - \frac{e^{-sT}}{s}\ [\, T^n + \frac{n}{s}\, T^{n-1} + \frac{n(n-1)}{s^2}\, T^{n-2}$$

$$+ \cdots + \frac{n!\, T}{s^{n-1}} + \frac{n!}{s^n}\,]\,, \tag{9.12}$$

after using integration by parts successively.

Example 9.7. $f(t) = t^a$, $a > -1$

Solution.

$$L_T[\, t^a;\ t \to s\,] = \int_0^T t^a.\, e^{-st}\, dt \quad , \text{ putting } u = st$$

$$= s^{-(a+1)} \int_0^{sT} u^a\, e^{-u}\, du$$

$$= s^{-(a+1)} \gamma\, (a+1\,,\ sT)\,, \tag{9.13}$$

by use of incomplete gamma function defined by

$$\gamma\, (\alpha, x) = \int_0^x e^{-u}\, u^{\alpha-1}\, du$$

Example 9.8. $f(t) =$ a periodic function of t of period w.

Solution. We have $f(t+nw) = f(t)$, for all integer value of n.

Now,

$$L_T[f(t)\ ;\ t \to s] = \int_0^T f(t)\, e^{-st}\, dt$$

$$- \int_0^{nw} f(t)\, e^{-st}\, dt$$

$$= \int_0^w f(t)e^{-st}dt + \int_w^{2w} f(t) e^{-st}dt + \cdots + \int_{(n-1)w}^{nw} f(t) e^{-st} dt$$

$$= \int_0^w f(u) e^{-su} du + e^{-sw} \int_0^w f(u) e^{-su} du + \cdots$$

$$+ e^{-s(n-1)w} \int_0^w f(u)e^{-su} du$$

$$= \frac{1 - e^{-nsw}}{1 - e^{-sw}} \int_0^w e^{-su} f(u) du$$

$$= \frac{1 - e^{-nsw}}{1 - e^{-sw}} \bar{f}(s,w) \tag{9.14}$$

When $n \to \infty$, the above result reduces to

$$\bar{f}(s) = \frac{\int_0^w f(u)e^{-su}du}{(1 - e^{-sw})},$$

which is a known result when $T \to \infty$(i, e $n \to \infty$) , for usual Laplace Transform.

Example 9.9. $f(t) = erf(at)$

Solution. $L_T[\, erf(at) ; t \to s\,] = \int_0^T erf(at) e^{-st} dt$

Defining the Error function by $erf(x) = \frac{2}{\sqrt{\pi}} \int_0^x e^{-\alpha^2} d\alpha$, we have

$$\int_0^T erf(at).e^{-st} dt$$

$$= -\left[\frac{e^{-st}}{s} erf(at) \right]_0^T + \frac{1}{s}\frac{2}{\sqrt{\pi}} \int_0^T exp\left[-\{\, st + a^2t^2 \,\}\right] dt$$

$$= -\frac{e^{-st}}{s} erf(aT) + \frac{1}{s} exp\left(\frac{s^2}{4a^2} \right) \frac{2}{\sqrt{\pi}} \int_{\frac{1}{2a}}^{aT+\frac{1}{2a}} e^{-u^2} du$$

$$= -\frac{e^{-sT}}{s} erf(aT) + \frac{1}{s} exp\left(\frac{s^2}{4a^2}\right) \left[erf\left(aT + \frac{s}{2a}\right) - erf\left(\frac{s}{2a}\right) \right]. \tag{9.15}$$

Example 9.10. $f(t) = H(t - a)$

Solution.

$$L_T[\, H(t - a) ; t \to s\,] = \int_0^T H(t - a) e^{-st} dt$$

$$= \int_a^T e^{-st} dt = \frac{1}{s}(e^{-as} - e^{-Ta}). \tag{9.16}$$

Example 9.11. Find finite Laplace Transform of

(i) $f(t) = \cosh at, a > 0$ (ii) $f(t) = t^4 e^{-at}, a > 0$

Solutions.

(i) $L_T [\cosh at \; ; \; t \to s] = \int_0^T \dfrac{1}{2} (e^{at} + e^{-at}) \cdot e^{-st} \, dt$

$= \dfrac{1}{2} \dfrac{1 - e^{-(s-a)T}}{s - a} + \dfrac{1}{2} \dfrac{1 - e^{-(s+a)T}}{s + a}$

$= \dfrac{s}{s^2 - a^2} - e^{-st} \left[\dfrac{e^{aT}}{s - a} + \dfrac{e^{-aT}}{s + a} \right]$

(ii) $L_T [t^4 e^{-at} \; ; \; t \to s] = \int_0^T t^4 e^{-at} e^{-st} \, dt$

$= \int_0^T t^4 e^{-(s+a)t} \, dt$

$= \dfrac{4!}{(s + a)^5} - \dfrac{e^{-(s+a)T}}{s + a} \left[T^4 + \dfrac{4 \, T^3}{s + a} + \dfrac{12 \, T^2}{(s + a)^2} + \dfrac{24 \, T}{(s + a)^3} + \dfrac{24}{(s + a)^4} \right]$

Example 9.12. If $f(x)$ has a discontinuity at $t = a$ where $0 < a < T$, prove that

$L_T [f'(t) \; ; \; t \to s] = s\bar{f} \, (s, T) + e^{-sT} f(T) - f(0) - e^{-sa} [f(a + 0) - f(a - 0)]$

Solution.

$L_T [f'(t) \; ; \; t \to s]$

$= \int_0^{a-0} f'(t) e^{-st} \, dt + \int_{a+0}^T f'(t) \, e^{-st} \, dt$

$= \left[e^{-st} f(t) \right]_0^{a-0} + s \int_0^{a-0} f(t) e^{-st} dt + \left[e^{-st} f(t) \right]_{a+0}^T + s \int_{a+0}^T e^{-st} f(t) dt$

$= s\bar{f}(s, T) + e^{-sT} f(T) - f(0) - e^{-sa} [f(a + 0) - f(a - 0)] .$

When $f(t)$ has a finite number k of finite discontinuities at $t = a_1, a_2, \cdots a_k$ this result can be generalised after application of the above method in a similar manner.

9.4   Operational Properties.

Theorem 9.1. If $L_T [f(t) \; ; \; t \to s] = \bar{f}(s, T)$, then

$$L_T \{ e^{-at} f(t) \; ; \; t \to s \} = \bar{f}(s + a \; , \; T)$$

This property of finite Laplace transform is called its shifting property.

Proof. It is given that $\bar{f}(s,T) = \int_0^T f(t)e^{-st}\, dt$.

Therefore,

$$
\begin{aligned}
L_T\left[e^{-at}\, f(t)\; ;\; t \to s\right] &= \int_0^T f(t)\,.\,e^{-at}.e^{-st}\, dt \\
&= \int_0^T f(t)\,.\,e^{-(s+a)t}.\, dt \\
&= \bar{f}\,(s+a,\; T)
\end{aligned}
\tag{9.17}
$$

Theorem 9.2. If $L_T\left[\; f(t)\; ;\; t \to s\;\right] = \bar{f}(s,T)$,

then $L_T\left[\; f(at)\; ;\; t \to s)\;\right] = \frac{1}{a}\bar{f}\,\left(\frac{s}{a}\,,aT\right)$

This property of finite Laplace transform is called its scaling property.

Proof. Since, $L_T\left[\; f(t)\; ;\; t \to s\;\right] = \bar{f}(s,T)$, we have

$$
\begin{aligned}
L_T[f(at)\; ;\; t \to s\;] &= \int_0^T f(at)\,.\,e^{-st}\, dt \\
&= \frac{1}{a} \int_0^{aT} f(x)\,.\,e^{-\frac{s}{a}x}.\, dx, \quad \text{where } at = x \\
&= \frac{1}{a}\bar{f}\left(\frac{s}{a}\,,aT\,\right).
\end{aligned}
\tag{9.18}
$$

Theorem 9.3. (Finite Laplace transform of Derivatives)

If $\quad L_T[f(t)\; ;\; t \to s\;] = \bar{f}\,(s,T)$,

then $\quad L_T[f'(t)\; ;\; t \to s\;] = s\bar{f}\,(s,T)\; -\; f(0) + e^{-sT}\, f(T)$

$L_T[f''(t)\; ;\; t \to s\;] = s^2\bar{f}\,(s,T)\; -\; sf(0) - f'(0) + sf(T)e^{-sT} + f'(T)e^{-sT}$

and generally, $\; L_T[\; f^{(n)}(t)\; ;\; t \to s\;] = s^n\bar{f}(s,T) - \displaystyle\sum_{k=1}^{n} s^{n-k}\, f^{(k-1)}(0)$

$$
+e^{-sT}\sum_{k=1}^{n} s^{n-k}\, f^{(k-1)}\,(T)\; .
$$

Proof. Applying integration by parts, we have

$$
\begin{aligned}
L_T\left[\; f'\,(t)\; ;\; t \to s\;\right] &= \int_0^T f'\,(t)\, e^{-st}\, dt \\
&= s\bar{f}\,(s,T) - f(0) + e^{-sT}\, f(T)
\end{aligned}
\tag{9.19}
$$

Let $\quad f'(t) \equiv \varphi(t)$. Then $f''(t) = \varphi'\,(t)$ and hence

$$LT\ [f''(t)\ ;\ t \to s] = LT\ [\varphi'(t)\ ;\ t \to s]$$
$$= s\ LT\ [\varphi(t)\ ;\ t \to s] - \varphi(0) + e^{-sT}\ \varphi(T)$$
$$= s[s\bar{f}(s,T) - f(0) + e^{-sT}\ f(T)] - f'(0) + e^{-sT}\ f'(T)$$
$$= s^2\ \bar{f}(s,T) - s\ f(0) - f'(0) + e^{-sT}[s\ f(T) + f'(T)]\ (9.20)$$

Repeating the above process $(n-1)$ times, the last formula can be proved easily.

Theorem 9.4. (Finite Laplace transform of Integrals).
If $LT[f(t)\ ;\ t \to s] = \bar{f}(s,T),$ then

$$LT\left[\int_0^t f(u)\ du\right] = \frac{1}{s}\left[\bar{f}(s,T) - e^{-sT}\int_0^T f(u)\ du\right]$$

Proof. Let the integral

$$\int_0^t f(u)\ du = \varphi(t),\ \text{so that}\ \varphi'(t) = f(t)$$

Now, $\quad LT[\varphi'(t)\ ;\ t \to s] = s\ LT[\varphi(t)\ ;\ t \to s] - \varphi(0) + e^{-sT}\ \varphi(T)$

$$\Rightarrow \quad LT[f(t)\ ;\ t \to s] = s\ LT\left[\int_0^t f(u)\ du\ ;\ t \to s\right] + e^{-sT}\ \varphi(T)$$

$$\Rightarrow \quad LT\left[\int_0^t f(u)\ du\ ;\ t \to s\right] = \frac{1}{s}\left[\bar{f}(s,T) - e^{-sT}\int_0^T f(u)\ du\right]$$

$$(9.21)$$

Theorem 9.5. (Derivative of finite Laplace transform).

If $\quad LT\ [f(t)\ ;\ t \to s] = \bar{f}(s,T),$ then

$$\frac{d}{ds}\ [\bar{f}(s,T)] = LT\ [(-t)f(t)\ ;\ t \to s]$$

$$\frac{d^2}{ds^2}\ [\bar{f}(s,T)] = LT\ [(-t)^2 f(t)\ ;\ t \to s]$$

and generally $\quad \dfrac{d^n}{ds^n}\ [\bar{f}(s,T)] = LT\ [(-t)^n\ f(t)\ ;\ t \to s]\ .$

Proof.

Since, $\quad \bar{f}(s,T) = \int_0^T e^{-st}\ f(t)\ dt$

we have $\quad \dfrac{d}{ds}\ \bar{f}(s,T) = \int_0^T \dfrac{\partial}{\partial s}\ e^{-st} \cdot f(t)\ dt$

$$= \int_0^T e^{-st} (-t) \, f(t) \, dt$$

$$= L_T[(-t) \, f(t) \, ; \, t \to s] \qquad (9.22)$$

$$\frac{d^2}{ds^2} \bar{f}(s,t) = \int_0^T \frac{\partial^2}{\partial s^2} e^{-st} \cdot f(t) \, dt$$

$$= \int_0^T e^{-st} \cdot (-t)^2 f(t) dt$$

$$= L_T[(-t)^2 f(t) \, ; \, t \to s] \, . \qquad (9.23)$$

In general, similarly we get

$$\frac{d^n}{ds^n} \bar{f}(s,T) = \int_0^T \frac{\partial^n}{\partial s^n} e^{-st} \cdot f(t) \, dt$$

$$= \int_0^T e^{-st} (-t)^n \, f(t) \, dt$$

$$= L_T \left[(-t)^n f(t) \, ; \, t \to s \right] . \qquad (9.24)$$

Theorem 9.6. Finite Laplace transform of the product of t^n and m th derivatives of $f(t)$ is

$$(-1)^n \frac{d^n}{ds^n} \left[s^m \bar{f}(s,t) - \sum_{k=1}^m s^{m-k} f^{k-1}(0) + e^{-sT} \sum_{k=1}^m s^{m-k} f^{(k-1)}(T) \right] .$$

Proof.

We have $\qquad L_T \left[t^n f'(t) \, ; \, t \to s \right] = (-1)^n \frac{d^n}{ds^n} \left[L_T \{ f'(t) \, ; \, t \to s \} \right]$

$$= (-1)^n \frac{d^n}{ds^n} \left[s \, \bar{f}(s,T) - f(0) + e^{-sT} \, f(T) \right]$$

$$= (-1)^n \left[\frac{d^n}{ds^n} \{ s \, \bar{f}(s,T) \} \right] + (T)^n \, e^{-sT} \, f(T) \qquad (9.25)$$

Similarly, $\qquad L_T \left[t^n f^{(m)}(t) \, ; \, t \to s \right]$

$$= (-1)^n \frac{d^n}{ds^n} \left[s^m \bar{f}(s,T) - \sum_{k=1}^m s^{m-k} \, f^{(k-1)} (0) + e^{-sT} \sum_{k=1}^m s^{m-k} f^{(k-1)}(T) \right]$$

$$\qquad\qquad (9.26)$$

Theorem 9.7. Integral of finite Laplace transform of a function is given by

$$\int_s^T \bar{f}(s,T) ds = \int_0^T \frac{f(t)}{t} e^{-st} dt - \int_0^T \frac{f(t)}{t} e^{-tT} dt.$$

Proof.

Let $\quad \bar{f}(s,T) = L_T[f(t) \; ; \; t \to s] = \int_0^T f(t) \cdot e^{-st} \, dt$

Then $\displaystyle \int_s^T \bar{f}(s,T) \, ds = \int_s^T ds \int_0^T f(t) \, e^{-st} \, dt$

$$= \int_0^T f(t) \, dt \int_s^T e^{-st} \, ds$$

$$= \int_0^T \frac{f(t)}{t} \, e^{-st} \, dt - \int_0^T \frac{f(t)}{t} \, e^{-tT} \, dt \, , \qquad (9.27)$$

provided that both the integrals on the right hand side of eqn. (9.27) exist.

9.5   The Initial Value and the Final Value Theorem .

These theorems give the behavior of the object functions in terms of the behavior of the transformed functions

Theorem 9.8. (The Initial Value Theorem).

If $f(t)$ be at most a piecewise continuous function, for $0 \leq t \leq T$ and if $L_T[f(t) \; ; \; t \to s] = \bar{f}(s,T)$ then,

$$\lim_{s \to \infty} [s\bar{f}\,(s,T)\,] = \lim_{t \to 0} f(t) \qquad (9.28)$$

Proof. The proof of the above theorem can be developed like that in the theorem of section 3.15.

Theorem 9.9. (The Final value Theorem)

If $f(t)$ be at most a piecewise continuous function for $0 \leqslant t \leqslant T$ and if $L_T[f(t) \; ; \; t \to s\,] = \bar{f}(s,T)$ then,

$$\lim_{t \to \infty} f(t) = \lim_{\substack{s \to 0}} s\bar{f}\,(s,T) \qquad (9.29)$$

Proof. The proof of this theorem can also be developed as that in the Theorem of section 3.16

Theorem 9.10. The solution of an initial value problem is identical with that of the final value problem.

Proof. Suppose $f_{in}(t)$ is the solution of the initial value problem, and it is given by

$$f_{in}(t) = \frac{1}{2\pi i} \int_\Gamma F\,(s,0) \, e^{st} \, ds \qquad (9.30)$$

where Γ is the Bromwich contour extending from $c - iR$ to $c + iR$ as $R \to \infty$.

Again suppose $f_{fi}(t)$ is the solution of the finial value problem and is defined by

$$f_{fi}(t) = \frac{1}{2\pi i} \int_\Gamma F(s, T) e^{s(t-T)} ds \qquad (9.31)$$

where Γ lies to the left of the singularities of $F(s, t)$ or of $F(s, 0)$. Then from eqns. (9.27) and (9.28) we get

$$
\begin{aligned}
f_{in}(t) - f_{fi}(t) &= \frac{1}{2\pi i} \int_C [\, F(s, 0) - F(s, T) \, e^{-sT} \,] \, e^{st} \, ds \\
&= \frac{1}{2\pi i} \int_C \bar{f}(s, T) \, e^{st} \, ds \qquad (9.32)
\end{aligned}
$$

where C is a closed contour which contains all the singularities of $F(s, 0)$ or of $F(s, T)$ and $F(s, t)$ is defined by

$$-F(s, t) e^{-st} = \int f(t) \, e^{-st} \, dt$$

The integrand in eqn. (9.32) is an entire function of s and hence by Cauchy's theorem, the integral must vanish.

Hence.

$$f_{in}(t) = f_{fi}(t) = f(t) \qquad (9.33)$$

9.6 Applications

Example 9.13. Verify Theorems 9.8 and 9.9 when $f(t) = \exp(-t)$.

Solution.

Let $\quad L_T[f(t) \; ; \; t \to s] = \bar{f}(s, T)$. Then for $f(t) = \exp(-t)$,

we have $\quad \bar{f}(s, T) = \dfrac{1 - e^{-(1+s)T}}{(1 + s)}$. Therefore,

$$\lim_{s \to \infty} \bar{f}(s, T) = 0, \quad \text{by L'Hospital's rule. Again,}$$

$$\lim_{s \to \infty} s \, \bar{f}(s, T) = \lim_{s \to \infty} (1 + s - 1) \frac{(1 - e^{-(s+1)T})}{1 + s}$$

$$= \lim_{s \to \infty} [1 - e^{-(s+1)T}] - \lim_{s \to \infty} \frac{1 - e^{-(s+1)T}}{1 + s}$$

$$= 1 - 0 = 1$$

Also, $\quad \lim_{t \to 0} f(t) = \lim_{t \to 0} e^{-t} = 1$

Thus $\quad \lim_{s \to \infty} s \ \bar{f}(s, T) = \lim_{t \to 0} f(t)$ is verified.

Also, $\quad \lim_{s \to 0} \bar{f}(s, T) = \lim_{s \to 0} \dfrac{1 - e^{-(1+s)T}}{1 + s} = 1 - e^{-T}$

and $\quad \displaystyle\int_{0}^{T} f(t) \ dt = 1 - e^{-T}$

$\therefore \quad \lim_{s \to 0} \bar{f}(s, T) = \displaystyle\int_{0}^{T} f(t) \ dt \quad$ is verified

Further, $\quad \lim_{s \to 0} s \ \bar{f}(s, T) = \lim_{s \to 0} s \ \dfrac{\left[1 - e^{-(1+s)T}\right]}{s + 1} = 0$

Thus both the theorems 9.8 and 9.9 are verified.

Example 9.14. Use finite Laplace transform to solve the initial value problem

$$\frac{di}{dt} = A \ , \ 0 \le t \le T \ ; \ i(0) = 0 \ .$$

Solution. Finite Laplace transform of the ODE under the initial condition $x(0) = 0$ gives

$$s \ \bar{i} \ (s, T) + i(T) \ e^{-sT} = \frac{A}{s} \left[1 - e^{-sT}\right]$$

or $\quad \bar{i} \ (s, T) + \dfrac{e^{-sT}}{s} \ i(T) = A \left[\dfrac{1 - e^{-sT}}{s^2}\right]$

Here $\bar{i} \ (s, T)$ is not an entire function unless we choose $i(T) = AT$. Then we have

$$\begin{aligned}
\bar{i} \ (s, T) &= \frac{-AT}{s} \left[e^{-sT} \right] + \frac{A}{s^2} \left(1 - e^{-sT}\right) \\
&= A \left[\frac{1}{s^2} - \frac{e^{-sT}}{s} \left(\frac{1}{s} + T\right) \right]
\end{aligned}$$

Inverting finite Laplace transform the above equation gives

$$i(t) = At.$$

Example. 9.15. The equation for vertical displacement of horizontal taut string between two fixed points $x = 0$ and $x = L$ caused by a

concentrated or distributed load $W(x)$ normalised with the tension of
the string is

$$\frac{d^2y}{dx^2} = W(x) , \ 0 \le x \le L$$

Find $y(x)$.

Solution. The above boundary value problem under the above statement is subjected to the boundary conditions

$$y(0) = y(L) = 0$$

Let us assume here that the given load is a concentrated load of unit
magnitude at $x = a$ and therefore.

$$W(x) = \delta(x - a) , \ 0 < a < L$$

Taking finite Laplace transform over $x \in (0, L)$ the above differential
equation under the loading $W(x) = \delta(x - a)$ gives

$$\bar{y}(s, L) = \frac{1}{s^2} \left[e^{-sa} + y'(0) - e^{-sL} \, y'(L) \right]$$

This function $\bar{y}(s, L)$ is not an entire function of s unless we choose
$y'(0) = y'(L) - 1$. Using this condition the above equation can be
expressed as

$$\bar{y}(s, L) = \frac{y'(0)}{s^2} \left[1 - e^{-sL} - sL \, e^{-sL} \right] + \frac{e^{-sa}}{s^2} \left[1 - e^{-s(L-a)} \right.$$
$$\left. + y'(0) \, sL \, e^{-s(L-a)} \right]$$

To complete the inversion of finite Laplace transform in the above equation we assume $Ly'(0) = a - L$ so that we get

$$y(x) = xy'(0) + (x - a) \, H \, (x - a)$$

Example 9.16. In an $L - R$ electrical circuit the current $I(t)$ satisfied
the ODE

$$L \frac{dI}{dt} + RI = E_0 \cos \, wt , \ 0 \leqslant t \leqslant T,$$

under the emf $E_0 \cos \, wt$ and $I(t) = 0$ at $t = 0$; L, R, E_0 being constants.
Find $I(t)$ using finite Laplace transform.

Solution.

Applying finite Laplace transform the given ODE becomes

$$s\bar{I}(s,T) + e^{-sT}I(T) + \frac{R}{L}\bar{I}(s,T) = \frac{E_0}{L}\left[\frac{s}{s^2+w^2}\right.$$
$$\left. + \frac{e^{-sT}}{s^2+w^2}(w\sin wT - s\cos wT)\right]$$

$$\bar{I}(s,T) = -\frac{e^{-sT}I(T)}{s+\frac{R}{L}} + \frac{E_0}{L(s+\frac{R}{L})}\left[\frac{s}{s^2+w^2}\right.$$
$$\left. + \frac{e^{-sT}}{s^2+w^2}(w\sin wT - s\cos wT)\right]$$

Since $\bar{I}(s,T)$ is not an entire function of s, we make it entire by choosing

$$I(T) = \frac{E_0}{L}\frac{w}{w^2+\frac{R^2}{L^2}}\left[\frac{R}{wL}\cos wT + \sin wT - \frac{R\,e^{-\frac{RT}{L}}}{wL}\right]$$

Using this result we get

$$\bar{I}(s,T) = \frac{w\,E_0}{L\left(w^2+\frac{R^2}{L^2}\right)}\left[\frac{R}{wL}\left\{\frac{s}{s^2+w^2} + \frac{e^{-sT}}{s^2+w^2}(w\sin wT - s\cos wT)\right\}\right]$$
$$+ \frac{w\,E_0}{L\left(w^2+\frac{R^2}{L^2}\right)}\left[\left\{\frac{w}{s^2+w^2} - \frac{e^{-sT}}{s^2+w^2}(s\sin wT + w\cos wT)\right\}\right]$$
$$- \frac{w\,E_0}{L\left(w^2+\frac{R^2}{L^2}\right)}\left[\frac{R}{wL(s+\frac{R}{L})}\left\{1 - e^{-(s+\frac{R}{L})T}\right\}\right]$$

Inverting finite Laplace transformed equation, the above relation gives

$$I(t) = \frac{w\,E_0}{L\left(w^2+\frac{R^2}{L^2}\right)}\left\{\frac{R}{wL}\cos wt + \sin wt - \frac{R}{wL}e^{-\frac{Rt}{L}}\right\}$$

It may be noted that the first two terms on the right hand side of the above equation represent the steady-state current and the third term there represents the transient part of the current field $I(t)$ at any time t.

Exercises.

(1) Find finite Laplace transform of

$$f(t) = \sinh \, at$$

$$\left[\text{Ans.} \quad \frac{a}{s^2-a^2} + \frac{e^{-sT}}{2}\left(\frac{e^{aT}}{a-s} + \frac{e^{-aT}}{a+s}\right)\right]$$

(2) Use finite Laplace transform to solve the initial value problem

$$\frac{dx}{dt} + \alpha \, x = At \, , \, 0 \leqslant t \leqslant T \, ; \, x(0) = a$$

to prove that

$$x(t) = \left(a + \frac{A}{\alpha^2}\right) e^{-\alpha t} + \frac{At}{\alpha} - \frac{A}{\alpha^2}$$

(3) Solve the simple harmonic oscillator governed by

$$\frac{d^2x}{dt^2} + w^2x = F \, , \, x(0) = a \, , \, \dot{x}(0) = u$$

where F, a and u are constants, to prove that

$$x(t) = \left(a - \frac{F}{w^2}\right) \cos \, wt + \frac{u}{w^2} \sin \, wt + \frac{F}{w^2} \, .$$

Chapter 10

Legendre Transforms

10.1 Introduction.

As an aid to solve some special type of boundary value problems this chapter deals with the Legendre transform whose kernel is a Legendre polynomial. Initially, definition and basic operational properties of the transform are discussed based on the papers of Churchill (J.Math. and Physics, 33. 165, 1954) and Churchill and Dolph (Proc. Amer. Math, Sec, 5, 93, 1954).

10.2 Definition of Legendre Transform.

Legendre transform of a function $f(x)$ over the interval $-1 \leqslant x \leqslant 1$ is defined by

$$\bar{f}_l(n) \equiv l\,[\,f(x)\,;\,x \to n\,] = \int_{-1}^{1} f(x)\,P_n(x)\,dx\;, \qquad (10.1)$$

where $P_n(x)$ is the Legendre polynomial of integral degree $n(\geqslant 0)$ provided that the integral in (10.1) exists.

The orthogonal property of Legendre polynomial is well-known as

$$\int_{-1}^{1} P_m(x)\,P_n(x)\,dx - \frac{2}{2n+1}\,\delta_{mn} \qquad (10.2)$$

and this can also be expressed as

$$l\,[\,P_m(x)\,;\,x \to n\,] = \frac{2}{2n+1}\,\delta_{mn} \qquad (10.3)$$

From eqns. (10.2) and (10.3) Fourier-Legendre expansion theorem of a piecewise continuous function $f(x)$ for $-1 \leqslant x \leqslant 1$ is given by

$$\sum_{n=0}^{\infty} \left(n + \frac{1}{2}\right) P_n(x) \bar{f}_l(n) = f(x) ,$$

if $f(x) \in C\ (-1,1) = \frac{1}{2}\ [f(x+0) + f(x-0)]$, if $f(x) \in P^1(-1,1)$

$$(10.4)$$

We, therefore, have the inversion formula for Legendre transform is

$$l^{-1}\left[\ \bar{f}_l\ (n)\ ;\ n \to x\ \right] = \sum_{n=0}^{\infty} \left(n + \frac{1}{2}\right) \bar{f}_l(n)\ P_n(x) \qquad (10.5)$$

10.3   Elementary properties of Legendre Transforms.

(1) Since $P_0(x) = 1$ and $P_1(x) = x$ we get

$$\bar{f}_l(0) = \int_{-1}^{1} f(x)\ dx \qquad\qquad (10.6)$$

and $\quad \bar{f}_l(1) = \int_{-1}^{1} x f(x)\ dx \qquad\qquad (10.7)$

Again, $\quad l[f'(x)\ ;\ x \to 0] = \int_{-1}^{1} f'(x) dx = f(1) - f(-1)$

$$(10.8)$$

$$l[f'(x)\ ;\ x \to 1] = \int_{-1}^{1} x f'(x)\ dx = f(1) + f(-1)$$

$$- \int_{-1}^{1} f(x)\ dx \qquad (10.9)$$

Also, $\quad l[f''(x)\ ;\ x \to 1] = f'(1) + f'(-1) - \int_{-1}^{1} f'(x) dx$

$$= f'(1) + f'(-1) - f(1) + f(-1) \quad (10.10)$$

and $\quad l[C\ ;\ x \to 0] = \int_{-1}^{1} C\ dx = 2C \qquad (10.11)$

$$l[C\ ;\ x \to n] = \int_{-1}^{1} C\ P_n(x) dx$$

$$= C \int_{-1}^{1} P_n(x) dx = C \int_{-1}^{1} P_0(x)\ P_n(x)\ dx$$

$\Rightarrow \quad l[C\ ;\ x \to n] = 0$, when $n = 1, 2, 3, \cdots$

(by putting $m = 0$ in eqn. (10.3)). $\qquad\qquad (10.12)$

(2) Again since, $P_n(-x) = (-1)^n\, P_n(x)$ we have

$$l[f(-x)\;;\; x \to n] = \int_{-1}^{1} f(-x)\,P_n(x)\,dx$$

$$= \int_{-1}^{1} f(x)\,P_n(-x)\,dx$$

$$= (-1)^n \int_{-1}^{1} f(x)\,P_n(x)\,dx$$

$$= (-1)^n\, l[f(x)\;;\; x \to n]$$

From this result we can write

$$l^{-1}\left[(-1)^n\,\bar{f}_l(n)\;;\; n \to x\right] = f(-x) \qquad (10.13)$$

(3) Legendre transform of $e^{i\alpha x}$ is given by

$$l_n\left[e^{i\alpha x}\;;\; x \to n\right] = \int_{-1}^{1} e^{i\alpha x}\,P_n(x)\,dx = \sqrt{\frac{2\pi}{\alpha}}\; i^n\, J_{n+\frac{1}{2}}(\alpha),$$

$$(10.14)$$

from table of integrals (Copson, 1935). Similarly,

$$l_n\left[e^{\alpha x}\;;\; x \to n\right] = \sqrt{\frac{2\pi}{\alpha}}\; I_{n+\frac{1}{2}}(\alpha) \qquad (10.15)$$

(4) Also

$$l\left[(1-x^2)^{-\frac{1}{2}}\;;\; x \to n\right] = \int_{-1}^{1} \frac{P_n(x)dx}{\sqrt{1-x^2}} = \pi\, P_n^2(0) \quad (10.16)$$

and $\quad l\left[\frac{1}{2(t-x)}\;;\; x \to n\right] = \frac{1}{2}\int_{-1}^{1} \frac{P_n(x)\,dx}{t-x} = Q_n(t)\;,\; |t| > 1$

$$(10.17)$$

These results are also obtained from "Tables of integral Transforms" by Harry Batemann (1954).

(5) From the generating relation of Legendre polynomial we have

$$(1 - 2rx + r^2)^{-\frac{1}{2}} = \sum_{n=0}^{\infty} r^n\, P_n(x)\;,\; |r| < 1$$

Multiplying both sides by $P_n(x)$ and integrating the result with respect to x from -1 to 1, we get

$$\int_{-1}^{1} (1 - 2rx + r^2)^{-\frac{1}{2}} P_n(x) \, dx = r^n \int_{-1}^{1} P_n^2(x) \, dx$$

$$= \frac{2r^n}{(2n+1)} \qquad (10.18)$$

When $\quad r = 1, \quad$ we get

$$\int_{-1}^{1} [2(1-x)]^{-\frac{1}{2}} P_n(x) \, dx = \frac{2}{2n+1}$$

$$\Rightarrow \quad \int_{-1}^{1} (1-x)^{-\frac{1}{2}} P_n(x) \, dx = \frac{2\sqrt{2}}{2n+1}$$

Or $\quad l \left[(1-x)^{-\frac{1}{2}} \, ; \, x \to n \, \right] = \frac{2\sqrt{2}}{2n+1} \qquad (10.19)$

Defferentiating eqn. (10.18) with respect to r and after multiplying the result by r, we get

$$\frac{1}{2} \int_{-1}^{1} (1 - 2rx + r^2)^{-\frac{3}{2}} (2rx - 2r^2) \, P_n(x) \, dx = \frac{2nr^n}{2n+1}$$

$$\Rightarrow \quad \frac{1}{2} \int_{-1}^{1} (1 - 2rx + r^2)^{-\frac{3}{2}} \left[-\{ 1 - 2rx + r^2 + r^2 - 1 \} \right] P_n(x) dx$$

$$= \frac{2nr^n}{2n+1}$$

$$\Rightarrow \quad -\frac{1}{2} \int_{-1}^{1} (1 - 2rx + r^2)^{-\frac{1}{2}} P_n(x)$$

$$+ \frac{(1 - r^2)}{2} \int_{-1}^{1} (1 - 2rx + r^2)^{-\frac{3}{2}} P_n(x) \, dx$$

$$= \frac{2nr^n}{2n+1}$$

$$\Rightarrow \quad -l \left[(1 - 2rx + r^2)^{-\frac{1}{2}} \, ; \, x \to n \, \right]$$

$$+ (1 - r^2) \, l \left[(1 - 2rx + r^2)^{-\frac{3}{2}} \, ; \, x \to n \right]$$

$$= \frac{4nr^n}{2n+1}$$

Now using eqn. (10.18), the above result reduces to

$$l[(1 - 2rx + r^2)^{-\frac{3}{2}} \, ; \, x \to n]$$

$$= \left[\frac{4nr^n}{2n+1} + \frac{2r^n}{2n+1} \right] \frac{1}{1 - r^2}$$

$$= \frac{2r^n}{1 - r^2} \tag{10.20}$$

(6) Legendre transform of $H(x)$ is given by

$$l\,[\,H(x)\;;\;x\to n\,]\;\;=\;\;\int_{-1}^{1} H\,(x)\;P_n(x)\;dx$$

$$=\;\;\int_{0}^{1} H(x)\;P_n(x)\;dx$$

$$=\;\;\int_{0}^{1} P_n(x)\;dx$$

If $n = 0$, $P_n(x) = P_0(x) = 1$ and hence $l[\,H(x)\;;\;x\to n\,] = 1$.

If $n > 1$, we know from the recurrence relation of Legendre polynomial that

$$(2n+1)\;P_n(x) = P'_{n+1}\,(x) - P'_{n-1}\,(x) \tag{10.21}$$

Therefore, in this case $\quad \displaystyle\int_{0}^{1} P_n\,(x)\;dx$

$$= \frac{1}{2n+1}\int_{0}^{1}[\,P'_{n+1}(x) - P'_{n-1}\,(x)\,]\;dx$$

$$= \frac{1}{2n+1}[P_{n-1}\,(0) - P_{n+1}\,(0)]$$

Thus, $l[\,H(x)\;;\;x\to n\;] = \begin{cases} 1\,, & \text{for } n = 0 \\ \frac{1}{2n+1}\,[P_{n-1}\,(0) - P_{n+1}\,(0)]\,, & \text{for } n \geqslant 1 \end{cases}$

$$\tag{10.22}$$

(7) From eqn. (10.21) we have

$$\int_{-1}^{1} g(x)\;P_n(x)\;dx = \frac{1}{2n+1}\left[\int_{-1}^{1} g(x)P'_{n+1}\,(x)\;dx \right.$$

$$\left. - \int_{-1}^{1} g(x)\;P'_{n-1}\,(x)\;dx\right]$$

$$= \frac{1}{2n+1}\left[\,g(x)\;P_{n+1}\,(x)|_{-1}^{1} - \int_{-1}^{1} g'(x)\;P_{n+1}\,(x)\;dx \right.$$

$$\left. -g(x)\;P'_{n-1}\,(x)|_{-1}^{1} + \int_{-1}^{1} g'(x)\;P_{n-1}(x)\;dx\right]$$

$$= \frac{1}{2n+1} \left[g(x) \left\{ P_{n+1}(x) - P_{n-1}(x) \right\}^1_{-1} \right.$$

$$\left. - \int_{-1}^1 g'(x) \left\{ P_{n+1}(x) - P_{n-1}(x) \right\} dx \right]$$

Thus if $g(x) = \int_{-1}^x f(t) dt$ is a continuous function in $(-1, 1)$

$$l\left[g(x) ; x \to n \right] = \frac{1}{2n+1} \left[l\{ g'(x) ; x \to (n-1) \} \right]$$

$$- \frac{1}{2n+1} \left[l\{ g'(x) ; x \to n+1 \} \right] \qquad (10.23)$$

provided the expression $[\cdots]^1_{-1}$ vanishes. Thus, in particular, we have the special case

$$l\left[\int_{-1}^x f(t) \, dt ; x \to n \right] = \frac{1}{2n+1} \left[\bar{f}_l(n-1) - \bar{f}_l(n+1) \right]$$

$$\qquad (10.24)$$

Therefore, by repeated application of (10.8) and (10.9) and by (10.23) we get

$$l\left[f'(x) ; x \to 2m \right]$$

$$= f(1) - f(-1) - \sum_{r=0}^{m-1} (4m - 4r - 1) \, \bar{f}_l (2m - 2r - 1)$$

$$\qquad (10.25)$$

$$l\left[f'(x) ; x \to 2m+1 \right]$$

$$= f(1) - f(-1) - \sum_{r=0}^{m-1} (4m - 4r + 1) \, \bar{f}_l (2m - 2r)$$

$$\qquad (10.26)$$

Also from the recurrence relation

$$(n+1) P_{n+1}(x) - (2n+1) \, x \, P_n(x) + n \, P_{n-1}(x) = 0$$

we can easily deduce that

$$l\left[xf(x) ; x \to n \right] = \frac{1}{2n+1} \left[(n+1) \, \bar{f}_l (n+1) + n \, \bar{f}_l (n) \right]$$

$$\qquad (10.27)$$

10.4 Operational Properties of Legendre Transforms

Theorem 10.1 If $f'(x)$ is continuous and $f''(x)$ is bounded and integrable in each subinterval of $-1 \leqslant x \leqslant 1$ and if Legendre transform of $f(x)$ exists and

$$\lim_{|x| \to 1} (1 - x^2) \left[f(x) , f'(x) \right] = 0$$

then $\quad l \left[\dfrac{d}{dx} \{ (1 - x^2) \dfrac{d}{dx} f(x) \} ; x \to n \right] = -(n+1) n \, \bar{f}_l \, (n).$

Proof. We have

$$l \left[\frac{d}{dx} \left\{ (1 - x^2) \frac{d}{dx} f(x) \right\} ; x \to n \right]$$

$$= \int_{-1}^{1} \frac{d}{dx} \left\{ (1 - x^2) \frac{d}{dx} f(x) \right\} P_n (x) \, dx.$$

$$= - \int_{-1}^{1} (1 - x^2) \frac{d}{dx} f(x) P_n'(x) \, dx ,$$

on integration by parts and using the given condition.

Again on integration by part the above result becomes

$$= - \left[(1 - x^2) P_n' (x) f(x) |_{-1}^{1} - \int_{-1}^{1} \frac{d}{dx} \left\{ (1 - x^2) P_n' (x) \right\} f(x) \, dx \right]$$

$$= \int_{-1}^{1} \frac{d}{dx} \left\{ (1 - x^2) \frac{d}{dx} P_n (x) \right\} f(x) \, dx , \text{ by using given condition.}$$

But it is known that

$$\frac{d}{dx} \left[(1 - x^2) \frac{d}{dx} P_n (x) \right] = -n(n+1) P_n(x)$$

and therefore,

$$l \left[\frac{d}{dx} \left\{ (1 - x^2) \frac{df(x)}{dx} \right\} ; x \to n \right] = -n (n+1) \int_{-1}^{1} f(x) P_n(x) \, dx$$

$$= -n(n+1) \, \bar{f}_l(n) \qquad (10.28)$$

Note 10.1. If we denote the operator R by

$$Rf(x) \equiv \frac{d}{dx} \left[(1 - x^2) \frac{d}{dx} f (x) \right]$$

we can extend the result of theorem 10.1 as

$$l\left[R^2 f(x) ; x \to n \right] = (-1)^2 n^2 (n+1)^2 \bar{f}_l(n)$$
$$l\left[R^3 f(x) ; x \to n \right] = (-1)^3 n^3 (n+1)^3 \bar{f}_l(n)$$

$$\cdots \qquad\qquad \cdots$$

$$l\left[R^k f(x) ; x \to n \right] = (-1)^k n^k (n+1)^k \bar{f}_l(n) \qquad (10.29)$$

Corollary 10.1. It is seen that $n(n+1) = \left(n+\frac{1}{2}\right)^2 - \frac{1}{4}$. Therefore we have

$$l\left[R f(x) ; x \to n\right] = -n(n+1)\, \bar{f}_l(n)$$
$$= -\left[\left(n+\frac{1}{2}\right)^2 - \frac{1}{4}\right] \bar{f}_l(n)$$
$$= -\left(n+\frac{1}{2}\right)^2 \bar{f}_l(n) + \frac{1}{4}\, \bar{f}_l(n)$$

Thus, $\quad l\left[\frac{1}{4}f(x) - R(f(x)) ; x \to n\right] = \left(n+\frac{1}{2}\right)^2 \bar{f}_l(n) \qquad (10.30)$

Extending this result, we get

$$(-1)^k\, l\left[R^k(f(x)) - 4^k ; x \to n\right] = \sum_{r=0}^{k-1} (-1)^r\, k_{c_r} \left[4^{-r}\left(n+\frac{1}{2}\right)^{2k-2r}\right] \bar{f}_l(n)$$

Theorem 10.2. (Convolution Theorem).

$$\text{If} \quad l\left[f(x) ; x \to n\right] = \bar{f}_l(n) \text{ and } l[g(x) ; x \to n] = \bar{g}_l(n)$$
$$\text{then} \quad l^{-1}\left[\bar{f}_l(n)\bar{g}_l(n) ; n \to x\right] = f(x) * g(x) \qquad (10.31)$$

where $f(x)*g(x)$ is called the convolution of $f(x)$ and $g(x)$ and is defined by

$$f(x) * g(x) = \frac{1}{\pi}\int_0^\pi f(\cos\lambda)\,\sin\,\lambda\, d\lambda \int_0^\pi g\,(\cos(\eta))\, d\beta \qquad (10.32)$$

with $x = \cos\mu$ and $\cos\eta = \cos\lambda\,\cos\mu + \sin\lambda\,\sin\mu\,\cos\beta$

Proof. We have

$$\bar{f}_l(n)\,\bar{g}_l(n) = \int_0^\pi f(\cos\lambda)P_n(\cos\lambda)(\sin\lambda)d\lambda \int_0^\pi g(\cos\eta)P_n(\cos\eta)\sin\eta\, d\eta$$
$$= \int_0^\pi f(\cos\lambda)\sin\lambda \left[\int_0^\pi g(\cos\eta)P_n(\cos\eta)P_n(\cos\lambda)\sin\eta\right] d\lambda$$

where $f(x) = f(\cos\lambda)$ and $g(x) = g\,(\cos\eta)$

Further, since

$$P_n(\cos \eta)\, P_n(\cos \lambda) = \frac{1}{\pi} \int_0^\pi P_n(\cos \mu)\, d\alpha$$

where $\cos \mu = \cos \eta\, \cos \lambda + \sin \eta\, \sin \lambda\, \cos \alpha$, we can rewrite

$$\bar{f}_l(n)\bar{g}_l(n) = \frac{1}{\pi} \int_0^\pi f(\cos \lambda) \sin \lambda \left[\int_0^\pi \int_0^\pi g(\cos \lambda) P_n(\cos \lambda) \sin \eta\, d\alpha\, d\eta \right] d\lambda$$

Now the double integral on the right hand of above equation is given by

$$\int_0^\pi \int_0^\pi g(\cos \lambda\, \cos \mu + \sin \lambda\, \sin \mu\, \cos \beta)\, P_n(\cos \mu) \sin \mu\, d\mu$$

Using this result and changing the order of integration, we get

$$\bar{f}_l(n)\bar{g}_l(n) = \frac{1}{\pi} \int_0^\pi P_n(\cos \mu) \sin \mu \left[\int_0^\pi \int_0^\pi f(\cos \lambda) \sin \lambda\, g(\cos \eta) d\lambda\, d\beta \right] d\mu$$

$$= \int_0^\pi h(\cos \mu)\, P_n(\cos \mu)\, \sin \mu\, d\mu,$$

where $\cos \eta = \cos \lambda\, \cos \mu + \sin \lambda\, \sin \mu\, \cos \beta$

and $h(\cos \mu) = \dfrac{1}{\pi} \displaystyle\int_0^\pi f(\cos \lambda)\, \sin \lambda\, d\lambda \int_0^\pi g(\cos \eta)\, d\beta$

Thus the theorem is proved.

In particular, if $\mu = 0$ we get from (10.32) that

$$h(1) = \int_{-1}^1 f(t)\, g(t)\, dt \tag{10.33}$$

and when $\mu = \pi$, $h(-1) = \displaystyle\int_{-1}^1 f(t)\, g(-t)\, dt$ (10.34)

10.5 Application to Boundary Value Problems.

As an illustration we consider application of finite Legendre transform in the solution of interior Dirichlet problem for the potential $u(r, \theta)$ inside a unit sphere $(r = 1)$ satisfying the partial differential equation

$$\frac{\partial}{\partial r}\left[r^2 \frac{\partial u}{\partial r} \right] + \frac{\partial}{\partial x}\left[(1 - x^2)\frac{\partial u}{\partial x} \right] = 0\ ,\ \ 0 < r \leqslant 1 \tag{10.35}$$

with boundary condition

$$u(1, x) = f(x)\ ,\ \ -1 < x < 1 \tag{10.36}$$

where $x = \cos \theta$.

Applying finite Legendre transform over the variable x, the PDE (10.35) and the boundary condition (10.36) reduce to

$$r^2 \frac{d^2 \bar{u}_l(r,n)}{dr^2} + 2r \frac{d\bar{u}_l(r,n)}{dr} - n(n+1) \, \bar{u}_l(r,n) = 0 \qquad (10.37)$$

and $\bar{u}_l(1,n) = \bar{f}_l(n)$ (10.38)

respectively, where $\bar{u}_l(r,n)$ is a continuous function of r in $0 < r \leqslant 1$

The bounded solution of (10.37) and (10.38) is

$$\bar{u}_l(r,n) = \bar{f}_l(n) \, r^n \; , \; 0 < r \leqslant 1 \; , \; \text{where} \; n \; \text{is a non} - \text{negative integer.}$$

On inversion the solution of the problem is given by

$$u(r,x) = \sum_{n=0}^{\infty} \left(n + \frac{1}{2} \right) \bar{f}_l(n) \, r^n \, P_n(x) \; , \; 0 < r \leqslant 1 \; , \; |x| < 1 \qquad (10.39)$$

Another representation of this solution can also be obtained through application of convolution theorem and also the formula in equation (10.20) as

$$\begin{aligned} u(r, \cos \theta) &= l^{-1} \left[\, \bar{f}_l(n) \, r^n \; ; \; n \to x = \cos \theta \, \right] \\ &= \frac{1}{2\pi} \int_0^\pi f(\cos \lambda) \sin \lambda \, d\lambda \int_0^\pi \frac{(1 - r^2) d\eta}{(1 - 2r \cos \mu + r^2)^{\frac{3}{2}}} \end{aligned}$$
$$(10.40)$$

where $\cos \mu = \cos \lambda \, \cos \theta + \sin \lambda \, \sin \theta \, \cos \eta$

Integral in (10.40) is called the Poisson integral formula of the potential inside the unit sphere.

The solution of exterior Dirichlet problem under boundary condition $w(1,x) = f(x)$, $x = \cos \theta$ of the unit sphere $r = 1$ is then given by

$$\bar{w}_l \, (r,n) = \frac{1}{r} \, \bar{f}_l(n) \, r^{-n} \; , \; n = 0, 1, 2, \cdots$$

On inversion it gives

$$\begin{aligned} w \, (r, \cos \theta) &= \frac{1}{r} \, w \left(\frac{1}{r}, \cos \theta \right) \; , \; r > 1 \\ &= \frac{1}{2\pi} \int_0^\pi f(\cos \lambda) \sin \lambda \, d\lambda \int_0^\pi \frac{(r^2 - 1) \, d\eta}{(1 - 2r \, \cos \mu + r^2)^{\frac{3}{2}}} \end{aligned}$$
$$(10.41)$$

where $\cos\mu = \cos\lambda\,\cos\theta + \sin\lambda\,\sin\theta\,\cos\eta$ and $w(r\,,\,\cos\theta)$ is the solution of the exterior problem.

Exercises.

(1) (a) Show that

$$l\,[\,x^n\,;\,x\to n\,] = \frac{2^{n+1}\,(n!)^2}{(2n+1)!}$$

(b) Using the result in eqn. (10.27) find $l\,[\,x^2 f(x)\,;\,x\to n\,]$

(2) Using the definition of odd Legendre transform as

$$\bar{f}_{2n+1} \equiv l_0\,[\,f(x)\,;\,x\to n\,] = \int_0^1 f(x)\,P_{2n+1}\,(x)\,dx$$

and the even Legendre transform as

$$\bar{f}_{2n} \equiv l_e\,[\,f(x)\,;\,x\to n\,] = \int_0^1 f(x)\,P_{2n}\,(x)\,dx$$

and their respective inversion formulae

$$\bar{l}_0\,[\,\bar{f}_{2n+1}\,;\,n\to x\,] \equiv f(x) = \sum_{n=0}^{\infty}(4n+3)\,\bar{f}_{2n+1}\,P_{2n+1}(x),\ 0\leqslant x\leqslant 1,$$

$$l_e^{-1}\,[\,\bar{f}_{2n}\,;\,n\to x\,] \equiv f(x) = \sum_{n=0}^{\infty}(4n+1)\,\bar{f}_{2n}\,P_{2n}(x)\,,\ 0\leqslant x\leqslant 1$$

and denoting the operator $L = \frac{\partial}{\partial x}\left[\,(1-x^2)\,\frac{\partial}{\partial x}\,\right]$

prove that

$$l_0\,[\,L\,f(x)\,;\,x\to n\,] = (2n+1)\,P_{2n}(0)\,f(0)$$
$$-(2n+1)(2n+2)\,\bar{f}_{2n+1}$$

and $\quad l_e\,[\,L\,f(x)\,;\,x\to n\,] = -P_{2n}(0)\,f'(0) - 2n(2n+1)\,\bar{f}_{2n}$

(3) Prove that the Legendre transformed solution of the Dirichlet boundary value problem for $u(r,\theta)$ defined by

$$\frac{\partial^2 u}{\partial r^2} + \frac{1}{r}\frac{\partial u}{\partial r} + (1-x^2)\frac{\partial^2 u}{\partial x^2} - 2x\,\frac{\partial u}{\partial x} = 0\,,\ 0\leqslant r\leqslant a\,,\ 0\leqslant\theta\leqslant\pi$$

$$u\,(a,\theta) = f(x)\,,\ 0\leqslant\theta\leqslant\pi$$

where $x = \cos\theta$, is given by

$$\bar{u}_l\,(r,n) = \left(\frac{r}{a}\right)^n\,\bar{f}_l(n)\,.$$

Chapter 11

The Kontorovich-Lebedev Transform

11.1 Introduction.

In discussing the variable separable solution of Boundary value Problems in cylindrical co-ordinates for wage type region Kontorovich-Lebedev transform has been found application. To understand the applicability of the transform some elementary properties of the modified Bessel function of the second kind, also called Macdonald function, is needed. We present below only operational proof of the results at various stages of required theorems for easy understanding of the readers in engineering and physics.

11.2 Definition of Kontorovich - Lebedev Transform.

Kontorovich-Lebedev transform of a function $f(x)$ over the positive real line is denoted by $\bar{f}(\tau)$ and is defined by

$$K[f(x) \; ; \; x \to \tau] \equiv \bar{f}(\tau) = \int_0^\infty x^{-1} \, f(x) K_{i\tau}(x) \, dx, \quad (x \geqslant 0) \quad (11.1)$$

The corresponding inversion theorem of Kontorovich-Lebedev transform is given by

$$K^{-1} \left[\bar{f}(\tau) \; ; \; \tau \to x \right] \equiv f(x) = \frac{2}{\pi^2} \int_0^\infty K_{i\tau}(x) \, \tau \, \sinh(\pi\tau) \bar{f}(\tau) \, d\tau, \quad (11.2)$$

if $f(x)$ is a function defined on the positive real line such that $\frac{1}{x} f(x)$ is continuously differentiable and $x f(x)$ and $x \frac{d}{dx}[\frac{1}{x} f(x)]$ are absolutely integrable over the positive real line.

If $f(x)$ is a piecewise continuous function over $(0 \leqslant x < \infty]$ having a finite discontinuity at x, then (11.2) becomes

$$
\begin{aligned}
K^{-1} \left[\bar{f}(\tau) \; ; \; \tau \to x \right] &= \frac{2}{\pi^2} \int_0^\infty K_{i\tau}(x) \tau \, \sinh(\pi\tau) \bar{f}(\tau) \, d\tau \\
&= \frac{1}{2} \left[f(x+0) + f(x-0) \right] \quad (11.3)
\end{aligned}
$$

We present below an operational proof of the above inversion theorem, since the rigorous proof is a bit complicated.

We know that the cosine transform of $K_{i\tau}(x)$ is given by

$$F_c[\ K_{i\tau}(x)\ ;\ \tau \rightarrow t\ ] = \sqrt{\frac{\pi}{2}}\ .\ e^{-x\,\cosh t} \qquad (11.4)$$

Therefore, we have

$$F_c\left[K\left\{f(x)\ ;\ x \rightarrow \tau\right\}\ ;\ \tau \rightarrow t\right]$$
$$= \sqrt{\frac{\pi}{2}} \int_0^\infty e^{-x\,\mathrm{ch}\,t}\ \frac{f(x)}{x}\ dx \qquad (11.5)$$

This result can also be expressed as Laplace transform of $\frac{1}{x}\ f(x)$ as

$$\mathfrak{L}\left[\frac{f(x)}{x}\ ;\ x \rightarrow p\right] = \frac{2}{\pi} \int_0^\infty \bar{f}(\tau)\ \cos\left(\tau\ \cosh^{-1} p\ \right)\ d\tau \qquad (11.6)$$

But since, $\quad \mathfrak{L}\left[\ f(x)\ ;\ x \rightarrow p\ \right] = \frac{-\partial}{\partial p}\ \mathfrak{L}\left[\frac{f(x)}{x}\ ;\ x \rightarrow p\right]$

we have from eqn. (11.6) that

$$\mathfrak{L}[f(x)\ ;\ x \rightarrow p\ ] = \frac{2}{\pi} \int_0^\infty \tau \bar{f}(\tau)\frac{\sin\left(\tau\,\cosh^{-1} p\right)}{\sqrt{p^2 - 1}}\ d\tau \qquad (11.7)$$

Taking Laplace inversion of both sides of the above equation we obtain

$$K^{-1}[\bar{f}(\tau)\ ;\ \tau \rightarrow x\ ] \equiv f(x) = \frac{2}{\pi^2} \int_0^\infty \tau \sinh\left(\pi\tau\right)\ K_{i\tau}\ (x)\ \bar{f}\ (\tau)\ d\tau$$
$$(11.8)$$

as the inversion formula for Kontorovich-Lebedev transform.

11.3 Parseval Relation for Kontorovich-Lebedev Transforms.

Theorem 11.1. If $\bar{f}(\tau)$ and $\bar{g}(\tau)$ are respectively Kontorovich-Labedev transforms of $f(x)$ and $g(x)$ respectively, $x \geqslant 0$ then

$$\frac{2}{\pi^2} \int_0^\infty \tau \sinh\left(\pi\tau\right)\ \bar{f}(\tau)\bar{g}(\tau)\ d\tau = \int_0^\infty x^{-1}f(x)\ g(x)\ dx \qquad (11.9)$$

Proof. We have

$$\frac{2}{\pi^2} \int_0^\infty \tau \sinh (\pi\tau) \, \bar{f} \, (\tau) \bar{g} \, (\tau) \, d\tau$$

$$= \frac{2}{\pi^2} \int_0^\infty \tau \sinh (\pi\tau) \, \bar{f} \, (\tau) \, d\tau \int_0^\infty x^{-1} \, g(x) \, K_{i\tau} \, (x) \, dx$$

$$= \int_0^\infty x^{-1} \, g(x) \, dx . \frac{2}{\pi^2} \int_0^\infty \tau \sinh (\pi\tau) \, \bar{f} \, (\tau) \, K_{i\tau} \, (x) \, d\tau$$

$$= \int_0^\infty x^{-1} g(x) \, f(x) \, dx \ , \ \text{by (11.8)}$$

This proves the Parseval relation of Kontorovich-Lebedev transform.

11.4   Illustrative Examples.

Example 11.1. Prove that

$$K \left[te^{-t \cos x} \; ; \; t \to \tau \right] = \frac{\pi \sinh(x\tau)}{\sinh(\pi\tau) \sin x}$$

and hence deduce the values of $K[t \; ; \; t \to \tau]$ and $K[te^{-t} \; ; \; t \to \tau]$.

Solution. By definition in (11.1) we have

$$K \left[t \, e^{-t \cos x} \; ; \; t \to \tau \right] = \int_0^\infty \frac{1}{t} . \, t \, e^{-t \cos x} \, K_{i\tau} \, (t) \, dt$$

$$= \int_0^\infty K_{i\tau} \, (t) \, e^{-pt} \, dt \ , \ \text{where} \ \ p = \cos x$$

$$= \mathcal{L} \left[\, K_{i\tau}(t) \; ; \; t \to p \, \right] = \frac{\pi \, \sinh \, (x\tau)}{\sinh \, (\pi\tau) \, \sin \, x} \ ,$$

using table of integral transforms.

Thus, in particular, if $x = \frac{\pi}{2}$ and $x = 0$ we get respectively
$K[t \; ; \; t \to \tau] = \frac{\pi}{2} \operatorname{sech} \frac{\pi\tau}{2}$ and $K[t \, e^{-t} \; ; \; t \to \tau] = \pi\tau \operatorname{cosech} (\pi\tau)$

Example 11.2. Prove that

$$K \left[x \, \sin \alpha \, e^{-x \cos \alpha} \; ; \; x \to \tau \right] = \frac{\pi sh(\alpha\tau)}{sh(\pi\tau)}$$

$$\text{and} \quad K \left[e^{-xcht} \; ; \; t \to \tau \right] = \frac{\pi \cos (\tau t)}{\tau \, sh \, (\pi\tau)}$$

Solution. Proceeding as the solution of Example 11.1, the above results can also be proved easily.

Example 11.3. Using the results of Example 11.2 deduce that

(a) $$\int_0^\infty e^{-x} f(x) \, dx = \frac{2}{\pi} \int_0^\infty \tau^2 \, \bar{f}(\tau) \, d\tau$$

(b) $$\int_0^\infty e^{-x(1+cht)} \, dx = 2 \int_0^\infty \frac{\cos(\tau t) \, d\tau}{sh(\pi\tau)}$$

and hence deduce that

$$F_c \, [\, \tau \, \text{cosech} \, (\pi\tau) \, ; \, \tau \to t \,] = \frac{1}{2\sqrt{2\pi}} \, \text{sech}^2 \left(\frac{t}{2} \right)$$

Solutions.

(a) Taking $g(x) = x \, \sin \, \alpha \, e^{-x \cos \alpha}$, we have from the table of integral transform of Copson that

$$\bar{g}(\tau) = K[\, g(x) \, ; \, x \to \tau \,] = \frac{\pi \, sh \, (\alpha\tau)}{sh \, (\pi\tau)}$$

Therefore, Parseval relation for Kontorovich-Lebedev transforms

$$\int_0^\infty x^{-1} \, f(x) \, g(x) \, dx = \frac{2}{\pi^2} \int_0^\infty \tau \, sh(\pi\tau) \, \bar{f}(\tau) \, \bar{g}(\tau) \, d\tau \, ,$$

for this $g(x)$, gives

$$\int_0^\infty x^{-1} f(x) \cdot x \sin \alpha \, e^{-x \cos \alpha} \, dx = \frac{2}{\pi^2} \int_0^\infty \tau \sinh(\pi\tau) \, \bar{f}(\tau) \cdot \frac{\pi sh(\alpha\tau)}{sh(\pi\tau)} d\tau$$

$$\Rightarrow \int_0^\infty e^{-x \cos \alpha} \, f(x) \, dx = \frac{2}{\pi} \int_0^\infty \tau \frac{sh(\alpha\tau)}{\sin \alpha} \, \bar{f}(\tau) \, d\tau$$

Now letting $\alpha \to 0$, we get

$$\int_0^\infty e^{-x} \, f(x) \, dx = \frac{2}{\pi} \int_0^\infty \tau^2 \bar{f}(\tau) \, d\tau$$

(b) Now taking $f(x) = e^{-xcht}$ we get

$$\bar{f}(\tau) = \frac{\pi}{\tau} \frac{\cos(\tau t)}{sh \, (\pi\tau)}$$

Then using these relations in the final result of (a), we get

$$\int_0^\infty e^{-x(1+cht)} dx = \frac{2}{\pi} \int_0^\infty \tau^2 \, \frac{\pi}{\tau} \frac{\cos(\tau t)}{sh(\pi\tau)} \, d\tau$$

$$\Rightarrow \quad \frac{1}{2} \operatorname{sech}^2 \frac{t}{2} = 2 \int_0^\infty \frac{\tau \cos (\tau t)}{\operatorname{sh} (\pi \tau)} \, d\tau$$

$$= \sqrt{2\pi} \, F_c \left[\, \tau \operatorname{cosech} (\pi \tau) \; ; \; \tau \to t \, \right]$$

$$\Rightarrow \quad F_c \left[\, \tau \operatorname{cosech} (\pi \tau) \; ; \; \tau \to t \, \right] = \frac{1}{2\sqrt{2\pi}} \operatorname{sech}^2 \left(\frac{t}{2} \right) .$$

11.5   Boundary Value Problem in a wedge of finite thickness.

Consider the boundary value problem of determining a harmonic function $u(\rho, \varphi, z)$ in a finite wedge defined by

$$\rho \geqslant 0 \; , \quad 0 \leqslant \varphi \leqslant \alpha \; , \quad 0 \leqslant z \leqslant a \tag{11.10}$$

when the function u satisfies the boundary conditions

$$\left. \begin{array}{c} u \, (\rho \, , \; \alpha \, , \; z) = f \, (\rho \, , \; z) \, , \\ u \, (\rho \, , \varphi \, , \; 0) = u \, (\rho \, , \; \varphi \, , \; a) = 0 \\ \text{and} \quad u \, (\rho \, , \; 0 \, , \; z) = 0 \, . \end{array} \right\} \tag{11.11}$$

Since $u \, (\rho \, , \; \varphi \, , \; z)$ is a harmonic function, it satisfies the PDE

$$\frac{\partial^2 u}{\partial \rho^2} + \frac{1}{\rho} \frac{\partial u}{\partial \rho} + \frac{1}{\rho^2} \frac{\partial^2 u}{\partial \varphi^2} + \frac{\partial^2 u}{\partial z^2} = 0 \tag{11.12}$$

We seek a variable separable non-zero solution of (11.12) in the form

$$u \, (\rho \, , \; \varphi \, , \; z) = R(\rho) \, \Phi(\varphi) \, Z(z) \neq 0$$

so that after substituting the above value of u, differential equation (11.12) becomes

$$\frac{1}{R} \frac{d^2 R}{d\rho^2} + \frac{1}{\rho R} \frac{dR}{d\rho} + \frac{1}{\Phi \, \rho^2} \frac{d^2 \Phi}{d\varphi^2} = -\frac{1}{Z} \frac{d^2 Z}{dz^2} = \sigma^2, \text{ say} \tag{11.13}$$

Therefore, $Z(z) = A(\sigma) \, \cos \, \sigma z + B(\sigma) \, \sin \, \sigma z$ \hfill (11.14)

and $$\frac{\rho^2}{R} \frac{d^2 R}{d\rho^2} + \frac{\rho}{R} \frac{dR}{d\rho} - \sigma^2 \rho^2 = \frac{-1}{\Phi} \frac{d^2 \Phi}{d\phi^2} = -\tau^2, \text{ say} \tag{11.15}$$

Therefore, $\Phi \, (\phi) = C(\tau) \, \operatorname{ch} \, (\tau \phi) + D(\tau) \, \operatorname{sh}(\tau \phi)$ \hfill (11.16)

Also from (11.15) we have

$$\rho^2 \frac{d^2 R}{d\rho^2} + \rho \frac{dR}{d\rho} - (\sigma^2 \rho^2 - \tau^2) R = 0 \tag{11.17}$$

Putting $\tau = -i\lambda \Rightarrow \lambda = i\tau$ eqn. (11.17) becomes

$$\rho^2 \frac{d^2 R}{d\rho^2} + \rho \frac{dR}{d\rho} - (\lambda^2 + \sigma^2 \rho^2) R = 0$$

One solution of this equation is $K_\lambda\,(\rho\sigma) = K_{i\tau}\,(\rho\sigma)$

Hence,

$$u(\rho, \phi, z) = L\,(\tau)\,K_{i\tau}(\rho\sigma)\,[A(\sigma)\ \cos\sigma z + B(\sigma)\ \sin\sigma z]$$
$$[C\,(\tau)\ ch\ \tau\phi + D\,(\tau)\ sh\ \tau\varphi\,].$$

From (11.11) since $u\,(\rho, \phi, 0) = u\,(\phi, \phi, a) = 0$, we have

$$A(\sigma) = 0 \text{ and } \sin\ \sigma a = 0 = \sin n\pi,\ n = 1, 2, \cdots$$

Also, since $u(\rho, 0, z) = 0$ we have $C\,(\tau)\ = 0$. Therefore, by method of linear superposition the solution under the last three boundary conditions in (11.11) of the problem can be expressed as

$$u(\rho, \phi, z) = \sum_{n=1}^{\infty} L(\tau)\,B(n)\,D(\tau)\sin\ \frac{n\pi z}{a}\ sh\ \tau\phi . K_{i\tau}\ \left(\frac{n\pi\rho}{a}\right)$$

for any variable value of the positive constant τ. After little adjustment of the constants $L(\tau)B(n)D(\tau)$, on linear superposition of solutions over τ the above solution $u(\rho, \phi, z)$ can also be expressed as

$$u(\rho, \phi, z) = \frac{2}{\pi^2} \sum_{n=1}^{\infty} \sin\ \frac{n\pi z}{a} \int_0^\infty \tau F_n(\tau) \frac{sh\ (\tau\phi)}{sh\ (\tau\,a)}\ K_{i\tau}\ \left(\frac{n\pi\rho}{a}\right)\ sh\ \pi\tau\ d\tau$$

$$(11.18)$$

This final form of $u\,(\rho, \phi, z)$ in (11.18) will provide a solution of the boundary value problem satisfying all the boundary conditions in (11 11), provided that we can find $F_n(\tau)$ such that

$$f(\rho, z) = \sum_{n=1}^{\infty} \sin\ \frac{n\pi z}{a}\ \left[K^{-1} \left\{\ F_n(\tau)\ ;\ \tau \rightarrow \frac{n\pi\rho}{a}\ \right\} \right]$$

Inverting as the finite Fourier transform we get

$$K^{-1} \left[\ F_n\,(\tau)\ ;\ \tau \rightarrow \frac{n\pi\rho}{a} \right] = \frac{2}{a} \int_0^a f\,(\rho, y)\ \sin\ \frac{n\pi y}{a}\ dy$$

Now replacing ρ by $\frac{ar}{n\pi}$, we see that the above equation is equivalent to

$$K^{-1}\left[\,F_n\left(\tau\right)\,;\,\tau\to r\,\right] = \frac{2}{a}\int_0^a f\left(\frac{ar}{n\pi},y\right)\sin\frac{n\pi y}{a}\,dy$$

and hence

$$
\begin{aligned}
F_n(\tau) &= \frac{2}{a}\int_0^a \sin\frac{n\pi y}{a}\,K\left[\,f\left(\frac{ar}{n\pi},\,y\right)\,;\,r\to\tau\,\right]dy \\
&= \frac{2}{a}\int_0^a \sin\frac{n\pi y}{a}\,dy\int_0^\infty r^{-1}\,f\left(\frac{ar}{n\pi},\,y\right)K_{i\tau}\,(r)\,dr
\end{aligned}
$$

Changing the variable of integration in last integration we can express it as

$$F_n(\tau) = \frac{2}{a}\int_0^a \sin\frac{n\pi y}{a}\,dy\left\{\int_0^\infty x^{-1}f(x,y)\,K_{i\tau}\left(\frac{n\pi x}{a}\right)\,dx\right\}(11.19)$$

Substituting this value of $F_n(\tau)$ in eqn. (11.18), the solution of the boundary value is obtained.

Exercises.

(1) Prove that

$$K\left[\,1\,;\,\tau\to x\,\right] = \frac{\pi}{2}\,\frac{1}{x\,\sinh\left(\frac{\pi}{2}x\right)}$$

(2) Prove that

(a) $$\int_0^\infty \left\{\tau^2\,K_{it}\,(x)\,dt\right\} = \frac{\pi}{2}\,x\,e^{-x}$$

(b) $$\int_0^\infty K_{it}\,(x)\,dt = \frac{\pi}{2}\,e^{-x}$$

(3) Prove that

$$K\left[\,e^{-x\cos t}\,;\,x\to\tau\,\right] = \frac{\pi\,\operatorname{ch}\,(t\tau)}{\tau\,\operatorname{sh}\,(\pi\tau)}$$

Hence deduce that

(a) $$K\left[\,e^{-x}\,;\,x\to\tau\,\right] = \frac{\pi}{\tau}\,\frac{1}{\operatorname{sh}(\pi\tau)}$$

(b) $$K\left[\,e^{x}\,;\,x\to\tau\,\right] = \frac{\pi}{\tau}\,\coth\,(\pi\tau)$$

(c) $$K\left[\,e^{-x\operatorname{cht}}\,;\,x\to\tau\,\right] = \frac{\pi\cos\,(t\tau)}{\tau\,\operatorname{sh}\,(\pi\tau)}$$

Chapter 12

The Mehler-Fock Transform

12.1 Introduction.

In solving some of the boundary value problems associated with the mathematical theory of elasticity application of the Mehler-Fock transform has been found. Specially in problems concerning analysis of stress field in the vicinity of external crack in elastic region these transforms have found significant applications . In this chapter our main objective is to develop some basic concepts of this transform in an elegant manner together with some useful basic properties of the associated Legendre functions of first kind.

12.2 Fock's Theorem (with weaker restriction).

Theorem 12.1 If a function $g(x)$ defined on $1 \leqslant x < \infty$ is such that the integral

$$\int_1^\infty \frac{|g(x)| \ dx}{\sqrt{(x+1)}}$$

exists, then at every point x in whose neighbourhood $g(x)$ has a bounded variation

$$\int_0^\infty P_{-\frac{1}{2}+i\tau} (x)g_0^*(\tau) \ d\tau = \frac{1}{2} [\ g(x+0) + g(x-0) \ ] \qquad (12.1)$$

where $g_0^*(\tau) = \tau \tanh (\pi\tau) \int_1^\infty P_{-\frac{1}{2}+i\tau} (x) \ g(x) \ dx \qquad (12.2)$

However, if $g(x)$ is continuous at $x \in [1, \infty)$ we have (12.1) is equivalent to

$$\int_0^\infty P_{-\frac{1}{2}+i\tau} (x)g_0^*(\tau) \ d\tau = g(x) \qquad (12.3)$$

Proof. To supply a formal proof of this form of Fock's theorem we replace $g(\cosh \alpha) = f(\alpha)$ and proceed as below.

Let us introduce a pair of operators Φ_0 and Φ_0^{-1} such that

$$\Phi_0 \left[f(\alpha) \; ; \; \alpha \to \tau \right] = f_0^*(\tau)$$
$$= \int_0^\infty f(\alpha) P_{-\frac{1}{2}+i\tau} (\cosh \alpha) \; \sinh \alpha \; d\alpha \tag{12.4}$$

and $\Phi_0^{-1} \left[f_0^*(\tau) \; ; \; \tau \to \alpha \right] = f(\alpha)$

$$= \int_0^\infty \tau \, \tanh (\pi \tau) \, P_{-\frac{1}{2}+i\tau}(\cosh \alpha) \, f_0^* (\tau) \, d\tau \tag{12.5}$$

We begin by finding the forms of Φ_0^{-1} and Φ_0. Determining these operators is equivalent to finding a solution

$$f_0^* (\tau) = \Phi_0 \left[f(\alpha) \; ; \; \alpha \to \tau \right]$$

of the integral equation (12.4). Now introducing the formula

$$P_{-\frac{1}{2}+i\tau} (\cosh \alpha) = \frac{\sqrt{2} \, cth \, (\pi\tau)}{\pi} \int_\alpha^\infty \frac{\sin (\tau t)}{\sqrt{cht - ch\alpha}} \, dt \tag{12.6}$$

in (12.5) we find

$$
\begin{aligned}
f(\alpha) &= \frac{\sqrt{2}}{\pi} \int_0^\infty \tau f_0^* (\tau) \, d\tau \int_\alpha^\infty \frac{\sin (\tau t)}{\sqrt{(cht - ch\alpha)}} \, dt \\
&= \frac{1}{\sqrt{\pi}} \int_\alpha^\infty \frac{dt}{\sqrt{ch \, t - ch \, \alpha}} \sqrt{\frac{2}{\pi}} \int_0^\infty \tau \, f_0^* (\tau) \, \sin (\tau \, t) \, d\tau \\
&= \frac{1}{\sqrt{\pi}} \int_\alpha^\infty F_s \left[\tau f_0^* (\tau) \; ; \; \tau \to t \right] \frac{dt}{\sqrt{(ch \, t - ch \, \alpha)}} \tag{12.7}
\end{aligned}
$$

where $F_s [\cdots]$ is the Fourier sine transform of $\tau f_0^* (\tau)$. Treating (12.7) as an Abel integral equation, its solution is given by

$$F_s[\tau f_0^* (\tau) \; ; \; \tau \to t] = -\frac{1}{\sqrt{\pi}} \frac{d}{dt} \int_t^\infty \frac{f(\alpha) \, d\alpha}{\sqrt{ch \, \alpha - \, cht}} \tag{12.8}$$

Again inverting (12.8) by means of Fourier sine inversion theorem we get

$$\tau \, f_0^* (\tau) = -\frac{1}{\sqrt{\pi}} F_s \left[\frac{d}{dt} \int_t^\infty \frac{f(\alpha) \, sh \, \alpha \, d\alpha}{\sqrt{ch \, \alpha - ch \, t}} \; ; \; t \to \tau \right]$$

$$= \frac{\tau}{\sqrt{\pi}} \; F_c \left[\int_t^\infty \frac{f(\alpha)\; sh\; \alpha\; d\alpha}{\sqrt{ch\; \alpha - ch\; t}} \; ; \; t \to \tau \right]$$

Therefore,

$$f_0^*(\tau) = \frac{\sqrt{2}}{\pi} \int_0^\infty \cos(\tau t) \; d\tau \int_t^\infty \frac{f(\alpha) sh\; \alpha\; d\alpha}{\sqrt{(ch\; \alpha - ch\; t)}}$$

$$= \int_0^\infty f(\alpha) \; sh\; \alpha \; d\alpha \; \frac{\sqrt{2}}{\pi} \int_0^\alpha \frac{\cos(\tau\; t)}{\sqrt{(ch\; \alpha - ch\; t)}} \; dt$$

$$\Rightarrow f_0^*(\tau) = \Phi_0[f(\alpha); \alpha \to \tau] = \int_0^\infty f(\alpha) P_{-\frac{1}{2}+i\tau}(ch\alpha) sh\alpha \; d\alpha$$

$$\tag{12.9}$$

since, $\quad P_{-\frac{1}{2}+i\tau}(ch\alpha) = \frac{\sqrt{2}}{\pi} \int_0^\alpha \frac{\cos(\tau\; t)}{\sqrt{(ch\; \alpha - ch\; t)}} \; dt$

Thus (12.4) is obtained after starting from (12.5).

If we replace $f(\alpha)$ by $g(ch\alpha)$ and if

$$g(x) = \int_0^\infty P_{-\frac{1}{2}+i\tau}(x) \; g_0(\tau) \; d\tau \; , \; x \in [1, \infty] \tag{12.10}$$

then eqn. (12.9) implies

$$g_0^*(\tau) = \tau \; th(\pi\tau) \int_1^\infty P_{-\frac{1}{2}+i\tau}(x) \; g(x) \; dx \tag{12.11}$$

This result proves the theorem for the case when $g(x)$ is a continuous function.

There are other two forms of the Fock's theorem which we need not persue here.

12.3 Mehler-Fock Transform of zero order and its properties.

The Mehler-Fock integral transform of order zero of a function $f(\alpha)$, $\alpha \in [0, \infty)$ is denoted by $f_0^*(\tau)$ and is defined by

$$f_0^*(\tau) = \Phi_0 \; [f(\alpha); \alpha \to \tau] = \int_0^\infty f(\alpha) \; P_{-\frac{1}{2}+i\tau}(ch\alpha) sh\alpha \; d\alpha \tag{12.12}$$

The inverse of this transform is given by

$$f(\alpha) = \Phi_0^{-1} \; [f_0^*(\tau) \; ; \; \tau \to \alpha] - \int_0^\infty \tau \; f_0^*(\tau) \; th(\pi\tau) \; P_{-\frac{1}{2}+i\tau}(ch\alpha) d\tau$$

$$\tag{12.13}$$

Properties.

Considering the special case $\nu = -\frac{1}{2} + i\tau$, $\theta = i\alpha$ and $m = 0$ in the F.G. Mehler's formula [Math.Ann. 18,161,1881] we find that

$$P_{-\frac{1}{2}+i\tau}\,(ch\alpha) \;=\; \frac{\sqrt{2}}{\pi}\int_0^\alpha \frac{\cos(\tau t)\,dt}{\sqrt{(ch\,\alpha - ch\,t)}} \tag{12.14}$$

$$=\; \frac{1}{\sqrt{\pi}}\sqrt{\frac{2}{\pi}}\int_0^\infty \frac{\cos(\tau t)H(\alpha - t)}{\sqrt{(ch\,\alpha - ch\,t)}}\,dt \tag{12.15}$$

and in the Lagrange formula [Mem. des. Sci. Math. Fasic 47, 1939] that

$$P_{-\frac{1}{2}+i\tau}\,(ch\alpha) \;=\; \frac{\sqrt{2}}{\pi}\,\text{cth}\,(\pi\tau)\int_\alpha^\infty \frac{\sin(\tau t)\,dt}{\sqrt{(ch\,t - ch\,\alpha)}} \tag{12.16}$$

$$=\; \frac{1}{\sqrt{\pi}}\sqrt{\frac{2}{\pi}}\,\text{cth}\,(\pi\tau)\int_0^\infty \frac{\sin(\tau t)H(t - \alpha)}{\sqrt{(ch\,t - ch\,\alpha)}}\,dt \tag{12.17}$$

Treating eqn. (12.14) as an Abel integral equation, we get after its inversion that

$$\frac{d}{dt}\left[\frac{\sin\tau t}{\tau}\right] = \cos(\tau t) = \frac{1}{\sqrt{2}}\frac{d}{dt}\int_0^t P_{-\frac{1}{2}+i\tau}(ch\alpha)\frac{sh\,\alpha\cdot d\,\alpha}{\sqrt{(ch\,t - ch\,\alpha)}}$$

$$\Rightarrow\quad \frac{\sin(\tau t)}{\tau} = \frac{1}{\sqrt{2}}\int_0^t P_{-\frac{1}{2}+i\tau}(ch\alpha)\frac{\sinh\alpha\,d\alpha}{\sqrt{(ch\,t - ch\,\alpha)}} \tag{12.18}$$

Also treating (12.16) as another Abel integral equation we get after its inversion that

$$\frac{\sqrt{2}}{\pi}\sin(\tau t)\,\text{cth}(\pi\tau) = -\frac{1}{\pi}\frac{d}{dt}\int_t^\infty \frac{P_{-\frac{1}{2}+i\tau}(ch\alpha)sh\alpha\,d\alpha}{\sqrt{\{(ch\,\alpha - ch\,t)\}}}$$

$$\Rightarrow -\frac{d}{dt}\left[\frac{\cos\tau t}{\tau}\right] = \sin\tau t = -\frac{1}{\sqrt{2}}\frac{d}{dt}\int_t^\infty P_{-\frac{1}{2}+i\tau}(ch\alpha)\cdot th(\pi\tau)\frac{sh\alpha\,d\alpha}{\sqrt{(ch\,\alpha - ch\,t)}}$$

$$\Rightarrow \frac{\cos\tau t}{\tau} = \frac{1}{\sqrt{2}}\,th\,\pi\tau\int_t^\infty \frac{P_{-\frac{1}{2}+i\tau}(ch\alpha)\cdot sh\alpha\,d\alpha}{\sqrt{(ch\,\alpha - ch\,t)}} \tag{12.19}$$

Thus from (12.18) we can have

$$\frac{1}{\sqrt{2}}\int_0^\infty H(t - \alpha)\,P_{-\frac{1}{2}+i\tau}(ch\alpha)\frac{sh\alpha\,d\alpha}{\sqrt{(ch\,t - ch\,\alpha)}} = \frac{\sin(\tau t)}{\tau}$$

The last equation can be expressed as

$$\Phi_0 \left[\frac{H(t-\alpha)}{\sqrt{(ch\ t - ch\ \alpha)}} \ ; \ \alpha \to \tau \right] = \frac{\sqrt{2}\ \sin\ \tau t}{\tau} \tag{12.20}$$

Again (12.19) gives

$$\int_0^\infty H(\alpha - t) P_{-\frac{1}{2}+i\tau}(ch\alpha) \frac{sh\alpha\ d\alpha}{\sqrt{(ch\ \alpha - ch\ t)}} = \frac{\sqrt{2}\cos(\tau t)}{\tau}\ cth\ (\pi\tau)$$

This equation can also be expressed as

$$\Phi_0 \left[\frac{H(\alpha - t)}{\sqrt{(ch\ \alpha - ch\ t)}} \ ; \ \alpha \to \tau \right] = \frac{\sqrt{2}}{\tau}\ cth\ (\pi\tau)\ \cos(\tau t) \tag{12.21}$$

12.4   Parseval type relation.

From Fourier transform the corresponding parseval type relation of Mehler-Fock transform of order zero can be deduced easily [N.N. Lebedev, Dokl. Akad. Nauk. SSSR, 68,445, 1949] if we define alternatively that

$$\phi_0[f(x) \ ; \ x \to \tau] = \bar{f}_0(\tau) = \int_1^\infty f(x) P_{-\frac{1}{2}+i\tau}(x)\ dx \tag{12.22}$$

and $\quad f(x) = \int_0^\infty \tau\ th\ (\pi\tau)\ \bar{f}_0(\tau)\ d\tau \equiv \phi_0^{-1}\ [\bar{f}_0(\tau) \ ; \ \tau \to x]$

$$\tag{12.23}$$

For convenience if we define

$$\bar{f}_0(\tau) = \phi_0[f(x) \ ; \ x \to \tau]$$
$$\bar{g}_0(\tau) = \phi_0[g(x) \ ; \ x \to \tau]$$

where $\quad \bar{f}_0(\tau) = \int_1^\infty f(x)\ P_{-\frac{1}{2}+i\tau}\ (x)\ dx, \tag{12.24}$

and $\quad \bar{g}_0(\tau) = \int_1^\infty g(x)\ P_{-\frac{1}{2}+i\tau}(x)\ dx \tag{12.25}$

then $\quad \int_0^\infty \tau\ th(\pi\tau)\ \bar{f}_0(\tau)\ \bar{g}_0(\tau)\ d\tau$

$$= \int_0^\infty \tau\ th\ (\pi\tau)\ \bar{g}_0\ (\tau)\ d\tau \int_1^\infty f(x) P_{-\frac{1}{2}+i\tau}(x)\ dx$$

$$= \int_1^\infty f(x)dx \int_0^\infty \tau\ th\ (\pi\tau)\ \bar{g}_0(\tau) P_{-\frac{1}{2}+i\tau}\ (x)\ d\tau$$

$$= \int_1^\infty f(x)\ g(x)\ dx, \tag{12.26}$$

by using inversion theorem in the inner integral.

This is the required Parseval type relation.

For example, if

$$f(x) = \frac{1}{x+s} \ , \quad g(x) = \frac{1}{x+t}$$

we have $\bar{f}_0(\tau) = \int_1^\infty \frac{1}{x+s} P_{-\frac{1}{2}+i\tau}(x) \, dx$

$$= \pi \mathrm{sech}\,(\pi\tau)\, P_{-\frac{1}{2}+i\tau}(s) \ , \ |s| < 1$$

and $\bar{g}_0(\tau) = \pi \mathrm{sech}\,(\pi\tau)\, P_{-\frac{1}{2}+i\tau}(t) \ , \ |t| < 1$

If $s \ne t$, $\displaystyle\int_1^\infty \frac{dx}{(x+s)(x+t)} = \frac{1}{t-s} \log \frac{1+t}{1+s}$

while $\displaystyle\int_1^\infty \frac{dx}{(x+s)^2} = \frac{1}{s+1}$

Thus from (12.26) we have

$$\int_0^\infty \frac{\tau\ sh\ \pi\tau}{ch^3\ \pi\tau} P_{-\frac{1}{2}+i\tau}(s) P_{-\frac{1}{2}+i\tau}(t)\, d\tau = \begin{cases} \frac{1}{\pi^2(t-s)} & \log \frac{1+t}{1+s} \ , \quad t \ne s \\ \frac{1}{\pi^2(1+s)} & , \qquad\qquad t = s \end{cases}$$

where $|s| < 1$, $|t| < 1$. $\hspace{3cm}$ (12.27)

As another example, let

$$f(x) = e^{-ax} \ , \quad g(x) = e^{-bx}$$

Then $\bar{f}_0(\tau) = \sqrt{\dfrac{2}{\pi a}}\, K_{i\tau}(a) \ $ and $\ \bar{g}_0(\tau) = \sqrt{\dfrac{2}{\pi b}}\, K_{i\tau}(b)$

Also, $\displaystyle\int_1^\infty e^{-(a+b)x}\, dx = \frac{e^{-(a+b)}}{(a+b)} \ .$

Therefore, by (12.26) we get

$$\int_0^\infty \tau\ th\ (\pi\tau)\, K_{i\tau}(a) K_{i\tau}(b)\, d\tau = \frac{\pi\sqrt{ab}}{2(a+b)} e^{-(a+b)} \hspace{1cm} (12.28)$$

As a third example, let

$$f(x) = x^{-\frac{3}{2}} \ , \quad g(x) = x^{-\frac{5}{2}}$$

and hence $\bar{f}_0(\tau) = \sqrt{2}\,\mathrm{sech}\left(\dfrac{\pi\tau}{2}\right) \ $ and

$$\bar{g}_0(\tau) = \frac{2\sqrt{2}}{3} \, \tau \, \text{cosech} \, \frac{\pi\tau}{2}$$

Again, since $\quad \displaystyle\int_1^\infty f(x) \, g(x) \, dx = \int_1^\infty x^{-4} \, dx = \frac{1}{3} \, ,$

we have from (12.26) that

$$\int_0^\infty \tau^2 \, \text{sech} \, \pi \, \tau \, d\tau = \frac{1}{8} \tag{12.29}$$

12.5 Mehler-Fock Transform of order m

The Mehler-Fock transform of order m is denoted and defined by

$$f_m^*(\tau) = \Phi_m \left[f(\alpha) \, ; \, \alpha \to \tau \right] = \int_0^\infty f(\alpha) P_{-\frac{1}{2}+i\tau}^m (ch\alpha) \, d\alpha \tag{12.30}$$

The corresponding inversion formula for the transform is

$$f(\alpha) = (-1)^m \int_0^\infty \tau \, \text{th}(\pi\tau) P_{-\frac{1}{2}+i\tau}^m (ch\alpha) f_m^*(\tau) \, d\tau$$

$$\equiv \Phi_m^{-1} \left[f_m^* (\tau) \, ; \, \tau \to \alpha \right] \tag{12.31}$$

where $\quad P_\nu^m(x) = (x^2 - 1)^{\frac{m}{2}} \dfrac{d^m}{dx^m} P_\nu(x) \, , \ m > 0$

Let $ch\alpha = x$. Then eqn. (12.31) can be expressed as

$$f(ch^{-1}x) = (-1)^m (x^2 - 1)^{\frac{m}{2}} \frac{d^m}{dx^m} \int_0^\infty \tau \, \text{th} \, (\pi\tau) P_{-\frac{1}{2}+i\tau} (x) \, f_m^* (\tau) \, d\tau \tag{12.32}$$

Also, if we define the Mehler-Fock transform of order m by the equation

$$\Phi_m \left[f(\alpha) \, ; \, \alpha \to \tau \right] = f_m^* (\tau) = \int_0^\infty f(\alpha) \, P_{-\frac{1}{2} \mid i\tau}^{-m} (ch\alpha) \, sh\alpha \, d\alpha \tag{12.33}$$

then the corresponding inversion formula will be

$$\Phi_m^{-1} \left[f_m^*(\tau) \, ; \, \tau \to \alpha \right] = f(\alpha) = (-1)^m \int_0^\infty \tau \, \text{th} \, (\pi\tau) \, f_m^*(\tau) P_{-\frac{1}{2}+i\tau}^m (ch\alpha) \, d\tau, \tag{12.34}$$

since the associated Legendre equation does not change after replacing m by $-m$ and therefore the corresponding Legendre functions.

Also, for some class of functions f for convenience, if we define altarnatively

$$\phi_m \left[f(x) ; x \to \tau \right] \equiv \bar{f}_m(\tau) = \int_1^\infty f(x) \, P^m_{-\frac{1}{2}+i\tau}(x) \, dx \qquad (12.35)$$

the corresponding inversion formula will be

$$\phi_m^{-1} \left[\bar{f}_m(\tau) ; \tau \to x \right] \equiv f(x) = (-1)^m \int_0^\infty \tau \, \text{th} \, (\pi\tau) \, f_m^*(\tau) \, P^m_{-\frac{1}{2}+i\tau}(x) \, dx \qquad (12.36)$$

12.6 Application to Boundary Value Problems.

12.6.1 First Example

At first we consider to determine a harmonic function, say, ψ in cylindrical co-ordinates (ρ, ϕ, z) under certain given boundary conditions like

$$\psi = f(\rho) , \; 0 \leqslant \rho \leqslant a \qquad (12.37)$$

and $\quad \dfrac{\partial \psi}{\partial z} = 0 , \; \rho > a$ on the boundary plane $z = 0$ $\qquad (12.38)$

If we introduce a toroidal co-ordinate system in the same region as (α, β, ϕ) the equations of transformations are

$$\rho = \frac{a \, sh\alpha}{ch\alpha + \cos\beta} \; , \quad z = \frac{a \sin\beta}{ch\alpha + \cos\beta} \; , \quad \phi = \phi \qquad (12.39)$$

implying $\quad \rho + iz = a \, \text{th} \, \dfrac{1}{2}(\alpha + i\beta) \equiv f(\alpha + i\beta), \; \text{say} ,$ $\qquad (12.40)$

where f is a holomorphic function of $(\alpha + i\beta)$. It may be noticed from (12.39) that

$$\rho^2 + (z + a \cot\beta)^2 = a^2 \, \text{cosec}^2 \, \beta \qquad (12.41)$$

and $\quad (\rho - a \, \text{th} \, \alpha)^2 + z^2 = a^2 \, \text{cosech}^2 \, \alpha \qquad (12.42)$

representing surfaces $\beta = \beta_1$, say and $\alpha = \alpha_1$, say, as toruses respectively about z-axis .

In (12.41), if $0 < \beta_1 < \pi$, $\beta = \beta_1$ is the part of the sphere with centre at $(0, 0, -\alpha \cot\beta_1)$ and radius a cosec β_1 lying above the xy-plane i.e, $z > 0$. Again if $-\pi < \beta_1 < 0$, the surface is that part of the above sphere which lies below the xy-plane i.e, where $z < 0$. In cylindrical system the Laplace equation in ψ

$$\frac{\partial^2 \psi}{\partial \rho^2} + \frac{1}{\rho}\frac{\partial \psi}{\partial \rho} + \frac{1}{\rho^2}\frac{\partial^2 \psi}{\partial \phi^2} + \frac{\partial^2 \psi}{\partial z^2} = 0 \qquad (12.43)$$

then takes the form

$$\frac{\partial}{\partial \alpha}\left[\rho\frac{\partial \psi}{\partial \alpha}\right] + \frac{1}{\partial \beta}\left[\rho\frac{\partial \psi}{\partial \beta}\right] + \frac{|f'|^2}{\rho}\frac{\partial^2 \psi}{\partial \phi^2} = 0 \qquad (12.44)$$

Again from (12.40) we have

$$|f'|^2 = a^2(ch\alpha + \cos \beta)^{-2}$$

and so the above PDE in (12.44) can be expressed as

$$\frac{\partial}{\partial \alpha}\left[\frac{sh\alpha}{ch\alpha + \cos \beta}\frac{\partial \psi}{\partial \alpha}\right] + \frac{\partial}{\partial \beta}\left[\frac{sh\alpha}{ch\alpha + \cos \beta}\frac{\partial \psi}{\partial \beta}\right]$$
$$+ \frac{1}{sh\alpha(ch\alpha + \cos \beta)}\frac{\partial^2 \psi}{\partial \phi^2} = 0 \quad (12.45)$$

Let us seek a solution of (12.45) in toroidal co-ordinates in the form

$$\psi(\alpha, \beta, \phi) = \sqrt{(ch\alpha + \cos \beta)} \cdot v(\alpha, \beta, \phi) \qquad (12.46)$$

Then (12.45) leads to a PDE in $v(\alpha, \beta, \phi)$ as

$$sh\alpha\frac{\partial^2 v}{\partial \alpha^2} + ch\alpha\frac{\partial v}{\partial \alpha} + sh\alpha\frac{\partial^2 v}{\partial \beta^2} + \frac{1}{4}v\, sh\alpha + \frac{1}{sh\alpha}\frac{\partial^2 v}{\partial \phi^2} = 0 \quad (12.47)$$

Eqn. (12.47) has the separable solutions of the form

$$v(\alpha, \beta, \phi) = A(\alpha) \cdot e^{\pm \tau \beta} \cdot e^{\pm im\phi} \ ,$$

if $A(\alpha)$ satisfy the equation

$$(1 - \mu^2)\frac{d^2 A}{d\mu^2} - 2\mu\frac{dA}{d\mu} - \left[\frac{m^2}{1 - \mu^2} + \left(\frac{1}{4} + \tau^2\right)\right]A = 0 \qquad (12.48)$$

where $\mu = ch\alpha$.

Now representing $\nu = -\frac{1}{2} + i\tau$, we have $\nu(\nu + 1) = -\left(\frac{1}{4} + \tau^2\right)$ and hence eqn, (12.48) can be expressed as

$$(1 - \mu^2)\frac{d^2 A}{d\mu^2} - 2\mu\frac{dA}{d\mu} + \left[\nu(\nu + 1) - \frac{m^2}{1 - \mu^2}\right]A = 0 \qquad (12.49)$$

which is an associated Legendre differential equation. This equation (12.49) has a solution of the form

$$A(\mu) = c\, P_\nu^{-m}(\mu) = c\, P_{-\frac{1}{2}+i\tau}^{-m}(ch\alpha)$$

implying $\qquad \nu(\alpha,\beta,\phi) = c\, P_{-\frac{1}{2}+i\tau}^{-m}(ch\alpha)\, e^{\pm\tau\beta} \cdot e^{\pm im\phi}$

Therefore, $\qquad \psi(\alpha,\beta,\phi) = \sqrt{(ch\alpha + \cos\beta)} \cdot P_{-\frac{1}{2}+i\tau}^{-m}(ch\alpha)\, e^{\pm\tau\beta} \cdot e^{\pm im\phi}$

$$(12.50)$$

are the forms of the general solution of eqn. (12.44)

If we are interested to find the separable solutions of the Laplace equation in the axisymmetric case that is in the ϕ-independent case, we have

$$\psi(\alpha,\beta) = \sqrt{(ch\alpha + \cos\beta)} \cdot P_{-\frac{1}{2}+i\tau}(ch\alpha)\, e^{\pm\tau\beta} \qquad (12.51)$$

Thus by the principle of superposition the general solution of the Laplace equation in axisymmetric case is expressed as

$$\psi(\alpha,\beta) = \sqrt{(ch\alpha + \cos\beta)} \cdot \Phi_0^{-1}\, [A(\tau)\, ch\,\beta\,\tau + B(\tau)\, sh\,\beta\,\tau\ ;\ \tau \to \alpha]$$

$$(12.52)$$

where $\qquad \Phi_0^{-1}\,[\chi(\tau)\ ;\ \tau \to \alpha] = \int_0^\infty \tau\, th\,(\pi\tau)\, P_{-\frac{1}{2}+i\tau}(ch\alpha)\, \chi(\tau)\, d\tau$

$$(12.53)$$

For the general case this solution similarly can be expressed as

$$\psi(\alpha,\beta,\phi) = \sqrt{(ch\alpha + \cos\beta)} \sum_{m=-\infty}^{\infty} e^{im\phi}\, \Phi_m^{-1}\,[A_m(\tau)ch(\beta\tau)$$

$$+ B_m(\tau)\, sh(\beta\tau)\ ;\ \tau \to \alpha]\ , \qquad (12.54)$$

where $\qquad \Phi_m^{-1}\,[\chi_m(\tau);\tau \to \alpha] = (-1)^m \int_0^\infty \tau\, th(\pi\tau)\, P_{-\frac{1}{2}+i\tau}^{-m}(ch\alpha)\chi_m(\tau)d\tau$

$$(12.55)$$

12.6.2   Second Example

Secondly for example, let us consider the particular problem of determination of the solutions ψ of the Laplace equation in a half-space $z \geqslant 0$ under the prescribed boundary conditions on $\psi(\rho, z, \phi)$ on a disk $0 < \rho < a$, $z = 0$ and $\frac{\partial\psi}{\partial z} = 0$ in the region of the boundary outside that disk in cylindrical co-ordinate system. In toroidal co-ordinate system (α, β, ϕ), as discussed above in 12.6.1, the given boundary conditions are transformed to

$$\psi\,(\alpha, 0, \phi) = f(\alpha, \phi)\,,\; 0 < \phi < 2\pi\,,\; \alpha > 0 \tag{12.56}$$

and $$\frac{\partial \psi}{\partial \beta}\,(\alpha, \pi, \phi) = 0\,. \tag{12.57}$$

Following the result in eqn. (12.50), we take a separable solution of the Laplace equation as

$$\psi(\alpha, \beta, \phi) = \sqrt{(ch\alpha + \cos\beta)} \cdot \sum_{m=-\infty}^{\infty} \Phi_m^{-1}\left[A_m(\tau)\frac{ch(\pi\tau - \beta\tau)}{ch(\pi\tau)}; \tau \to \alpha\right] e^{im\phi} \tag{12.58}$$

Here we may note that this form of solution identically satisfied the condition $\frac{\partial \psi}{\partial \beta}(\alpha, \pi, \phi) = 0$. Now, to satisfy the boundary condition in eqn. (12.56) we have to find the functions $A_m(\tau)$ for m ranging from $m \to -\infty$ to $m \to +\infty$ through integral values such that

$$\sqrt{2}\,sh\left(\frac{\alpha}{2}\right) \sum_{m=-\infty}^{m=+\infty} \Phi_m^{-1}\,[A_m(\tau)\,;\; \tau \to \alpha]e^{im\phi} = f(\alpha, \phi) \tag{12.59}$$

implying that

$$\Phi_m^{-1}\,[A_m(\tau)\,;\; \tau \to \alpha] = \frac{2^{-\frac{3}{2}}}{\pi}\,\mathrm{cosech}\left(\frac{\alpha}{2}\right)\int_0^{2\pi} f(\alpha, \phi)e^{-im\phi}\,d\phi$$

Therefore, $A_m(\tau) = 2^{-\frac{3}{2}}\pi^{-1}\int_0^{2\pi} e^{-im\phi}\Phi_m\left[\mathrm{cosech}\left(\frac{\alpha}{2}\right)f(\alpha, \phi); \alpha \to \tau\right]d\phi$

$$\tag{12.60}$$

12.6.3 Third Example

Thirdly another important problem of interest is to find the solution $\psi(\rho, z)$ of Laplace equation in a half space $z \geqslant 0$ for the axisymmetric case under the boundary conditions

$$\psi(\rho, 0) = F(\rho)\;\; \text{for}\;\; 0 \leqslant \rho \leqslant a \tag{12.61}$$

and $$\left[\frac{\partial \psi}{\partial z}\right] = 0,\;\; \text{for}\;\; \rho > a \tag{12.62}$$

Then in the toroidal co-ordinate system the ϕ independent solution is obtained from eqn. (12.58) with a little variation after putting $\phi = 0$ and $m = 0$ in the form

$$\psi(\alpha, \beta) = \sqrt{(ch\alpha + \cos\beta)}\ \Phi_0^{-1}\left[A_0(\tau)\frac{ch(\pi\tau - \beta\tau)}{ch(\pi\tau)}\ ;\ \tau \to \alpha\right] \quad (12.63)$$

for identically satisfying the condition in eqn. (12.62). Also the condition in eqn. (12.61) will be satisfied if we have found out $A_0(\tau)$ such that

$$A_0(\tau) = \Phi_0\left[\frac{1}{\sqrt{2}}\ \text{sech}\left(\frac{\alpha}{2}\right)F\left(a\ th\frac{\alpha}{2}\right)\ ;\ \alpha \to \tau\right] \quad (12.64)$$

$$= \text{Mehler} - \text{Fock transform of } \frac{1}{\sqrt{2}}\ \text{sech}\left(\frac{\alpha}{2}\right)F\left(a\tan\frac{\alpha}{2}\right)$$

Thus eqns. (12.63) together with (12.64) will constitute the required solution of the given boundary value problem in the axi-symmetric case.

This boundary value problem corresponds to the axisymmetric contact problem for a half-space in the theory of elasticity.

As a particular case of the above problem, let us suppose that $F(\alpha) = k$, a constant

implying $\quad \dfrac{1}{\sqrt{2}}\ \text{sech}\left(\dfrac{\alpha}{2}\right)F\left(a\ th\dfrac{\alpha}{2}\right) = \dfrac{k}{\sqrt{2}}\ \text{sech}\left(\dfrac{\alpha}{2}\right)$,

Now, $\quad A_0(\tau) = \Phi_0\left[\dfrac{k}{\sqrt{2}}\ \text{sech}\ \dfrac{\alpha}{2}\ ;\ \alpha \to \tau\right] = \dfrac{k}{\sqrt{2}}\cdot 2\ \tau^{-1}\ \text{cosech}\ (\pi\tau)$

$$= \sqrt{2}\ k\ \tau^{-1}\ \text{cosech}\ (\pi\tau)$$

$$\therefore\ \psi(\alpha, \beta) = k\sqrt{[2(ch\alpha + \cos\beta)]}\ \Phi_0^{-1}\left[\tau^{-1}\frac{2\ ch(\pi\tau - \beta\tau)}{sh(2\pi\tau)}\ ;\ \tau \to \alpha\right]$$

$$= \frac{2k}{\pi}\tan^{-1}\left[\frac{ch\alpha + \cos\beta}{1 - \cos\beta}\right]^{\frac{1}{2}}, \quad (12.65)$$

from table of integrals.

A quantity of physical interest is the number

$$P = -2\pi\int_0^a \rho\left(\frac{\partial\psi}{\partial z}\right)_{z=0} d\rho$$

to be evaluated now.

We have, $\quad a\dfrac{\partial\psi}{\partial z} = -sh\,\alpha\sin\beta\,\dfrac{\partial\psi}{\partial\alpha} + (1 + ch\alpha\cos\beta)\dfrac{\partial\psi}{\partial\beta}$

$\therefore\quad \left[\dfrac{\partial\psi}{\partial z}\right]_{z=0} = \dfrac{1 + ch\alpha}{a}\left(\dfrac{\partial\psi}{\partial\beta}\right)_{\beta=0}$

Now, $\quad \dfrac{\partial\psi}{\partial\beta} = \dfrac{\sqrt{2}\,k}{\pi}\cos\left(\dfrac{\beta}{2}\right)(ch\alpha + \cos\beta)^{-\frac{1}{2}}$

So, $\quad \left[\dfrac{\partial\psi}{\partial z}\right]_{z=0} = -\dfrac{2k}{\pi a}\,ch\left(\dfrac{\alpha}{2}\right)$

$\therefore\quad P = 2\,k\,a\displaystyle\int_0^\infty th\,\dfrac{\alpha}{2}\,\mathrm{sech}\,\dfrac{\alpha}{2}\,d\alpha\ ,\ \text{since } \rho = a\,th\,\dfrac{\alpha}{2}$

$$= 4\,a\,k \tag{12.66}$$

12.6.4 Fourth Example

Finally we consider the problem of determining the axisymmetric stresses in an elastic body weakened by a penny-shaped crack . This problem is reduced to determining a harmonic function ψ in the half-space $z \geqslant 0$ satisfying the conditions

$$\left[\dfrac{\partial\psi}{\partial z}\right]_{z=0} = p\,(\rho)\ ,\ 0 \leqslant \rho \leqslant a \tag{12.67}$$

$$[\Psi]_{z=0} = 0\ ,\ \rho > a \tag{12.68}$$

As before, introducing toroidal co-ordinates (α, β, ϕ) under transformation

$$\rho = \dfrac{a\,sh\,\alpha}{ch\alpha + \cos\beta}\ ,\ z = a\,\dfrac{\sin\beta}{ch\alpha + \cos\beta}\ ,\ \phi = \phi$$

the boundary conditions in eqns. (12.67) and (12.68) reduce to

$$\dfrac{\partial\psi\,(\alpha, 0)}{\partial\beta} = -2a\,\mathrm{sech}^2\left(\dfrac{\alpha}{2}\right)p\left(a\,th\,\dfrac{\alpha}{2}\right) \tag{12.69}$$

and $\quad \psi(\alpha, \pi) = 0 \quad$ respectively. $\tag{12.70}$

So, we choose $\psi(\alpha, \beta)$ satisfying (12.70) identically as

$$\psi(\alpha, \beta) = \sqrt{(ch\alpha + \cos\beta)}\ \Phi_0^{-1}\left[\tau^{-1}\,f^*(\tau)\dfrac{sh(\pi\tau - \beta\tau)}{ch(\pi\tau)}\ ;\ \tau \to \alpha\right]$$

To satisfy (12.69) we take

$$f^*(\tau) = \Phi_0\,[f(\alpha)\ ;\ \alpha \to \tau]$$

where $\quad f(\alpha) = \sqrt{2}\,a\,p\left(a\,\tanh\dfrac{\alpha}{2}\right)\mathrm{sech}^3\dfrac{\alpha}{2} \tag{12.71}$

For a particular case, if $p(\rho) = k$, a constant, then

$$f(\alpha) = \sqrt{2}\, k\, a\, \text{sech}^3\left(\frac{\alpha}{2}\right)$$

and hence $f^*(\tau) = 8\sqrt{2}\, k\, a\, \tau\, \text{cosech}\,(\pi\tau)$,   from table of integrals.

Therefore, $\psi(\alpha, \beta) = 16\sqrt{\left(ch^2\dfrac{\alpha}{2} - \sin^2\dfrac{\beta}{2}\right)k\, a}$

$$\Phi_0^{-1}\left[\frac{2\,sh(\pi\tau - \beta\tau)}{sh(2\pi\tau)}\; ; \tau \to \alpha\right]$$

12.7 Application of Mehler-Fock Transform for solving dual integral equation.

Consider the dual integral equations for determination of $f(\tau)$ as

$$\int_0^\infty f(\tau)\cos\, x\,\tau\, d\tau = g_1(x)\,,\; 0 \leqslant x \leqslant a \qquad (12.72)$$

$$\int_0^\infty f(\tau)\,cth\,(\pi\tau)\sin\, x\,\tau\, d\tau = h_2(x)\,,\; x > a\,, \qquad (12.73)$$

where $g_1(x)$ and $h_2(x)$ are two prescribed functions in their given ranges.

Multiplying both sides of the eqn. (12.72) by $\dfrac{\sqrt{2}}{\pi\sqrt{(ch\alpha - chx)}}$ and

integrating the result with respect to x from 0 to α and finally using the result in eqn. (12.14) we get

$$\int_0^\alpha \frac{\sqrt{2}}{\pi\sqrt{(ch\alpha - ch\, x)}}\left[\int_0^\infty f(\tau)\cos x\tau\, d\tau\right] dx = \int_0^\alpha \frac{\sqrt{2}}{\pi}\frac{g_1(x)dx}{\sqrt{(ch\alpha - ch\, x)}}$$

Thus $\displaystyle\int_0^\infty f(\tau)\, P_{-\frac{1}{2}+i\tau}\,(ch\alpha)\, d\tau = \Omega_1(\alpha)$,   say

$$= \frac{\sqrt{2}}{\pi}\int_0^\alpha \frac{g_1(x)dx}{\sqrt{(ch\alpha - ch\, x)}}\,, 0 < \alpha < a$$

$$(12.74)$$

Similarly, multiplying eqn. (12.73) by $\frac{\sqrt{2}}{\pi}(chx - ch\alpha)^{-\frac{1}{2}}$ and integrating the result with respect to x from α to ∞ and finally using the result in eqn (12.16) we get

$$\int_0^\infty f(\tau)\, P_{-\frac{1}{2}+i\tau}\,(ch\alpha)\, d\tau = \frac{\sqrt{2}}{\pi}\int_\alpha^\infty \frac{h_2(x)dx}{\sqrt{(chx - ch\alpha)}} \equiv \Omega_2(\alpha)\text{, say}, \alpha > a$$

$$(12.75)$$

Let us now define

$$\Omega\left(\alpha\right) = \begin{cases} \Omega_1\left(\alpha\right) & , \quad 0 \leqslant \alpha \leqslant a \\ \Omega_2\left(\alpha\right) & , \quad \alpha > a \end{cases}$$

so that the eqns in (12.74) and (12.75) are equivalent to

$$\int_0^\infty f(\tau)\, P_{-\frac{1}{2}+i\tau}\left(ch\alpha\right)\, d\tau = \Omega\left(x\right)$$

$$\Rightarrow \quad \Phi_0^{-1}\left[\,\tau^{-1}\,cth\left(\pi\tau\right) f(\tau)\,;\,\tau \to \alpha\,\right] = \Omega\left(\alpha\right)$$

After Mehler-Fock inversion the above equation yields

$$
\begin{aligned}
f(\tau) \quad &= \quad \tau\,th\left(\pi\tau\right)\Phi_0\left[\,\Omega(\alpha)\,;\,\alpha \to \tau\,\right] \\
&= \quad \tau\,th\left(\pi\tau\right)\int_0^a \Omega_1(\alpha)\, P_{-\frac{1}{2}+i\tau}\left(ch\alpha\right)\,sh\,\alpha\,d\alpha \\
&\quad +\quad \tau\,th\left(\pi\tau\right)\int_a^\infty \Omega_2(x)\, P_{-\frac{1}{2}+i\tau}\left(ch\alpha\right)\,sh\,\alpha\,d\alpha \quad (12.76)
\end{aligned}
$$

Thus the dual integral equation is solved in closed form.

Exercises

(1) Considering the alternative notation of Mehler-Fock transform as

$$\bar{f}_0(\tau) = \varphi_0\left[f(x)\,;\,x \to \tau\right] = \int_1^\infty f(x)\, P_{-\frac{1}{2}+i\tau}(x)\,dx$$

and the inverse

$$f(x) = \varphi_0^{-1}\left[\bar{f}_0(\tau)\,;\,\tau \to x\right] = \int_0^\infty \tau\,th\left(\pi\tau\right)\bar{f}_0(\tau)\, P_{-\frac{1}{2}+i\tau}(x)\,d\tau$$

prove that

(a) $\varphi_0\left[e^{-ax}\,;\,x \to \tau\right] = \sqrt{\dfrac{2}{\pi a}}\, K_{i\tau}(a)$, $|arg\,a| < \dfrac{\pi}{2}$

(b) $\varphi_0\left[x^{-\frac{3}{2}}\,;\,x \to \tau\right] = \sqrt{2}\,\text{sech}\left(\dfrac{\pi\tau}{2}\right)$

(c) $\varphi_0\left[(ch\alpha + \cos\beta)^{-\frac{3}{2}}\,;\,\alpha \to \tau\right] = 2\sqrt{2}\,\text{cosec}\,\beta\dfrac{sh(\beta\tau)}{sh(\pi\tau)}, -\pi < \beta < \pi$

(d) $\varphi_0\left[\text{sech}^3\left(\dfrac{\alpha}{2}\right)\,;\,\alpha \to \tau\right] = 8\,\tau\,\text{cosech}\left(\pi\tau\right)$

(2) Following the dual integral equations discussed in section 12.7 prove that

$$\int_0^\infty f(\tau) \cos(x\,\tau)\,d\tau = \frac{1}{\sqrt{2}} \frac{d}{dx} \int_0^a \frac{\Omega_1(\alpha) sh\ \alpha\ d\alpha}{\sqrt{(ch\ x - ch\alpha)}}$$

$$+\frac{1}{\sqrt{2}} \frac{d}{dx} \int_a^x \frac{\Omega_2(\alpha)\ sh\ \alpha\ d\alpha}{\sqrt{(ch\ x - ch\alpha)}}\ , \qquad \text{for} \quad x > a$$

$\Omega_1(\alpha)$, $\Omega_2(\alpha)$ are functions defined in the section.

(3) Reduce the following dual integral equations

$$\int_0^\infty \tau^{-1} f(\tau) \sin x\tau\ d\tau = j_1(x)\ , \quad 0 \leqslant x \leqslant a$$

$$\int_0^\infty cth\ (\pi\tau)\ f(\tau) \sin x\tau\ d\tau = h_2(x)\ , \quad x > a$$

as a pair of dual integral equations discussed in section 12.7.

(4) Prove that

(a) $\quad \Phi_0 \left[sech \left(\frac{\alpha}{2} \right)\ ;\ \alpha \to \tau \right] = 2\tau^{-1} \operatorname{cosech} (\pi\tau)$

(b) $\quad \Phi_0 \left[\sqrt{(sech\ (\alpha))}\ ;\ \alpha \to \tau \right] = \frac{1}{\sqrt{2}\,\tau} \operatorname{cosech} \left(\frac{\pi\tau}{2} \right)$

Chapter 13

Jacobi, Gegenbauer, Laguerre and Hermite Transforms

13.1   Introduction.

In this section we present below a brief account of some important integral transforms like Jacobi, Gegenbauer, Laguerre and Hermite transforms. These transforms are applicable to a limited class of problems of physical science and therefore these are sparingly discussed in literature. We only persue here their definitions, basic operational properties and a few important applications.

13.2   Definition of Jacobi Transform.

In view of the orthogonal relation of Jacobi polynomials of orders $\alpha(> -1)$ and $\beta(> -1)$ and degrees m and n given by

$$\int_{-1}^{1} (1-x)^{\alpha} \, (1+x)^{\beta} \, P_n^{(\alpha,\beta)} \, (x) \, P_m^{(\alpha,\beta)} \, (x) \, dx = \delta_n \, \delta_{mn} \qquad (13.1)$$

where δ_{nm} is the Kronecker delta symbol and

$$\delta_n = \frac{2^{\alpha+\beta+1} \, \Gamma(n+\alpha+1) \, \Gamma(n+\beta+1)}{n! \, (\alpha+\beta+2n+1) \, \Gamma(n+\alpha+\beta+1)} \qquad (13.2)$$

there exist under certain restrictions a series expansion of $f(x)$ of the form

$$f(x) = \sum_{n=1}^{\infty} a_n \, P_n^{(\alpha,\beta)} \, (x) \qquad , -1 < x < 1 \qquad (13.3)$$

with $\quad a_n = \dfrac{1}{\delta_n} \int_{-1}^{1} f(x) \, P_n^{(\alpha,\beta)} \, (x) \, dx = \dfrac{1}{\delta_n} \, f^{(\alpha,\beta)}(n) \qquad (13.4)$

Accordingly, the Jacobi transform of degree n of a function $f(x)$ in $-1 < x < 1$ is defined by Debnath (Bull Cal Math Soe., 55, 1963) as

$$J[f(x); x \to n] \equiv f^{(\alpha,\beta)}(n) = \int_{-1}^{1} (1-x)^\alpha (1+x)^\beta P_n^{(\alpha,\beta)}(x) f(x) \, dx$$

$$(13.5)$$

and the inverse Jocobi transform is given by

$$J^{-1}\left[f^{(\alpha,\beta)}(n); n \to x \right] \equiv f(x) = \sum_{n=1}^{\infty} (\delta_n)^{-1} f^{(\alpha,\beta)}(n) P_n^{(\alpha,\beta)}(x)$$

$$(13.6)$$

Example 13.1. Prove that the Jacobi Transform of degree n of

(a) $f(x)$, a polynomial of degree $m < n$ is zero.

(b) $P_m^{(\alpha,\beta)}(x)$ is δ_{mn}

Solution.

(a) Since $f(x)$ is a polynomial of degree $m < n$, it admits of a series expansion in Jacobi polynomials as

$$f(x) = \sum_{r=1}^{m<n} a_r P_r^{(\alpha,\beta)}(x), \quad -1 < x < 1$$

and therefore,

$$
\begin{aligned}
J[f(x); x \to n] &= \int_{-1}^{1} (1-x)^\alpha (1+x)^\beta P_n^{(\alpha,\beta)}(x) \left\{ \sum_{r=1}^{m<n} a_r P_r^{(\alpha,\beta)}(x) \right\} dx \\
&= \sum_{r=1}^{m} a_r \int_{-1}^{1} (1-x)^\alpha (a+x)^\beta P_n^{(\alpha,\beta)}(x) P_r^{(\alpha,\beta)}(x) \, dx \\
&= \delta_n \, \delta_{nr} = 0, \text{ since } n \neq r(= 1, 2, \cdots m < n)
\end{aligned}
$$

with δ_n is as defined in (13.2).

(b) $J\left[P_m^{(\alpha,\beta)}(x); x \to n \right]$

$$= \int_{-1}^{1} (1-x)^\alpha (1+x)^\beta P_n^{(\alpha,\beta)}(x) P_m^{(\alpha,\beta)}(x) \, dx$$

$$= \delta_{nm}.$$

[Tables of Integral Transforms, Erdelyi et al, Mc Graw Hill, NY]

Theorem 13.1

Defining a differential operator R over a function $f(x)$ by

$$R\left[f(x)\right] = (1-x)^{-\alpha}(1+x)^{-\beta}\frac{d}{dx}\left[(1-x)^{\alpha+1}(1+x)^{\beta+1}\frac{d}{dx}f(x)\right]$$

and assuming

$$\lim_{|x|\to 1}\left[(1-x)^{\alpha+1}(1+x)^{\beta+1}f(x)\right] = 0 = \lim_{|x|\to 1}\left[(1-x)^{\alpha+1}(1+x)^{\beta+1}f'(x)\right]$$

$$(13.7)$$

if $\quad J[f(x)\,;\,x\to n] = f^{(\alpha,\beta)}(n)$, then

$$J\left\{R[f(x)]\,;\,x\to n\right\} = -n(n+\alpha+\beta+1)\,f^{(\alpha,\beta)}(n), n = 0, 1, 2, \cdots$$

$$(13.8)$$

Proof. By definition

$$J\{R[f(x)]\} = \int_{-1}^{1}(1-x)^{\alpha}(1+x)^{\beta}(1-x)^{-\alpha}(1+x)^{-\beta}$$

$$\frac{d}{dx}\left[(1-x)^{1+\alpha}(1+x)^{1+\beta}\frac{df(x)}{dx}\right]P_n^{(\alpha,\beta)}(x)\,dx$$

$$= \int_{-1}^{1}\frac{d}{dx}\left[(1-x)^{1+\alpha}(1+x)^{1+\beta}\frac{df(x)}{dx}\right]P_n^{(\alpha,\beta)}(x)\,dx$$

$$= \left[P_n^{(\alpha,\beta)}(x)(1-x)^{1+\alpha}(1+x)^{1+\beta}f'(x)\right]_{-1}^{1} - \int_{-1}^{1}(1-x)^{1+\alpha}(1+x)^{1+\beta}f'(x)$$

$$\frac{d}{dx}P_n^{(\alpha,\beta)}(x)\,dx$$

$$= -\int_{-1}^{1}(1-x)^{1+\alpha}(1+x)^{1+\beta}f'(x)\frac{d}{dx}P_n^{(\alpha,\beta)}(x)dx, \text{ after integrating by parts}$$

Again integrating by parts and using the orthogonal relation (13.1), the above relation further reduces to

$$J\{R[f(x)]\} \quad = \quad -n(n+\alpha+\beta+1)\int_{-1}^{1}(1-x)^{\alpha}(1+x)^{\beta}P_n^{(\alpha,\beta)}(x)f(x)dx$$

$$= \quad -n(n+\alpha+\beta+1)\,f^{(\alpha,\beta)}(n)$$

This completes the proof.

Corollary 13.1 If $f(x)$ and $R[f(x)]$ satisfy the conditions in (13.7) then $J\{R\{R[f(x)]\}\} \equiv J\{R^2[f(x)]\}$ exists and is given by

$$J\{R^2\left[f(x)\right]\} = (-1)^2\,n^2(n+\alpha+\beta+1)^2\,f^{(\alpha,\beta)}(n)$$

More generally, if $f(x)$ and $R^j[f(x)]$ satisfy the conditions in (13.7) for $j = 1, 2, \cdots$ then $J\{R^k[f(x)]\} = (-1)^k \, n^k (n + \alpha + \beta + 1)^k \, f^{(\alpha,\beta)}(n)$.

Example 13.2

Find the Jacobi Transform of the following functions :

(a) $f(x) = x^n$

(b) $f(x) = (1 + x)^{p-\beta}$

(c) $f(x) = (1 - x)^{q-\alpha}$

Solution.

(a) Since $P_n^{(\alpha,\beta)}(x) = \dfrac{(1 + \alpha + \beta)_{2n} \, x^n}{2^n \, n!(1 + \alpha + \beta)_n} + a$ polynomial of degree$(n - 1)$,

therefore

$$x^n = \frac{2^n n!(1 + \alpha + \beta)_n}{(1 + \alpha + \beta)_{2n}} P_n^{(\alpha,\beta)}(x) + a \text{ polynomial of}(n - 1)^{\text{th}}\text{degree}$$

So,

$$J\,[x^n;\ x \to n] = \int_{-1}^1 (1 - x)^\alpha \, (1 + x)^\beta P_n^{\alpha,\beta}(x) \, x^n \, dx$$

$$= \int_{-1}^1 (1 - x)^\alpha \, (1 + x)^\beta \cdot \left[P_n^{\alpha,\beta}(x)\right]^2 \cdot \frac{2^n n!(1 + \alpha + \beta)_n}{(1 + \alpha + \beta)_{2n}} \, dx$$

$$+ \int_{-1}^1 (1 - x)^\alpha \, (1 + x)^\beta P_n^{\alpha,\beta}.(x). \,[\text{a polynomial of } (n - 1)\text{th degree}] \, dx$$

$$= 2^{n+\alpha+\beta+1} \frac{\Gamma(n + \alpha + 1) \, \Gamma(n + \beta + 1)}{\Gamma(n + \alpha + \beta + 1)}, \text{as the second integral vanishes}$$

Here, $(\alpha)_n$ is the factorial function defined by

$$\begin{aligned}(\alpha)_n &= \alpha(\alpha + 1)(\alpha + 2)\cdots(\alpha + n - 1), \ n \geqslant 1 \\ &= \frac{\Gamma(\alpha + n)}{\Gamma(\alpha)}.\end{aligned}$$

(b) $J\left[(1 + x)^{p-\beta} \ ; \ x \to n\right]$

$$= \int_{-1}^1 (1 - x)^\alpha (1 + x)^p \, P_n^{(\alpha,\beta)}(x) \, dx$$

$$= \binom{n + \alpha}{n} 2^{n+p+1} \frac{\Gamma(p + 1) \, \Gamma(\alpha + 1) \, \Gamma(p - \beta + 1)}{\Gamma(\alpha + p + n + 2) \, \Gamma(p - \beta + n + 1)},$$

follows from the Tables of Integrals [Erdelyi et al, McGraw Hill, NY.] In particular, if $\alpha = \beta = 0$, then

$$
J\left[(1+x)^{p-\beta} \; ; \; x \to n\right]
$$

$$
= \int_{-1}^{1} P_n^{(0,0)}(x)(1+x)^p \; dx
$$

$$
= \int_{-1}^{1} (1+x)^p \; P_n(x) dx
$$

$$
= \frac{2^{p+1}\{\Gamma(1+p)\}^2}{\Gamma(p+n+2)\;\Gamma(p+n+1)}.
$$

This result again follows from the Tables of Integrals [Erdelyi et al, Mc Graw Hill, NY]

(c) $\quad J\left[(1-x)^{q-\alpha}; \; x \to n\right]$

$$
= \int_{-1}^{1} (1-x)^q \; (1+x)^\beta P_n^{\alpha,\beta}(x) dx
$$

$$
= \frac{2^{\,q+\beta+1}}{n!\Gamma(\alpha-q)} \frac{\Gamma(q+1)\;\Gamma(n+\beta+1)\Gamma(n+\alpha-q)}{\Gamma(n+\beta+q+2)}
$$

This result also follows from the tables of integrals [Erdelyi et al, Mc Graw Hill, NY.]

13.3 The Gegenbauer Transform.

The Gegenbauer Transform and its inversion is denoted and defined by

$$
G\left[f(x); x \to n\right] \equiv f^{(\nu)}(n) = \int_{-1}^{1} (1-x^2)^{(\nu-\frac{1}{2})} C_n^\nu(x) f(x) \; dx
$$

and $\quad G^{-1}\left[f^{(\nu)}(n); n \to x\right] = f(x) = \sum_{n=0}^{\infty} \delta_n^{-1} \; C_n^\nu(x) f^{(\nu)}(n)$

with $\quad \delta_n = \dfrac{2^{1-2\nu}\;\pi\Gamma(n+2\nu)}{n!(\nu+n)\;[\Gamma(\nu)]^2}$ $\qquad(13.9)$

When $\alpha = \beta = \nu - \frac{1}{2}$, the Jacobi Polynomial $P_n^{(\alpha,\beta)}(x)$ becomes the Gegenbauer polynomial $C_n^\nu(x)$. The corresponding orthogonal relation of Gegenbauer polynomial is

$$
\int_{-1}^{1} (1-x^2)^{\nu-\frac{1}{2}} \; C_m^\nu(x) C_n^\nu(x) dx = \delta_n \; \delta_{nm} \qquad(13.10)
$$

where $\quad \delta_n = \dfrac{2^{1-2\nu}\pi\Gamma(n+2\nu)}{n!(n+\nu)[\Gamma(\nu)]^2}$

Also the differential form R defined by

$$R[f(x)] = \left(1-x^2\right)\frac{d^2 f(x)}{dx^2} - (2\nu+1)\,x\,\frac{df(x)}{dx}$$

can be expressed as

$$R[f(x)] = (1-x^2)^{\frac{1}{2}-\nu}\frac{d}{dx}\left[(1-x^2)^{\nu+\frac{1}{2}}\frac{df}{dx}\right] \tag{13.11}$$

Then under the Gegenbauer transform it reduces to

$$G[R\{f(x)\}; x \to n] = -n(n+2\nu)\,f^{(\nu)}(n) \tag{13.12}$$

Similarly,

$$G[R^k\{f(x)\};\ x \to n] = (-1)^k\,n^k(n+2\nu)^k\,f^{(\nu)}(n),$$

where $k = 1, 2, 3, \cdots$ $\hspace{6cm}$ (13.13)

13.4 Convolution Theorem

If $G[\,f(x)\,;\,x \to n] = f^{(\nu)}(n)$ and $G\,[\,g(x)\,;\,x \to n] = g^{(\nu)}(n)$, then
$f^{\nu}(n)g^{(\nu)}(n) = G\,[\,h(x)\,;\,x \to n\,] = h^{(\nu)}(n)$,

where $h(x)$ is given by

$$h(\cos\psi) = A(\sin\psi)^{1-2\nu}\int_0^\pi\int_0^\pi f(\cos\theta)\,g\,(\cos\phi)\,(\sin\theta)^{2\nu}$$
$$(\sin\phi)^{2\nu-1}\,(\sin\lambda)^{2\nu-1}\,d\theta\,d\alpha$$

and α is defined by

$$\cos\phi = \cos\theta\,\cos\psi + \sin\theta\,\sin\psi\,\cos\alpha$$

Proof. From definition

$$f^{(\nu)}(n)\,g^{(\nu)}(n)$$
$$= \int_{-1}^1 f(x)\,(1-x^2)^{\nu-\frac{1}{2}}\,C_n^\nu\,(x)\,dx\int_{-1}^1 g(x)\,(1-x^2)^{\nu-\frac{1}{2}}\,C_n^\nu\,(x)\,dx$$
$$= \int_0^\pi f(\cos\theta)(\sin\theta)^{2\nu}\,C_n^\nu(\cos\theta)d\theta.\int_0^\pi g(\cos\phi)(\sin\phi)^{2\nu}\,C_n^\nu(\cos\phi)\,d\phi$$
$$= \int_0^\pi f(\cos\theta)(\sin\theta)^{2\nu}\left[\int_0^\pi g(\cos\phi)\,C_n^\nu(\cos\theta)\,C_n^\nu(\cos\phi)(\sin\phi)^{2\nu}d\phi\right]d\theta$$

$$\tag{13.14}$$

By the additional formula of Gegenbauer polynomial, we have

$$C_n^\nu(\cos\theta)\, C_n^\nu(\cos\phi) = A \int_0^\pi C_n^\nu(\cos\psi)(\sin\lambda)^{2\nu-1}\, d\lambda, \quad (13.15)$$

where $\quad A = \{\Gamma(n+2\nu)/\, n!\, 2^{2\nu-1}\, \Gamma^2(\nu)\}$ $\qquad\qquad$ (13.16)

and $\quad \cos\psi = \cos\theta\, \cos\phi + \sin\theta\, \sin\phi\, \cos\lambda$ $\qquad$ (13.17)

Accordingly we get

$$f^{(\nu)}(n)\, g^{(\nu)}(n) = \int_0^\pi f(\cos\theta)\, (\sin\theta)^{2\nu} \left[\int_0^\pi \int_0^\pi g\,(\cos\phi)\, C_n^\nu\,(\cos\psi)\right.$$
$$\left. (\sin\phi)^{2\nu}\, (\sin\lambda)^{2\nu-1}\, .\, d\lambda\, d\phi\,\right] d\theta \qquad (13.18)$$

If we make the variable transformation

$$\cos\phi = \cos\theta \cos\psi + \sin\theta \sin\psi \cos\alpha \qquad (13.19)$$

then from eqns (13.17) and (13.19) we get

$$d\lambda\, d\phi = (\sin\psi/\sin\phi)\, d\psi\, d\alpha$$

and so the double integral in eqn. (13.18) becomes

$$\int_0^\pi \int_0^\pi (\cos\phi)\, C_n^\nu\,(\cos\psi)\, (\sin\phi)^{2\nu-1}(\sin\lambda)^{2\nu-1} \sin\psi\, d\psi\, d\alpha \quad (13.20)$$

Then we get from (13.18) as

$$\begin{aligned}
f^{(\nu)}(n)\, g^{(\nu)}(n) &= \int_0^\pi (\sin\psi)^{2\nu}\, C_n^\nu(\cos\psi)\, h(\cos\psi)\, d\psi \\
&= G[h\,(\cos\psi)],
\end{aligned}$$

$$\text{where } h\,[\cos\psi] = A(\sin\psi)^{1-2\nu} \int_0^\pi \int_0^\pi f\,(\cos\theta)\, g\,(\cos\theta)\, (\sin\theta)^{2\nu}.$$
$$(\sin\phi)^{2\nu-1}\, (\sin\lambda)^{2\nu-1}\, d\theta\, d\alpha \qquad (13.21)$$

Thus the convolution theorem is proved.

13.5 Application of the Transforms

Case I We consider below the one-dimensional generalised heat conduction equation defined by

$$\frac{\partial}{\partial x}\left[(1-x^2)\,\frac{\partial u}{\partial x}\right] + [(1-x)\,\beta - (1+x)\alpha]\,\frac{\partial u}{\partial x} = \rho c\,\frac{\partial u}{\partial t} \qquad (13.22)$$

under the initial condition

$$u(x,0) = g(x) \ , \ \text{for} \ -1 < x < 1 \tag{13.23}$$

This equation can be expressed as

$$(1-x)^{-\alpha} (1+x)^{-\beta} \left[\frac{\partial}{\partial x} \left\{ (1-x)^{\alpha+1} (1+x)^{\beta+1} \frac{\partial u}{\partial x} \right\} \right] = \rho\alpha \frac{\partial u}{\partial t} \tag{13.24}$$

Or $\quad R[u(x,t)] = \dfrac{1}{c} \dfrac{\partial u}{\partial t} \ ,$

where $c = \dfrac{\alpha}{\rho d}$ is a positive constant depending on the density, specific heat and thermal conductivity of the material.

We solve this boundary value problem by using the Jacobi transform. Taking Jacobi transform, the eqns. (13.24) and (13.23) give rise to

$$\frac{d}{dt} u^{(\alpha,\beta)}(n,t) = -cn(n+\alpha+\beta+1) \, u^{(\alpha,\beta)}(n,t) \tag{13.25}$$

and $\quad u^{(\alpha,\beta)}(n,0) = g^{(\alpha,\beta)}(n) \tag{13.26}$

respectively. Solving these equations we get

$$u^{(\alpha,\beta)}(n,t) = g^{(\alpha,\beta)}(n) \, exp \, [-n(n+\alpha+\beta+1) \, ct]$$

Therefore, the formal solution of the problem after inversion of Jacobi transform then becomes

$$u(x,t) = \sum_{n=0}^{\infty} \delta_n^{-1} \, g^{(\alpha,\beta)}(n) \, P_n^{(\alpha,\beta)}(x) \, exp \, [-n(n+\alpha+\beta+1) \, ct \,] \tag{13.27}$$

Case II If the heat conduction equation for a homogeneous solid beam be

$$\frac{\partial}{\partial x} \left[(1-x^2) \frac{\partial u}{\partial x} \right] - (2\nu + 1) \, x \, \frac{\partial u}{\partial x} = \frac{1}{c} \frac{\partial u}{\partial t} \tag{13.28}$$

under the initial condition $\ u(x,0) \ = \ g(x), \ -1 < x < 1 \tag{13.29}$

we can apply Gegenbauer transform to solve it. After application of the Gegenbauer transform the PDE (13.28) becomes

$$\frac{d}{dt} u^{(\nu)}(n,t) = -dn(n+2\nu) \, u^{(\nu)}(n,t) \tag{13.30}$$

and the initial condition (13.29) becomes

$$u^{(\nu)}(n,0) = g^{(\nu)}(n) \qquad (13.31)$$

The solution of equation (13.30) under (13.31) is then given by

$$u^{(\nu)}(n,t) = g^{(\nu)}(n) \; exp \left[-n(n+2\nu) \; ct \; \right] \qquad (13.32)$$

On inversion of (13.32) we get the formal solution of the given boundary value problem as

$$u(x,t) = \sum_{n=0}^{\infty} \delta_n^{-1} \; C_n^{\nu}(x) \; exp \left[-n(n+2\nu) \; ct \right] \qquad (13.33)$$

13.6 The Laguerre Transform

Debnath and McCully introduced this transform through their pioneering works. Laguerre transform of a function $f(x)$ over $0 \leqslant x < \infty$ is denoted and defined by the integral

$$L[f(x); \; x \to n] \equiv \bar{f}_a(n) = \int_0^{\infty} e^{-x} x^{\alpha} \; L_n^{\alpha}(x) \; f(x) \; dx \qquad (13.34)$$

where $L_n^{\alpha}(x)$ is the Laguerre polynomial of degree $n (\geqslant 0)$ and order $\alpha \; (> -1)$ which satisfy the ODE

$$\frac{d}{dx} \left[e^{-x} \; x^{\alpha+1} \frac{d}{dx} L_n^{\alpha}(x) \right] + n e^{-x} \; x^{\alpha} \; L_n^{\alpha}(x) = 0 \qquad (13.35)$$

These Laguerre polynomials satisfy the orthogonal property

$$\int_0^{\infty} e^{-x} \; x^{\alpha} \; L_n^{\alpha}(x) \; L_m^{\alpha}(x) \; dx = \binom{n+\alpha}{n} \Gamma(n+1) \; \delta_n = \delta_n \; \delta_{mn}$$

$$(13.36)$$

where δ_{mn} is the Kronecker delta function and δ_n is given by

$$\delta_n = \binom{n+\alpha}{n} \Gamma(n+1) \qquad (13.37)$$

Accordingly, the inverse Laguerre transform is defined by

$$f(x) \equiv L^{-1} \left[\bar{f}_a(n) \; ; \; n \to x \right] = \sum_{n=0}^{\infty} (\delta_n)^{-1} \; \bar{f}_\alpha(n) \; L_n^{\alpha}(x) \qquad (13.38)$$

If $\alpha = 0$, according to McCully, Laguerre transform pair takes the form

$$L[f(x);\ x \to n] \equiv \bar{f}_0(n)\ =\ \int_0^\infty e^{-x} L_n(x)\ f(x)\ dx \qquad (13.39)$$

$$L^{-1}[\bar{f}_0(n);\ n \to x] \equiv f(x) = \sum_{n=0}^\infty \bar{f}_0\ (n)\ L_n\ (x) \qquad (13.40)$$

where $L_n(x)$ is the Laguerre polynomial of degree n and order zero.

Example 13.3. If $f(x) = L_m^\alpha(x)$ then prove that $L[f(x)] = \delta_n \delta_{mn}$

Solution. We have from definition

$$L\ [f(x);\ x \to n\ ] = L[L_m^\alpha(x);\ x \to n\ ]$$

$$= \int_0^\infty e^{-x} x^\alpha L_n^\alpha\ (x)\ L_m^\alpha\ (x)\ dx$$

$$= \left(\begin{array}{c} n+\alpha \\ n \end{array} \right) \Gamma\ (\alpha+1)\ \delta_{mn}\ ,$$

by the orthogonal property of the Laguerre polynomial

$$= \delta_n\ \delta_{mn}$$

Example 13.4 Find the Laguerre transform of $f(x) = e^{-\alpha x}, \alpha > -1$

Solution. We have $L[e^{-\alpha x}\ ;\ x \to n\ ]$

$$= \int_0^\infty e^{-(\alpha+1)x}\ x^\alpha\ L_n^\alpha\ (x)\ dx$$

$$= \frac{\Gamma(n+\alpha+1)\ \alpha^n}{n!(\alpha-1)^{n+\alpha+1}}\ ,$$

from the tables of integrals [Erdelyi et al (1954).]

Example 13.5. Prove that $\displaystyle\int_0^\infty e^{-x}\ x^k\ L_n(x)\ dx = \left\{ \begin{array}{l} 0,\ k \neq n \\ (-1)^n\ n!,\ k = n \end{array} \right.$

where $L_n(x)$ is a Laguerre polynomial due to McCully.

Solution. We can express x^k as linear combination of polynomials $L_0(x),\ L_1\ (x), \cdots L_k(x)$. Let it be

$$x^k = \sum_{r=0}^k C_r L_r(x)$$

Then $\displaystyle\int_0^\infty e^{-x} x^k L_n(x)\, dx = \sum_{r=0}^{k} C_r \int_0^\infty e^{-x} L_r(x)\, L_n(x)\, dx$

$$= \sum_{r=0}^{k} C_r \times 0 \ ,$$

by the orthogonality relation since $k = r \neq n$.

But, if $\qquad k = n$, then $\displaystyle x^k = x^n = \sum_{r=0}^{n} C_r\, L_r(x)$

$$= \sum_{r=0}^{n-1} C_r\, L_r(x) + C_n\, L_n(x)$$

Therefore, $\displaystyle\int_0^\infty e^{-x}\, x^n L_n(x)\, dx$

$$= \int_0^\infty e^{-x} L_n^2(x) . C_n\, dx$$

$$= C_n . 1$$

Now, in the expansion of x^n

$$x^n = C_0\, L_0(x) + C_1 L_1(x) + \cdots + C_n\, L_n(x)$$

$$= [C_0 L_0(x) + C_1\, L_1(x) + \cdots + C_{n-1}\, L_{n-1}]$$

$$+ C_n \left[\frac{(-1)^n x^n}{n!} + \text{a polynomial of degree } (n-1) \right]$$

as x^n occurs only on the single term $C_n \frac{(-1)^n}{n!} x^n$ on the right hand side of the above expansion we have by comparing the coefficients of x^n on both sides that

$$1 = C_n\, (-1)^n / n!$$

This implies that $C_n = (-1)^n n!$ and therefore

$$\int_0^\infty e^{-x}\, x^n\, L_n(x)\, dx = (-1)^n\, n!$$

13.7 Operational properties

(a) **Laguerre transform of the derivative of a function.**

Let $L[f(x) \ ; \ x \to n] = \bar{f}_\alpha(n)$, then

$$L[f'(x) \ ; \ x \to n] = \int_0^\infty e^{-x}\, x^\alpha L_n^\alpha(x) f'(x)\, dx$$

$$= \left[e^{-x} \, x^{\alpha} L_n^{\alpha}\left(x\right) f(x)\right]_0^{\infty} + \int_0^{\infty} e^{-x} x^{\alpha} L_n^{\alpha}\left(x\right) f(x) \, dx$$

$$-\alpha \int_0^{\infty} e^{-x} \, x^{\alpha-1} \, L_n^{\alpha}(x) \, f(x) \, dx - \int_0^{\infty} e^{-x} x^{\alpha} \left\{ \frac{d}{dx} L_n^{\alpha}(x) \right\} f(x) \, dx$$

$$= \bar{f}_\alpha(n) - \alpha \sum_{k=0}^{n} \bar{f}_{n-1}(k) + \sum_{k=0}^{n} \bar{f}_\alpha(k) \, , \tag{13.41}$$

from Higher Transandental function, [Erdelyi, 1953.]

(b) If $L\left[f(x)\,;\,x \to n\right] = \bar{f}_a(n)$, then

$$L\left[R\{f(x)\}\,;x \to n\right] = -n\bar{f}_a(n) \tag{13.42}$$

where R is the differential operator defined by

$$R[f(x)] = e^x \, x^{-\alpha} \frac{d}{dx} \left[e^{-x} \, x^{\alpha+1} \frac{df(x)}{dx}\right].$$

Solution. We have by definition

$$L\left[R\{f(x)\}\,;\,x \to n\right] = \int_0^{\infty} L_n^{\alpha}\left(x\right) \frac{d}{dx} \left[e^{-x} \, x^{\alpha+1} \frac{df(x)}{dx}\right] dx$$

$$= \left[L_n^{\alpha}(x) \left\{ e^{-x} x^{\alpha+1} \frac{df(x)}{dx} \right\}\right]_0^{\infty} - \int_0^{\infty} e^{-x} x^{\alpha+1} \frac{dL_n^{\alpha}(x)}{dx} \cdot \frac{df(x)}{dx} dx$$

$$= -\left[e^{-x} x^{\alpha+1} \frac{d}{dx} L_n^{\alpha}\left(x\right) f(x)\right]_0^{\infty} + \int_0^{\infty} f(x) \frac{d}{dx} \left\{ e^{-x} x^{\alpha+1} \frac{d}{dx} L_n^{\alpha}(x) \right\} dx$$

$$= \int_0^{\infty} f(x) \cdot \left\{ -n e^{-x} \, x^{\alpha} \, L_n^{\alpha}\left(x\right) \right\} \, dx \quad , \text{by eqn. (13.35)}$$

$$= -n \int_0^{\infty} e^{-x} \, x^{\alpha} \, L_n^{\alpha}(x) \, f(x) dx$$

$$= -n\bar{f}_\alpha(n) \tag{13.43}$$

On extendingthis proof, we can have

$$L\left[R^2\{f(x)\,;\,x \to n\}\right] = (-1)^2 \, n^2 \bar{f}_\alpha(n)$$

More generally,

$$L\left[R^k\{f(x)\}\,;\,x \to n\right] = (-1)^k \, n^k \bar{f}_\alpha(n), \ k = 1, 2, 3, \cdots \tag{13.44}$$

(c) The convolution theorem of Laguerre transform of order zero $(\alpha = 0)$ is given by

$$L^{-1}[\bar{f}_0(n) \, \bar{g}_0(n)] = h(x) \tag{13.45}$$

where $h(x)$ is given by

$$h(x) = \frac{1}{\pi} \int_0^\infty e^{-t} f(t) \, dt \int_0^\pi exp \left(\sqrt{xt} \cos \theta \right) \cos \left(\sqrt{xt} \, \sin \theta \right)$$

$$g \left(x + t - 2\sqrt{xt} \cos \theta \right) d\theta, \quad (13.46)$$

if $\quad L[f(x) \; ; \; x \rightarrow n] = \bar{f}_0(n)$ and $L[g(x) \; ; \; x \rightarrow n] = \bar{g}_0(n).$

Proof. We have by definition

$$\bar{f}_0(n) \, \bar{g}_0(n) = \int_0^\infty e^{-x} L_n (x) \, f(x) dx \int_0^\infty e^{-y} L_y(y) g(y) dy$$

$$= \int_0^\infty e^{-x} f(x) \, dx \left[\int_0^\infty e^{-y} L_n (x) \, L_y(y) \, g(y) \, dy \right]$$

Now from a formula of the product of two Laguerre functions we have

$$L_n(x) L_y(y) = \frac{1}{\pi} \int_0^\pi e^{\sqrt{xy} \cos \theta} \cos(\sqrt{xy} \sin \theta) \, L_n(x+y-2\sqrt{xy} \cos \theta) \, d\theta$$

[Bateman, Partial differential equation, 1944]. Therefore we can express

$$\pi \bar{f}_0(n) \bar{g}_0(n) = \int_0^\infty e^{-x} f(x) \, dx \left[\int_0^\infty e^{-y} g(y) \right.$$

$$\left. \left\{ \int_0^\pi exp(\sqrt{xy} \cos \theta) \cos(\sqrt{xy} \sin \theta). L_n(x + y - 2\sqrt{xy} \cos \theta) \, d\theta \right\} dy \right]$$

The integral within the square bracket is

$$\int_0^\infty e^{-t} L_n (t) \, dt \int_0^\pi exp \left(\sqrt{xt} \cos \phi \right) \cos(\sqrt{xt} \sin \phi)$$

$$g \left(x + t - 2\sqrt{xt} \cos \phi \right) d\phi$$

So, we can write, $\quad \bar{f}_0(n) \bar{g}_0(n) = L[h(t); t \rightarrow n]$

$$= \int_0^\infty e^{-t} L_n(t) h(t) dt$$

where, $h(x) = \frac{1}{\pi} \int_0^\infty e^{-t} f(t) dt \left[\int_0^\pi exp \left(\sqrt{xt} \cos \theta \right) \cos(\sqrt{xt} \sin \theta) \right.$

$$\left. g(x + t - 2\sqrt{xt} \cos \theta) d\theta \right]$$

This is the convolution theorem of the Laguere transform and the proof of it is complete.

13.8   Hermite Transform.

In 1964 Debnath introduced Hermite transform through his paper entitled "on Hermite transform", Mathematicki Vesnik, 1,(16)[]. He denoted and defined it by

$$H[f(x); \ x \to n] \equiv f_H(n) = \int_{-\infty}^{+\infty} exp\ (-x^2)\ H_n(x)\ f(x)\ dx,$$

(13.47)

where $H_n(x)$ is the Hermite polynomial of degree n defined through the Rodrigues formula

$$H_n(x) = (-1)^n\ exp\ (x^2)\ \frac{d^n}{dx^n}\left\{exp\ (-x^2)\right\}$$

(13.48)

The inverse Hermite transform denoted by $H^{-1}[f_H(n); n \to x]$ and is defined by

$$H^{-1}[f_H(n)\ ;\ n \to x] \equiv f(x) = \sum_{n=0}^{\infty} \delta_n^{-1}\ f_H(n)\ H_n(x)$$

(13.49)

where δ_n is given by

$$\delta_n = \sqrt{\pi}\ n!\ 2^n,$$

(13.50)

Since any function $f(x)$ can be expanded in a series of Hermite polynomials like

$$f(x) = \sum_{n=0}^{\infty} a_n\ H_n(x)$$

where the coefficients a_n of the $(n+1)$th term is given by

$$a_n = \delta_n^{-1}\ f_H(n)$$

after using the orthogonality relation of Hermite polynomials

$$\int_{-\infty}^{\infty} exp\ (-x^2)\ H_n(x)\ H_m(x)\ dx = \delta_{nm}\delta_n$$

(13.51)

Example 13.6. Find Hermite transform of each of the following functions :

(a) $f(x) = $ a polynomial of degree $m < n$

(b) $f(x) = H_m(x)$

(c) $f(x) = exp\ (2xt - t^2)$

(d) $f(x) = exp\ (ax)$

Solutions.

(a) Let $f(x) = P_r(x), r < n$

Then

$$H[f(x) \; ; \; x \to n] = \int_{-\infty}^{+\infty} exp\,(-x^2) \sum_{k=0}^{r} a_k \; H_k(x).$$
$$H_n(x) \; dx, \; r < n$$

$$= \sum_{k=0}^{\infty} a_k \, \{0\} = 0 \; ,$$

from the orthogonality relation of Hermite polynomials.

(b) $H\left[H_m\,(x) \; ; \; x \to n\right]$

$$= \int_{-\infty}^{+\infty} exp(-x^2) H_m\,(x) H_n(x) \; dx$$

$$= \delta_n \, \delta_{mn}$$

(c) To find $H\left[exp(2xt - t^2) \; ; \; x \to n\,\right]$ we consider the generating function of Hermite polynomial as

$$exp\,(2xt - t^2) = \sum_{n=0}^{\infty} \frac{t^n}{n!} \; H_n(x)$$

Then $H\left[exp(2xt - t^2) \; ; \; x \to n\right]$

$$= \int_{-\infty}^{+\infty} exp(-x^2) \; H_n(x). \sum_{n=0}^{\infty} \frac{t^n}{n!} \; H_n\,(x) \; dx$$

$$= \sum_{n=0}^{\infty} \frac{t^n}{n!} \int_{-\infty}^{+\infty} exp(-x^2) \; H_n^2\,(x) \; dx$$

$$= \sum_{n=0}^{\infty} \frac{t^n}{n!} \, \delta_n = \sqrt{\pi} \sum_{n=0}^{\infty} (2t)^n \quad , \quad |t| < \frac{1}{2}.$$

(d) We have from definition

$$H[exp(ax) \; ; \; x \to n]$$

$$= \int_{-\infty}^{+\infty} exp(-x^2) \; exp\,(ax) \; H_n(x) \; dx$$

Now since ,

$$\int_{-\infty}^{+\infty} exp(-x^2 + 2bx) \; H_n\,(x) \; dx$$

$$= \sqrt{\pi} \sum_{n=0}^{\infty} (2b)^n \ exp \ (b^2),$$

the above relation gives

$$H \left[exp \ (ax) \ ; \ x \to n \right] = \sqrt{\pi} \sum_{n=0}^{\infty} a^n \ exp \ \left(\frac{a^2}{4} \right)$$

13.9   Operational Properties.

(a) **Theorem 13.2.** If $f'(x)$ is continious and $f''(x)$ is integrable in $-\infty < x < \infty$ and if $H[f(x); \ x \to n] = f_H(n)$, then

$$H[R\{f(x)\} \ ; \ x \to n] = -2nf_H(n),$$

where $R[f(x)]$ is defined as

$$R[f(x)] = exp(x^2) \ \frac{d}{dx} \left[exp(-x^2) \frac{df}{dx} \right].$$

Proof. We have by definition,

$$
\begin{aligned}
H\left[R\{f(x)\} \ ; \ x \to n\right] &= \int_{-\infty}^{+\infty} H_n \ (x). \ \frac{d}{dx} \left[exp \ (-x^2) \ \frac{df}{dx} \right] dx \\
&= - \int_{-\infty}^{+\infty} \frac{df}{dx} \ (x) \ e^{-x^2} \ H_n'(x) \ dx \\
&= \int_{-\infty}^{+\infty} f(x) \left[e^{-x^2} H_n''(x) - 2xe^{-x^2} H_n'(x) \right] dx \\
&= \int_{-\infty}^{+\infty} e^{-x^2} \ f(x) \left[-2n \ H_n \ (x) \right] \ dx \\
&= -2n \ f_H \ (n) \quad\quad\quad\quad\quad (13.52)
\end{aligned}
$$

If $R \ [f(x)]$ also satisfies conditions of the theorem, then
$$H \left[R^2 \{f(x)\} \ ; \ x \to n \right] = (-1)^2 \ (2n)^2 f_H(n).$$
In general, when R^{k-1} satisfies the conditions of the theorem, then
$$H \left[R^k \{f(x)\} \ ; \ x \to n \right] = (-1)^k \ (2n)^k f_H(n),$$
where $k = 1, 2, \cdots$

Theorem 13.3. If $f(x)$ is bounded and integrable and $f_H(0) = 0$, then for each constant C

$$H^{-1} \left[-\frac{f_H(n)}{2n} \right] = R^{-1}[f(x)] = \int_{-\infty}^{x} exp \ (s^2) \int_{-\infty}^{s} exp \ (-t^2) \ f(t) \ dt \ + C,$$

where R^{-1} is the inverse of the differential operator R and n is a positive integer.

Proof.

$$\text{Let } R^{-1}[f(x)] = y(x) \Rightarrow R\,[y(x)] = f(x).$$

Thus $y(x)$ is the solution of a second order ODE. Since,

$f_H(0) = 0$ and $H_0(x) = 1$, we have

$$\int_{-\infty}^{+\infty} e^{-x^2}.\, R\,[y\,(x)]\, dx = \int_{-\infty}^{+\infty} e^{-x^2}\, R\,[y\,(x)]\, H_o(x)\, dx$$

$$= \int_{-\infty}^{+\infty} e^{-x^2}\, f(x)\, H_0(x)\, dx = f_H(0) = 0 \qquad (13.53)$$

Therefore, the first integral of $R\,[y(x)] = f(x)$ is

$$exp(-x^2)\, \frac{dy}{dx} = \int_{-\infty}^{x} exp(-t^2)\, f\,(t)\, dt$$

As $|x| \to \infty$, the right hand side integral in above tends to zero by virtue of eqn. (13.53). Then the second integral is

$$y(x) = \int_{0}^{x} exp\,(s^2) \left\{ \int_{-\infty}^{s} [exp\,(-t^2) f(t)]\, dt \right\} ds + C \quad (13.54)$$

Therefore ,

$$\lim_{|x|\to\infty} exp\,(-x^2)\, y(x) = 0 \text{ and } H\,[y(x)] \text{ exists.}$$

So, $\quad H\,[R\{y\,(x)\}\,;\, x \to n] = -2n\,H\,[y(x)]$ implying

$$H[f\,(x)\,;\, x \to n] = -2n\,H\,[y\,(x)]$$

$$\Rightarrow \quad f_H(n) = -2n\,H\,[R^{-1}\{f(x)\}]$$

$$\Rightarrow \quad H\,[R^{-1}\{f\,(x)\}] = -\frac{f_H\,(n)}{2n} \qquad (13.55)$$

13.10   Hermite Transform of derivative of a function.

Theorem 13.4. If $H[f\,(x)\,;\, x \to n\,] = f_H\,(n)$ and m is a positive integer, then $H[f^{(m)}\,(x)\,;\, x \to n\,] = f_H\,(n+m)$

Proof. we have by definition

$$H[f'\,(x)\,;\, x \to n\,] = \int_{-\infty}^{+\infty} exp\,(-x^2)\, H_n\,(x)\, f'\,(x)\, dx$$

$$= \left[exp\ (-x^2)\ f(x)\ H_n\ (x) \right]_{-\infty}^{+\infty} - \int_{-\infty}^{+\infty} f(x)\ \frac{d}{dx} \left\{ e^{-x^2} H_n(x) \right\}\ dx$$

$$= 2 \int_{-\infty}^{+\infty} x e^{-x^2}\ H_n\ (x)\ f(x)\ dx - \int_{-\infty)}^{+\infty} f(x)\ e^{-x^2} H'_n\ (x)\ dx$$

$$= \int_{-\infty}^{+\infty} e^{-x^2}\ [H_{n+1}\ (x) + 2n\ H_{n-1}\ x]\ f(x)\ dx$$

$$-2n \int_{-\infty}^{+\infty} e^{-x^2} H_{n-1}\ (x)\ f(x)\ dx,$$

since $H_{n+1}(x) = 2x\ H_n\ (x) - 2n\ H_{n-1}\ (x)$ (13.56)

and $H'_n\ (x) = 2n\ H_{n-1}\ (x)$ (13.57)

$$= \int_{-\infty}^{+\infty} exp\ (-x^2)\ H_{n+1}\ (x)\ f(x)\ dx$$

$$= f_H(n+1).$$ (13.58)

Proceeding in the similar manner we can arrive at

$$H\left[f^{(m)}(x)\ ;\ x \to n \right] = f_H(n+m)$$

Thoerem 13.5. If Harmite transforms of $f(x)$ and of $x\ f^{m-1}(x)$ exists, then

$$H\left[x\ f^{(m)}(x)\ ;\ x \to n \right] = n\ f_H(m+n-1) + \frac{1}{2} f_H(m+n+1)$$

Proof. We have by definition

$$H[x\ f^{(m)}(x)\ ;\ x \to n] = \int_{-\infty}^{+\infty} exp\ (-x^2) H_n(x) \left\{ x\ f^{(m)}(x) \right\} dx$$

$$= \left[x\ exp\ (-x^2)\ H_n(x)\ f^{(m-1)}(x) \right]_{-\infty}^{+\infty}$$

$$- \int_{-\infty}^{+\infty} \frac{d}{dx} \left\{ x\ exp(-x^2)\ H_n(x) \right\} f^{m-1}(x)\ dx$$

$$= \int_{-\infty}^{+\infty} 2x^2\ exp(-x^2)\ H_n(x) f^{m-1}(x) dx$$

$$- \int_{-\infty}^{+\infty} exp(-x^2) H_n(x) f^{(m-1)}(x) dx$$

$$-n \int_{-\infty}^{+\infty} 2x\ exp\ (-x^2) H_{n-1}(x) f^{(m-1)}(x)\ dx$$

Using eqns. (13.56) and (13.57), we therefore have

$$H[x f^m(x); x \to n] = \int_{-\infty}^{+\infty} x\ e^{-x^2} [H_{n+1}(x) + 2n\ H_{n-1}(x)] f^{(m-1)}(x) dx$$

$$-n \int_{-\infty}^{+\infty} e^{-x^2} [H_n(x) + 2(n-1)H_{n-2}(x)] f^{(m-1)}(x) dx - f_H(n+m+1)$$

$$= \frac{1}{2} \int_{-\infty}^{+\infty} e^{-x^2} [H_{n+2}(x) + 2(n+1)H_n(x)] \ f^{(m-1)}(x) \ dx$$

$$+n \int_{-\infty}^{+\infty} e^{-x^2} [H_n(x) + 2(n-1)H_{n-2}(x)] \ f^{(m-1)}(x) \ dx$$

$$-n \ f_H(n+m-1) - 2n(n-1)f_H(n+m-3) - f_H(n+m+1) \ ,$$
$$\text{since } H[f^{(m)}(x)] = f_H(n+m)$$

$$= \frac{1}{2} \ f_H(n+m+1) + (n+1)f_H(n+m-1) + n[f_H(n+m-1)$$

$$+2(n-1)f_H(n+m-3)] - nf_H(n+m-1) - 2n(n-1)f_H(n+m-3)$$
$$-f_H(n+m+1)$$

$$= n \ f_H(n+m-1) + \frac{1}{2} \ f_H \ (n+m+1)$$

Accordingly, for $m = 1$ and $m = 2$, we get

$$H\left[x \ f'(x) \ ; \ x \to n\right] = n \ f_H(n) + \frac{1}{2} \ f_H(n+2)$$

$$H\left[x \ f''(x) \ ; \ x \to n\right] = n \ f_H(n+1) + \frac{1}{2} \ f_H(n+3)$$

Exercises

(1)   Find Jacobi Transform of the following functions to prove that

(a) $J(1 \ ; \ n) = 0$

(b) $J[x^m \ ; \ x \to n] = \begin{cases} 0 \ , & \text{if} \ \ m \neq n \\ 1 \ , & \text{if} \ \ m = n \end{cases}$

(c) $J[(1-x)^{-1} P_n^{(\alpha,\beta)}(x) \ ; \ x \to n]$

$$= \frac{2^{\alpha+\beta} \ \Gamma(\alpha+n+1)\Gamma(\beta+n+1)}{n! \ \alpha\Gamma(\alpha+\beta+n+1)} \ , \quad \text{Re } \alpha > 0 \ , \ \text{Re } \beta > -1$$

(d) $J[(1-x)^\alpha \ P_n^{(\alpha,\beta)}(x) \ ; \ x \to n]$
$$= \frac{2^{4\alpha+\beta+1} \ \Gamma\left(\alpha+\frac{1}{2}\right) \{\Gamma(\alpha+n+1)\}^2 \ \Gamma(\beta+2n+1)}{\sqrt{\pi}(n!)^2 \ \Gamma(\alpha+1) \ \Gamma(2\alpha+\beta+2n+2)}, \quad \text{Re } \alpha > -\frac{1}{2}$$
$$\text{Re } \beta > -1$$

after using the Tables of integrals, if necessary.

(2)   Express $\dfrac{\partial}{\partial x}\left[(1-x^2)\dfrac{\partial u}{\partial x}\right]+[\beta(1-x)-\alpha(1+x)]\dfrac{\partial u}{\partial x}$

as $R\,[u(x,t)]$, if R is the differential operator defined by

$$(1-x)^{-\alpha}(1+x)^{-\beta}\left[\dfrac{\partial}{\partial x}\left\{(1-x)^{\alpha+1}\,(1+x)^{\beta+1}\,\dfrac{\partial}{\partial x}\right\}\right]$$

(3)   Prove that

$$\int_{-1}^{1}\left[(1-x)\,P_n^{(\alpha+1,\beta)}(x)+(1+x)\,P_n^{(\alpha,\beta+1)}(x)\right]$$
$$(1-x)^{-\alpha}(1+x)^{-\beta}\,P_n^{(\alpha,\beta)}\,(x)\,dx$$
$$=2\,\int_{-1}^{1}(1-x)^{-\alpha}\,(1+x)^{-\beta}\left[P_n^{(\alpha,\beta)}\,(x)\right]^2\,dx$$
$$=2.$$

(4)   Expressing x^β , $\beta>0$ as a series using the result from Erdelyi in the form

$$x^\beta=\Gamma(\alpha+\beta+1)\sum_{n=0}^{\infty}\dfrac{(-\beta)_n}{\Gamma(n+\alpha+1)}\,L_n^\alpha(x)\ ,\ x>0,\ \alpha>-1.$$

prove that $\quad L[x^\beta\ ;\ x\to n]=\Gamma(\alpha+\beta+1)\displaystyle\sum_{n=0}^{\infty}\dfrac{(-\beta)_n}{\Gamma(n+\alpha+1)}\delta_n$

where $\quad \delta_n=\begin{pmatrix} n+\alpha \\ n \end{pmatrix}\Gamma(\alpha+1)$

(5)   Prove that

$$\bar f_{\alpha+1}\,(n)=(n+\alpha+1)\,\bar f_\alpha(n)-(n+1)\,\bar f_\alpha(n+1)$$

by using the recurrence relation of Laguerre polynomial

$$x\,L_n^{\alpha+1}\,(x)=(n+\alpha+1)L_n^\alpha\,(x)-(n+1)\,L_{n+1}^\alpha\,(x)$$

(6)   Prove that the zero order Laguerre transfrom of

(a) $f(x)=H(x-a),a\geqslant0$ is exp $(-a)$ if $n=0$ and is $e^{-a}[L_n(a)-L_{n-1}(a)]$, if $n\geqslant1$

(b) $f(x)=e^{-ax}$, $a>-1$ is $a^n(1+a)^{n+1}$

(c) $f(x)=x^m$ is 0 if $n>m$ and is $(-1)^n\begin{pmatrix} m \\ n \end{pmatrix}$, if $m\geqslant n$

(d) $f(x)=L_n(x)$ is 1, if n is a non zero positive integer.

(7) If $L\left[f(x) ; x \to n\right] = \bar{f}_0(n) = \int_0^\infty e^{-x} L_n(x) f(x) \, dx,$

  prove that $L\left[e^x \dfrac{d}{dx} \{x \, e^{-x} \, f'(x)\} ; x \to n\right] = -n \, \bar{f}_0(n)$

(8) Prove that $H[x^m ; x \to n] = 0$, for $m = 0, 1, 2, \cdots (n-1)$

(9) Show that $H[x^n ; x \to n] = \sqrt{\pi}(n!) \, P_n(1)$, where $P_n(x)$
  is a Legendre Polynomial.

(10) Prove that $H[e^{ax}(1 + ax) ; x \to n] = H[x \, e^{ax} ; x \to n].$

Chapter 14

The Z-Transform

14.1 Introduction.

Analysing the behavior of functions whose values are known on a finite or infinite set of discrete points in a given domain cannot be achieved by using the existing Fourier or Laplace Transform techniques. Accordingly, some efficient procedure for the numerical evaluation of the system will be aimed at. For example, such necessity do arise for the determination of the output relation when the discrete input function in known in a signal processing system. The Z-transform technique is one such methods to answer the problem proposed above.

14.2 Z - Transform : Definition.

We have learnt that, under certain restrictions, the Laplace transform of a function $f(t)$ is defined as

$$F(s) = \int_0^\infty f(t)e^{-st} \, dt \tag{14.1}$$

After writing the integral on the right hand side of eqn. (14.1) as a Riemann sum we have

$$F(s) = T \sum_{k=0}^\infty f(kT)exp(-kTs), \tag{14.2}$$

where T is the length of the subinterval chosen to get a proper representation of $f(t)$ with its sampled values at the discrete set of points $t_k = kT$, given by $f(kT), k = 0, 1, 2, \ldots$ and the value of the integral under (14.1) with desired accuracy.

Then putting $z = exp(sT)$, we have $F(s)$ in (14.2) is expressed as

$$F(s) = T \sum_{k=0}^{\infty} f(kT) z^{-k} \qquad (14.3)$$

Here, it shows that $F(s)$ is expanded in a series in power of z. This relation in (14.3) can further be generalized by considering a sequence $\{x_k\}$ for $k = 0, 1, 2, \cdots$ and write a series in powers of z corresponding to the above as

$$\sum_{k=0}^{\infty} x_k z^{-k} = x(z) \qquad (14.4)$$

after conveniently setting $T = 1$ as the interval between samples. This setting and representations have no effect on the properties of this transform.

This relation in (14.4) defines the Z - transform of the sequence $\{x_k\}$. Symbolically, we write

$$Z[\{x_k\}] = X(z) = \sum_{k=0}^{+\infty} x_k z^{-k} \qquad (14.5)$$

as the Z-transform of $\{x_k\}$ and

$$Z^{-1}[X(z)] = \{x_k\} \qquad (14.6)$$

as the inverse Z-transform of $X(z)$. It means that with a given sequence we can associate a series in a domain of complex z-plane and vice versa.

To obtain the inversion of Z-transform in terms of a complex integral in the z-plane, we consider eqn. (14.5) as

$$X(z) = \sum_{k=0}^{\infty} x_k \, z^{-k} \equiv \sum_{k=0}^{\infty} x(k) \, z^{-k} \quad , \text{ say} \qquad (14.7)$$

Or $\quad X(z) = x(0) + x(1)z^{-1} + x(2)z^{-2} + \cdots + x(k)z^{-k} + \cdots \qquad (14.8)$

Multiplying both sides by z^{k-1} and integrating with respect to z along the closed positive contour C, which encloses all singularities of $X(z)$, we get

$$\frac{1}{2\pi i} \oint_C X(z) \, z^{k-1} \, dz = \frac{1}{2\pi i} \left[\oint_C x(0) \, z^{k-1} \, dz + \oint_C x(1) z^{k-2} dz + \cdots + \right.$$

$$\left. + \oint_C x(k) z^{-1} \, dz + \int_C x(k+1) z^{-2} \, dz + \cdots \cdots \right]$$

By Canchy's integral formula, the above relation gives

$$\frac{1}{2\pi i} \oint_C X(z) \, z^{k-1} \, dz = \frac{1}{2\pi i} \oint_C \frac{x(k)}{z} \, dz$$
$$= x(k) \,,$$

since all other integrals on the right hand side vanish.

Thus we get,

$$\frac{1}{2\pi i} \oint_C X(z) \, z^{k-1} \, dz = x(k) \qquad (14.9)$$

implying that

$$z^{-1}[X(z)] = x(k) = \{x_k\} = \frac{1}{2\pi i} \oint_C X(z) \, z^{k-1} \, dz \qquad (14.10)$$

Example 14.1.

If $x(n) = H(n) = \begin{cases} 1, & n \geqslant 0 \\ 0, & n < 0 \end{cases}$, find the $Z-$ transform of $x(n)$.

Solution.

$$Z[x(n)] = \sum_{n=0}^{\infty} [x(0) + x(1)z^{-1} + \cdots + x(n)z^{-n} + \cdots]$$
$$= \sum_{n=0}^{\alpha} \left(\frac{1}{z}\right)^n = \frac{1}{1 - \frac{1}{z}} = \frac{z}{z-1} \,, \qquad |z| > 1.$$

Example 14.2. If $x(n) = a^n$, then find its Z-transform.

Solution.

$$X(z) = Z[x(n)] = \sum_{n=0}^{\infty} \left(\frac{a}{z}\right)^n = \frac{z}{z-a} \,, \qquad |z| > a.$$

Example 14.3. Find the Z-transform of the following functions

(a) $x(n) = n$

(b) $x(n) = n^2$

(c) $x(n) = \frac{1}{n!}$

(d) $x(n) = e^{inx}$

(e) $x(n) = \cosh nx$

(f) $x(n) = H(n)$, where $H(n) = \begin{cases} 0, & \text{if } n < 0 \\ 1, & \text{if } n > 0 \end{cases}$ and n is an integer.

Solution.

(a) $$Z[n] = \sum_{n=0}^{\infty} n\, z^{-n} = z \sum_{n=0}^{\infty} n\, z^{-n-1} = -z\frac{d}{dz}\left[\sum_{n=0}^{\infty} z^{-n}\right]$$
$$= z/(z-1)^2, \quad |z| > 1$$

(b) $$Z[n^2] = Z[n.n] = \sum_{n=0}^{\infty} n.n z^{-n}$$
$$= z \sum_{n=0}^{\infty} n.n\, z^{-(n+1)}$$
$$= z \sum_{n=0}^{\infty} n.\left[-\frac{d}{dz}\, z^{-n}\right]$$
$$= -z\frac{d}{dz} \sum_{n=0}^{\infty} n\, z^{-n}$$
$$= -z\frac{d}{dz}\left[\frac{z}{(z-1)^2}\right], |z| > 1, \text{ by using result in } (a)$$
$$= \frac{z(z+1)}{(z-1)^3}, \ |z| > 1$$

(c) $$Z\left[\frac{1}{n!}\right] = \sum_{n=0}^{\infty} \frac{1}{n!}\, z^{-n} = \exp\left(\frac{1}{z}\right), \text{ for all } z$$

(d) $$Z\left[e^{inx}\right] = \sum_{n=0}^{\infty} \frac{e^{inx}}{z^n} = \sum_{n=0}^{\infty} \left(e^{ix}\right)^n z^{-n} = \sum_{n=0}^{\infty} \left(\frac{e^{ix}}{z}\right)^n$$
$$= \frac{1}{1 - \frac{e^{ix}}{z}} = \frac{z}{z - e^{ix}}$$

Similarly $Z\left[e^{-inx}\right] = \dfrac{z}{z - e^{-ix}}$

Therefore, $Z\left[\cos n\,x\right] = \dfrac{z(z - \cos x)}{z^2 - 2z \cos x + 1}$

and $Z\left[\sin nx\right] = \dfrac{z \sin x}{z^2 - 2z \cos x + 1}$

(e) $Z\left[\cosh nx\right] = \dfrac{1}{2} Z\left[e^{nx} + e^{-nx}\right]$

$= \dfrac{1}{2}\left[\dfrac{z}{z - e^x} + \dfrac{z}{z - e^{-x}}\right] = \dfrac{z(z - chx)}{z^2 - 2z\,chx + 1}$

(f) $[H(n)] = \displaystyle\sum_{n=0}^{\infty} H(n)\,z^{-n} = \sum_{n=0}^{\infty} z^{-n} = \dfrac{z}{z - 1}$, $|z| > 1$

Example 14.4. If $f(n)$ is a periodic sequence of integral period N, then prove that

$$F(z) = Z[f(n)] = \dfrac{z^N}{z^N - 1}\,F_1(z)$$

where $F_1(z) = \displaystyle\sum_{k=0}^{N-1} f(k)z^{-k}$

Solution. By definition,

$$F(z) = Z[f(n)] = \sum_{n=0}^{\infty} f(n)z^{-n}$$

$$= \left[f(0) + f(1)z^{-1} + f(2)z^{-2} + \cdots + f(N-1)z^{-N+1}\right]$$

$$+ \left[f(0)z^{-N}z^0 + f(1)z^{-N}.z^{-1} + \cdots + f(N-1)\,z^{-N}\,z^{-N+1}\right] + \cdots$$

$$= \sum_{k=0}^{N-1} f(k)z^{-k}.1 + z^{-N}\sum_{k=0}^{N-1} f(k)z^{-k} + z^{-2N}\sum_{k=0}^{N-1} f(k)z^{-k} + \cdots$$

$$= \sum_{k=0}^{N-1} f(k)z^{-k}\left[1 + \dfrac{1}{z^N} + \dfrac{1}{z^{2N}} + \cdots\right]$$

$$= \dfrac{1}{1 - \frac{1}{z^N}}\,F_1(z) = \dfrac{z^N}{z^N - 1}\,F_1(Z)$$

14.3   Some Operational Properties of Z-Transform.

Theorem 14.1 : (Translation). If $Z[f(n)] = F(z)$ and $m \geqslant 0$, then

$$Z\left[f(n-m)\right] = z^{-m}\left[F(z) + \sum_{k=-m}^{-1} f(k)z^{-k}\right], \qquad (14.11)$$

$$Z\left[f(n+m)\right] = z^m \left[F(z) - \sum_{k=0}^{m-1} f(k)z^{-k}\right]. \qquad (14.12)$$

In particular, if $m = 1, 2, 3, \cdots$, then from the first relation

$$\left.\begin{array}{l} Z[f(n-1)] = z^{-1} \; F(z) \\[2mm] Z[f(n-2)] = z^{-2} \; F(z) + \displaystyle\sum_{k=-2}^{-1} f(k)z^{-k} \end{array}\right\} \qquad (14.13)$$

Similarly, it follows from the second relation (14.12) that

$$\left.\begin{array}{l} Z[f(n+1)] = z[F(z) - f(0)] \\[2mm] Z[f(n+2)] = z^2[F(z) - f(0)] - zf(1) \\[2mm] Z[f(n+3)] = z^3[F(z) - f(0)] - z^2 f(1) - zf(2) \end{array}\right\} \qquad (14.14)$$

Result in (14.12) will be useful in solving difference and differential equation for initial value problems.

Proof. By definition

$$\begin{aligned} Z[f(n-m)] \;&=\; \sum_{m=0}^{\infty} f(n-m)\, z^{-n} \\[2mm] &=\; \sum_{k=-m}^{\infty} f(k)z^{-m-k} \\[2mm] &=\; z^{-m} \sum_{k=-m}^{\infty} f(k)\, z^{-k} \\[2mm] &=\; z^{-m}\left[\sum_{k=-m}^{-1} f(k)z^{-k} + \sum_{k=0}^{\infty} f(k)\, z^{-k}\right] \\[2mm] &=\; z^{-m}\left[\sum_{k=-m}^{-1} f(k)z^{-k} + F(z)\right] \end{aligned}$$

If $m = 1$, then $Z\left[f(n-1)\right] = z^{-1} \; F(z)$, when $f(k) = 0, \; k < 0$

Also, $\quad Z\left[f(n+m)\right] = \displaystyle\sum_{m=0}^{\infty} f(n+m)z^{-n}$

$$= z^m \sum_{k=m}^{\infty} f(k)z^{-k}$$

$$= z^m \left[\sum_{k=0}^{\infty} f(k)z^{-k} - \sum_{k=0}^{m-1} f(k) \ z^{-k} \right]$$

$$= z^m \left[F(z) - \sum_{k=0}^{m-1} f(k) \ z^{-k} \right]$$

For $m = 1, 2, 3, \cdots$ the results in (14.14) will follow immediately.

Theorem 14.2 : (Multiplication)

If $Z\left[f(n)\right] = F(z)$, then $Z\left[a^n f(n)\right] = F\left(\dfrac{z}{a}\right), |z| > |a|,$

$Z\left[e^{-bn} f(n)\right] = F(ze^b)$ and $Z\left[nf(n)\right] = -z\dfrac{d}{dz}F(z).$

Proof. From definition $\quad Z\left[a^n f(n)\right] = \sum_{n=0}^{\infty} a^n \ f(n) \ z^{-n}$

$$= \sum_{n=0}^{\infty} f(n) \left(\frac{z}{a}\right)^{-n} = F\left(\frac{z}{a}\right), \frac{|z|}{|a|} > 1 \qquad (14.15)$$

Again, $\quad Z\left[e^{-bn} \ f(n)\right] = \sum_{n=0}^{\infty} f(n)z^{-n}e^{-bn}$

$$= \sum_{n=0}^{\infty} f(n) \left(ze^b\right)^{-n}$$

$$= F\left(ze^b\right) \qquad (14.16)$$

Also, $\quad Z\left[nf(n)\right] = \sum_{n=0}^{\infty} nf(n) \ z^{-n}$

$$= z \sum_{n=0}^{\infty} f(n) \left\{-\frac{d}{dz}z^{-n}\right\}$$

$$= -z\frac{d}{dz} \left[\sum_{n=0}^{\infty} f(n)z^{-n}\right]$$

$$= -z\frac{d}{dz} \left[F(z)\right] \qquad (14.17)$$

Thus, all the results of the theorem is proved.

Theorem 14.3 : (Division)

If $Z\left[f(n)\right] = F(z),$ then $\quad Z\left[\dfrac{f(n)}{n+m}\right] = -z^m \displaystyle\int_0^z \frac{F(\xi)}{\xi^{m+1}} \ d\xi$

Proof. By definition, we have

$$Z\left[\frac{f(n)}{n+m}\right] = \sum_{n=0}^{\infty}\frac{f(n)}{n+m}z^{-n}, \quad m \geqslant 0$$

$$= \sum_{n=0}^{\infty}f(n)\left[-\int_{0}^{z}\xi^{-(m+n+1)}d\xi\right]$$

$$= -z^{m}\int_{0}^{z}\xi^{-(m+1)}\left[\sum_{n=0}^{\infty}f(n)\xi^{-n}\right]d\xi$$

$$= -z^{m}\int_{0}^{z}\xi^{-(m+1)}F(\xi)\ d\xi \qquad (14.18)$$

Theorem 14.4 : (The convolution Theorem).

If $Z[f(n)] = F(z)$ and $Z[g(n)] = G(z)$, then Z-transform of the convolution $f(n) * g(n)$ defined by

$$f(n) * g(n) = \sum_{m=0}^{\infty} f(n-m)g(m) \qquad (14.19)$$

is given by

$$Z[f(n) * g(n)]$$
$$= F(z)\ G(z)$$
$$= Z[f(n)]\ Z[g(n)]$$

Equivalently, $Z^{-1}[F(z)\ G(z)] = f(n) * g(n) = \sum_{m=0}^{\infty} f(n-m)\ g(m)$

Proof. From definition

$$Z[f(n) * g(n)] = \sum_{n=0}^{\infty}z^{-n}\left[\sum_{m=0}^{\infty}f(n-m)\ g(m)\right]$$

$$= \sum_{m=0}^{\infty}g(m)\sum_{n=0}^{\infty}f(n-m)z^{-n}$$

$$= \sum_{m=0}^{\infty}g(m)z^{-m}\sum_{k=-m}^{\infty}f(k)z^{-k}$$

$$= \sum_{m=0}^{\infty}g(m)z^{-m}\sum_{k=0}^{\infty}f(k)z^{-k} \ , \ \text{on the assumption that f(k)} = 0, \text{k} < 0$$

$$= Z[f(n)]\ Z[g(n)] \qquad (14.20)$$

Thus the theorem is proved.

Theorem 14.5 : (The initial value Theorem).

If $\quad Z[f(n)] = F(z)$, then $f(0) = \lim\limits_{z \to \infty} F(z)$.

Also if $\quad f(0) = 0$, then $f(1) = \lim\limits_{z \to \infty} zF(z)$.

Proof. we have by definition

$$F(z) = \sum_{n=0}^{\infty} f(n)z^{-n} = f(0) + \frac{f(1)}{z} + \frac{f(2)}{z^2} + \cdots$$

Then letting $z \to \infty$, we get from this relation that

$$\lim_{z \to \infty} F(z) = f(0) \tag{14.21}$$

If $f(0) = 0$, then we get

$$F(z) = \frac{f(1)}{z} + \frac{f(2)}{z^2} + \cdots$$

Therefore, $\quad \lim\limits_{z \to \infty} z\, F(z) = f(1) \tag{14.22}$

Theorem 14.6 : (The Final Value Theorem)

If $Z[f(n)] = F(z)$, then $\lim\limits_{n \to \infty} f(n) = \lim\limits_{z \to 1} [(z-1)F(z)]$,

provided the limits exist.

Proof. We have from (14.14)

$$Z[f(n+1) - f(n)] = z[F(z) - f(0)] - F(z)$$

This implies that

$$\sum_{n=0}^{\infty} [f(n+1) - f(n)]\, z^{-n} = (z-1)F(z) - zf(0)$$

In the limit as $z \to 1$, we obtain

$$\lim_{z \to 1} \sum_{n=0}^{\infty} [f(n+1) - f(n)]\, z^{-n} = \lim_{z \to 1}(z-1)\, F(z) - f(0)$$

Or $\ f(\infty) - f(0) = \lim\limits_{z \to 1}(z-1)\, F(z) - f(0)$

Thus , $\quad \lim\limits_{z \to \infty} f(n) = \lim\limits_{z \to 1}(z-1)\, F\,(z), \tag{14.23}$

provided the limits exist. Hence the result of this theorem is proved.

Theorem 14.7 : The Z-transform of the partial derivatives of a function.

The Z-transform of $\frac{\partial}{\partial a}[f(n,a)]$ is given by

$$Z\left[\frac{\partial}{\partial a} f(n,a)\right] = \frac{\partial}{\partial a}[Z\{f(n,a)\}]$$

Proof. We have by definition

$$Z\left[\frac{\partial}{\partial a} f(n,a)\right] = \sum_{n=0}^{\infty} \left[\frac{\partial}{\partial a} f(n,a)\right] z^{-n}$$

$$= \frac{\partial}{\partial a}\left[\sum_{n=0}^{\infty} f(n,a)z^{-n}\right]$$

$$= \frac{\partial}{\partial a} Z[f(n,a)] \qquad (14.24)$$

For example , if $\quad f(n,a) = ne^{na}$ then

$$Z[ne^{na}] = Z\left[\frac{\partial}{\partial a}e^{na}\right]$$

$$= \frac{\partial}{\partial a}[Z\{e^{na}\}]$$

$$= \frac{\partial}{\partial a}\left[\frac{z}{z - e^a}\right] = \frac{ze^a}{(z - e^a)^2}$$

Example 14.5 If $F(z) = \frac{z}{(z-a)(z-b)}$, find $f(0)$ and $f(1)$ by using the initial value theorem.

Solution. By initial value theorem we get from eqns. (14.21) and (14.22) that

$$f(0) = \lim_{z \to \infty} F(z) = \lim_{z \to \infty} \frac{z}{(z-a)(z-b)} = 0$$

Again,

$$f(1) = \lim_{z \to \infty} zF(z)$$

$$= \lim_{z \to \infty} \frac{z^2}{(z-a)(z-b)} = 1$$

Example. 14.6

Find the inverses of the following Z-transformed functions:

$$(a) \quad F(z) = \frac{z}{z - 2}$$

$$(b) \quad F(z) \;=\; exp\left(\frac{1}{z}\right)$$

$$(c) \quad F(z) \;=\; \frac{3z^2 - z}{(z-1)(z-2)^2}$$

$$(d) \quad F(z) \;=\; \frac{z(z+1)}{(z-1)^3}$$

Solution.

(a) Here $F(z) = \frac{z}{z-2}$

$$= \frac{1}{1 - \frac{2}{z}} = \left(1 - \frac{2}{z}\right)^{-1}$$

$$= 1 + \frac{2}{z} + \frac{2^2}{z^2} + \frac{2^3}{z^3} + \cdots$$

Therefore, $Z^{-1}[F(z)] = Z^{-1}\left[1 + \frac{2}{z} + \frac{2^2}{z^2} + \cdots\right]$

Here $\quad f(0) = 1, \;\; f(1) = 2, \;\; f(2) = 2^2, \;\; f(3) = 2^3, \cdots$

Therefore, $\quad f(n) = 2^n$

So, $\quad Z^{-1}[F(z)] = f(n) = 2^n$

(b) We have in this case

$$F(z) = 1 + \frac{1}{z} + \frac{1}{2!}\frac{1}{z^2} + \cdots$$

Therefore, $\quad f(0) = 1, \quad f(1) = 1, \quad f(2) = \frac{1}{2!}, \cdots$

$\therefore \quad f(n) = \frac{1}{n!}$

$\therefore \quad Z^{-1}[F(z)] = f(n) = \frac{1}{n!}$

Thus $\quad Z^{-1}\left[e^{\frac{1}{z}}\right] = \frac{1}{n!}$

(c) Here, by partial fraction method

$$\frac{3z^2 - z}{(z-1)(z-2)^2} \equiv 2\frac{z}{z-1} - 2\frac{z}{z-2} + \frac{5}{2}\frac{2z}{(z-2)^2}$$

Then, $\quad Z^{-1}\left[\frac{3z^2 - z}{(z-1)(z-2)^2}\right]$

$$= Z^{-1}\left[\frac{2z}{z-1}\right] - Z^{-1}\left[\frac{2z}{z-2}\right] + \frac{5}{2}Z^{-1}\left[\frac{2z}{(z-2)^2}\right]$$

$$= 2.1 - (2)^{n+1} + \frac{5}{2} \cdot n \, 2^n, \quad \text{by theorem (14.7)}$$

$$= 2 - 2^{n+1} + 5n \, 2^{n-1}$$

(d) We can express

$$\frac{z(z+1)}{(z-1)^3} \equiv \frac{z}{(z-1)^2} \left[\frac{z}{z-1} + \frac{1}{z-1} \right]$$

Let $\quad F(z) = \dfrac{z}{(z-1)^2} \quad$ and $\quad G(z) = \dfrac{z}{z-1} + \dfrac{1}{z-1}$

Then by inversion of Z-transform we have

$$f(n) = n \quad \text{and} \quad g(n) = H(n) + H(n-1)$$

where $H(n)$ is the unit function .

Therefore, by convolution theorem

$$Z^{-1}\left[\frac{z(z+1)}{(z-1)^3} \right] = f(n) * g(n)$$

$$= \sum_{m=0}^{n} m[H(n-m) + H(n-m-1)]$$

$$= n^2.$$

14.4   Application of Z-Transforms.

(a) **Solution of difference equations.**

Example 14.7

Solve the following initial value problems :

(i) $f(n+1) + 3f(n) = n, \quad f(0) = 1$

(ii) $f(n+2) - f(n+1) - 6f(n) = 0, \quad f(0) = 0, \; f(1) = 3$

(iii) $u_{n+1} = u_n + u_{n-1}, \; u_1 = u(0) = 1$

(iv) $u(n+2) - u(n+1) + u(n) = 0, u(0) = 1, \; u(1) = 2$

Solutions.

(i) $f(n+1) + 3\,f(n) = n$, $f(0) = 1$

The Z-transform of this difference equation yields

$$(z+3)\,F(z) = z + \frac{z}{(z-1)^2}$$

Or $F(z) = \dfrac{z}{z+3}\left(1 + \dfrac{1}{16}\right) + \dfrac{1}{4}\dfrac{z}{(z-1)^2} - \dfrac{1}{16}\dfrac{z}{z-1}$

Upon inverting this transformed equation by Z-transform we get

$$f(n) = \frac{17}{16}(-3)^n - \frac{1}{16}(1)^n + \frac{1}{4}n.$$

(ii) $f(n+2) - f(n+1) - 6\,f(n) = 0$, $f(0) = 0$, $f(1) = 3$.

The Z-transform of the difference equation under the initial conditions yeilds

$$F(z)[z^2 - z - 6] = 3z$$

Or $F(z) = \dfrac{3z}{5}\left[\dfrac{1}{z-3} - \dfrac{1}{z+2}\right]$

$$= \frac{3}{5}\frac{z}{z-3} - \frac{3}{5}\frac{z}{z+2}$$

Upon inversion the Z-transformed equation yeilds

$$f(n) = \frac{3}{5}3^n - \frac{3}{5}\,(-2)^n.$$

(iii) $u_{n+1} = u_n + u_{n-1}$, $u_1 = u(0) = 1$

This equation defines a sequence which is known as the **Fibonacci Sequence.**

The Z-transform of the above difference equation yeilds

$$U(z) = \frac{z^2}{z^2 - z - 1}$$

Upon inversion of the above transformed equation we get

$$u_n = Z^{-1}\left[\frac{z^2}{z^2 - z - 1}\right]$$

$$\equiv Z^{-1}\left[\frac{z^2}{(z-a)(z-b)}\right], \text{say}$$

$$\text{where } (a,b) = \left(\frac{1+\sqrt{5}}{2}\ ,\ \frac{1-\sqrt{5}}{2}\right)$$

Therefore,

$$
\begin{aligned}
u_n &= Z^{-1}\left[\frac{az}{z-a}\frac{1}{a-b}\right] - Z^{-1}\left[\frac{bz}{z-b}\frac{1}{a-b}\right] \\
&= \frac{a}{a-b}(a)^n - \frac{b}{a-b}(b)^n \\
&= \frac{a^{n+1}-b^{n+1}}{a-b}
\end{aligned}
$$

(iv) $u(n+2) - u(n+1) + u(n) = 0$, $u(0) = 1$, $u(1) = 2$

The Z-transform of this difference equation under the initial conditions yeilds

$$
U(z)\left[z^2 - z + 1\right] = (z^2 + z)
$$

Or $U(z) = \dfrac{z^2 - \frac{z}{2}}{z^2 - z + 1} + \dfrac{\sqrt{3}\left(\frac{\sqrt{3}}{2}z\right)}{z^2 - z + 1}$

Upon inversion of the Z-transformed equation we get

$$
u(n) = \cos\frac{n\pi}{3} + \sqrt{3}\sin\left(\frac{n\pi}{3}\right)
$$

as the solution of the given initial value problem.

(b) **Summation of Infinite Series**

Theorem 14.8   If $Z[f(n)] = F(z)$, then

$$
\sum_{k=1}^{n} f(k) = Z^{-1}\left[\frac{z}{z-1}F(z)\right]
$$

and $\displaystyle\sum_{k=1}^{\infty} f(k) = \lim_{z \to 1} F(z) = F(1).$

Proof. Let $h(n) = \displaystyle\sum_{k=1}^{n} f(k)$. Then clearly,

$$
h(n) - h(n-1) = f(n)
$$

Applying Z-transform this relation gives

$$
H(z) - z^{-1} H(z) = F(z) \text{ , by (14.13)}.
$$

Therefore,

$$
H(z) = \frac{z}{z-1} F(z)
$$

Inverting Z-transform we get

$$h(n) = \sum_{k=1}^{n} f(k)$$

$$= Z^{-1}\left[\frac{z}{z-1}F(z)\right]$$

Also, if $z \to 1$ we have from the Final Value Theorem that

$$\lim_{n\to\infty} h(n) = \lim_{n\to\infty} \sum_{k=1}^{n} f(k)$$

$$= \lim_{z\to 1}\left[\frac{(z-1).z}{z-1}F(z)\right]$$

Or $\quad \sum_{k=1}^{\infty} f(k) = \lim_{z\to 1} [zF(z)] = F(1).$

Thus the proof is complete.

Example 14.8 Find the sum of the following series using Z-transform:

(a) $\quad \sum_{n=0}^{\infty} a^n e^{inx}$

(b) $\quad \sum_{n=0}^{\infty} e^{-(2n+1)x}$

(c) $\quad \sum_{n=0}^{\infty} (-1)^n \frac{x^{n+1}}{n+1}$

(d) $\quad \sum_{n=0}^{\infty} a^n \cos nx$

Solutions.

(a) $\quad$ Let $f(n) = a^n e^{inx}$

We know that $Z\left[e^{inx}\right] = \dfrac{z}{z-e^{ix}}$

Therefore , $Z[f(n)] = Z\left[a^n e^{inx}\right] = \dfrac{z}{z-ae^{ix}}$

So, $\quad \sum_{n=0}^{\infty} a^n e^{inx} = \lim_{z\to 1} \dfrac{z}{z-ae^{ix}}$

$$= \left(1 - ae^{ix}\right)^{-1}$$

(b) $\quad \displaystyle\sum_{n=0}^{\infty} e^{-(2n+1)x} = e^{-x} \sum_{n=0}^{\infty} e^{-2nx}$

So, $\quad Z\left[e^{-(2n+1)x}\right] = Z\left[e^{-x}e^{-2nx}\right]$

$$= \frac{e^{-x}}{1 - ze^{-2x}}$$

Therefore , $\quad \displaystyle\sum_{n=0}^{\infty} e^{-(2n+1)x}$

$$= \lim_{z \to 1} \frac{e^{-x}}{1 - ze^{-2x}}$$

$$= \frac{1}{e^x - e^{-x}} = (2\sinh x)^{-1}$$

(c) We know that $\quad Z\left[x^{n+1}\right] = \dfrac{zx}{z - x}$

Hence , $\quad Z\left[\dfrac{x^{n+1}}{n+1}\right] = z \displaystyle\int_{z}^{\infty} \frac{zx}{z - x} \frac{dz}{z^2}$

$$= xz \int_{z}^{\infty} \frac{dz}{z(z - x)}$$

$$= xz \left[\frac{1}{x} \log \frac{z - x}{z}\right]_{z}^{\infty}$$

$$= -z \log \left[\frac{z - x}{z}\right]$$

Now replacing x by $-x$, this above result gives

$$Z\left[(-1)^n \frac{x^{n+1}}{n+1}\right] = z \log \left[\frac{z + x}{z}\right]$$

Therefore ,

$$\sum_{n=0}^{\infty} (-1)^n \frac{x^{n+1}}{n+1} = \lim_{z \to 1} z \log \left[\frac{z + x}{z}\right]$$

$$= \log(1 + x)$$

(d) We know that

$$Z[\cos nx] = \frac{z(z - \cos x)}{z^2 - 2z \cos x + 1}$$

$$\therefore Z\left[a^n \cos nx\right] = \frac{\frac{z}{a}\left(\frac{z}{a} - \cos x\right)}{\frac{z^2}{a^2} - \frac{2z}{a} \cos x + 1}$$

$$= \frac{z(z - a\cos x)}{z^2 - 2az \cos x + a^2}$$

$$\text{So },\sum_{n=0}^{\infty} a^n \cos nx = \lim_{z \to 1}\left[\frac{z(z - a\cos x)}{z^2 - 2az\cos x + a^2}\right]$$

$$= \frac{1 - a\cos x}{1 - 2a\cos x + a^2}$$

Exercises.

(1) Prove the following :

    (a) $Z\left[n^3\right] = (z^3 + 4z^2 + z)/(z - 1)^4$

    (b) $Z\left[H(n) - H(n - 2)\right] = 1 + \dfrac{1}{z}$

    (c) $Z\left[n^2 a^n\right] = \dfrac{z(z + a)}{(z - a)^2}$

    (d) $Z\left[\sinh na\right] = \dfrac{z\sinh a}{z^2 - 2z\cosh a + 1}$

    (e) $Z\left[na^n f(n)\right] = -z\dfrac{d}{dz}\left[F\left(\dfrac{z}{a}\right)\right]$

    (f) $Z\left[\dfrac{f(n)}{n + m}\right] = z^m \displaystyle\int_z^{\infty} \dfrac{F(\xi)d\xi}{\xi^{m+1}}$,

Hence duduce from the result of (f) that

$$Z\left[\frac{1}{n + 1}\right] = z\,\log\left(\frac{z}{z - 1}\right)$$

(2) Find the inverse Z transform of the following functions :

    (a) $\dfrac{1}{(z - a)^2}$ (b) $\dfrac{z^3}{(z^2 - 1)(z - 2)}$ (c) $\dfrac{z^2}{(z - 1)(z - \frac{1}{2})}$

    (d) $\dfrac{z}{(z - a)^{m+1}}$ (e) $z\,\log\dfrac{z + 1}{z}$ (f) $\dfrac{z}{z - 2}$

$$\left[\text{Ans. (a) } (n - 1)\,a^{n-2}\,H(n - 1) \quad \text{(b) } \frac{1}{6}\left[(-1)^n + 2^{n+3} - 3(1)^n\right]\right.$$

    (c) $(2 - 2^{-n})$ (d) $n(n - 1)\cdots(n - m + 1)\,a^{n-m}\,/m!$

    (e) $\dfrac{(-1)^n}{(n + 1)}$ (f) $\left.2^n\right]$

(3) Solve the following difference equations :

    (a) $f(n + 1) + 2f(n) = n,\, f(0) = 1$

(b) $f(n+2) - 3f(n+1) + 2f(n) = 0, f(0) = 1, f(1) = 2$

(c) $f(n+2) - 5f(n+1) + 6f(n) = 2^n, f(0) = 1, f(1) = 0$

(d) $f(n+2) - 2xf(n+1) + f(n) = 0$, $|x| \leq 1$,

$$f(0) = f_0, \ f(1) = 0, \ n \geqslant 2$$

(e) $f(n+3) - 3f(n+2) + 3f(n+1) - f(n) = 0,$

$$f(0) = 1, f(1) = 0 \ ; \ f(2) = 1$$

Ans. (a) $\dfrac{1}{9} [10(-2)^n + 3n - 1]$ (b) 2^n (c) $2^{n+1} - 3^n - n\,2^{n-1}$

(d) $\dfrac{f_0 \, z[(z-x) - 2x(1-x^2)^{-\frac{1}{2}}]}{(z^2 - 2xz + 1)}$ (e) $(n-1)^2$

(4) Show that

$$\sum_{n=0}^{\infty} \frac{(-1)^n \, e^{-n}}{n+1} = e \, \log \, (1 + e^{-1})$$

(5) (a) If $F(z) = \frac{z}{z-a}$, prove by using initial value theorem that $f(0) = 1$

(b) Use final value theorem to evaluate

$$\lim_{n \to \infty} f(n) \ \text{if} \ F(z) = \frac{(3z^2 - z)}{[(z-1)(z-2)^2]}$$

[Ans.2]

(6) Solve the difference equation

$$y_{k+2} - 3y_{k+1} + 2y_k = 0$$

to prove that $Y(z) = \dfrac{1}{1-2z} - \dfrac{1}{1-z}$

and hence deduce that $y_k = 2^k - 1$, if $y(0) = 0$, $y(1) = 1$.

Appendix : Tables of Integral Transforms

We collect here some important formulae derived in this text or elsewhere which are of common use. For an exhaustive list of transforms the reader should also consult G.A. Campbell and R.M. Foster (Bell Telephone-system, Technical Publication, 1931), W. Magnus and F. Oberhettinger (Springer, Berlin, 1948), R.V. Churchill (Modern Operational Mathematics in Engineering, Mc Graw-Hill, NY, 1944), Erdelyi et al. (Tables of Integral Transforms, Mc Graw-Hill, NY, 1954), Oberhettinger and Higgens (Boeing Sc. Res. Lab. Maths Note 246, Seattle, 1961), Ian N. Sneddon (Mc Graw-Hill Book Company, 1951), etc. Unless otherwise stated here a, t are assumed to be positive constants.

A1 : Fourier Transforms.

$f(t)$	$F[f(t) \; ; \; t \to \xi]$
1	$\sqrt{2\pi}\,\delta(\xi)$
$t^k H(t)$	$\sqrt{2\pi}\,i^k\left[\frac{1}{2}\delta^{(k)}(\xi) + \frac{(-1)^k k!}{2\pi i \xi^{k+1}}\right]$
$e^{-a|t|}$	$\sqrt{\frac{2}{\pi}}\,\frac{a}{a^2+\xi^2}$
$t e^{-\alpha|t|}$	$\sqrt{\frac{2}{\pi}}2ai\xi\,(a^2+\xi^2)^{-2}$
$|t| e^{-a|t|}$	$\sqrt{\frac{2}{\pi}}(a^2-\xi^2)\,(a^2+\xi^2)^{-2}$
$e^{-a^2 t^2}$	$\frac{1}{\sqrt{2}a}e^{-\frac{1}{4}\frac{\xi^2}{a^2}}$
$\delta(t)$	$\frac{1}{\sqrt{2\pi}}$
$\frac{1}{t}$	$-i\sqrt{\frac{\pi}{2}}\;\; sgn\;\;\xi$

A1 : Fourier Transforms. (Continued)

$f(t)$	$F[f(t) ; \ t \to \xi]$
$t^k \ \text{sgn} \ t$	$\sqrt{\frac{2}{k}}(-i\xi)^{-k-1}k!$
$\frac{\sin at}{t}$	$\begin{cases} \sqrt{\frac{\pi}{2}} , & \text{if} \ \ \lvert\xi\rvert < a \\ 0 , & \text{if} \ \ \lvert\xi\rvert > a \end{cases}$
$\frac{chat}{ch\pi t}, -\pi < a < \pi$	$\sqrt{\frac{2}{\pi}} \cos \frac{a}{2} \ ch\frac{\xi}{2}/[ch\xi + \cos a]$
$\frac{shat}{sh\pi t}, -\pi < a < \pi$	$\frac{1}{\sqrt{2\pi}} \frac{\sin a}{ch\xi + \cos a}$
$\begin{cases} \frac{1}{\sqrt{a^2-t^2}} , & \lvert t\rvert < a \\ 0 , & \lvert t\rvert > a \end{cases}$	$\frac{\pi}{2} \ J_0(a\xi)$
$\lvert t\rvert^{-1}$	$\frac{1}{\lvert\xi\rvert}$
$H(a - \lvert t\rvert)$	$\sqrt{\frac{2}{\pi}} \frac{\sin a\xi}{\xi}$
t^{-1}	$i\sqrt{\frac{\pi}{2}} \ \text{sgn} \ \xi$
$(1 - t^2)(1 + t^2)^{-2}$	$\sqrt{\frac{\pi}{2}}\xi \ e^{-\xi}$

A2 : Fourier Cosine Transforms.

$f(t)$	$f_c \ [f(t) ; \ t \to \xi]$
$\begin{cases} 1 , & 0 < t < a \\ 0 , & t > a \end{cases}$	$\sqrt{\frac{2}{\pi}} \frac{\sin a\xi}{\xi}$
$a(t^2 + a^2)^{-1}, a > 0$	$\sqrt{\frac{\pi}{2}}e^{-a\xi}$
e^{-at}	$\sqrt{\frac{2}{\pi}} \frac{a}{a^2+\xi^2}$
$t^{p-1}, \ 0 < p < 1$	$\sqrt{\frac{2}{\pi}} \ \Gamma(p)\xi^{-p} \cos \frac{p\pi}{2}$

A2 : Fourier Cosine Transforms. (Continued)

$f(t)$	$f_c\,[f(t)\ ;\ t \to \xi]$
$e^{-a^2 t^2}$	$\dfrac{1}{\sqrt{2}\,\lvert a\rvert}\,e^{-\frac{1}{4}\frac{\xi^2}{a^2}}$
$\cos\frac{t^2}{2}$	$\dfrac{1}{\sqrt{2}}\,[\cos\frac{\xi^2}{2} + \sin\frac{\xi^2}{2}]$
$\sin\frac{t^2}{2}$	$\dfrac{1}{\sqrt{2}}\,[\cos\frac{\xi^2}{2} - \sin\frac{\xi^2}{2}]$
$\dfrac{sh(\alpha t)}{sh(\beta t)}\,,\ 0<\alpha<\beta$	$\sqrt{\dfrac{\pi}{2}}\dfrac{1}{\beta}\dfrac{\sin(\frac{\pi\alpha}{\beta})}{ch(\frac{\pi\xi}{\beta})+\cos\left(\frac{\pi\alpha}{\beta}\right)}$
$\dfrac{ch(\alpha t)}{ch(\beta t)}\,,\ 0<\alpha<\beta$	$\sqrt{2\pi}\,\dfrac{1}{\beta}\dfrac{\cos\left(\frac{\pi\alpha}{2\beta}\right)ch\left(\frac{\pi\xi}{2\beta}\right)}{ch\left(\frac{\pi\xi}{\beta}\right)+\cos\left(\frac{\pi\alpha}{\beta}\right)}$
$t^{-\nu}J_\nu(at),\ \nu>-\frac{1}{2}$	$\dfrac{(a^2-\xi^2)^{\nu-\frac{1}{2}}\,H(a-\xi)}{2^{\nu-\frac{1}{2}}\,a^\nu\,\Gamma(\nu+\frac{1}{2})}$
$K_0(at)$	$\sqrt{\dfrac{\pi}{2}}\left(a^2+\xi^2\right)^{\frac{1}{2}}$
$2e^{-t}\,\dfrac{\sin t}{t}$	$\sqrt{\dfrac{2}{\pi}}\,\tan^{-1}\left(\dfrac{2}{\xi^2}\right)$
$\dfrac{(1-t^2)}{(1+t^2)^2}$	$\sqrt{\dfrac{\pi}{2}}\,\xi\,e^{-\xi}$

A3 : Fourier Sine Transforms.

$f(t)$	$F_s\,[f(t)\ ;\ t \to \xi]$
$\dfrac{1}{t}$	$\sqrt{\dfrac{\pi}{2}}\,sgn\,\xi$
e^{-at}	$\sqrt{\dfrac{2}{\pi}}\,\dfrac{\xi}{\xi^2+a^2}$
$t\,e^{-at}$	$\sqrt{\dfrac{2}{\pi}}\,\dfrac{2a\xi}{(\xi^2+a^2)^2}$

A3: Fourier Sine Transforms. (Continued)

$f(t)$	$F_s\,[f(t)\,;\,t \to \xi]$
$t^{-1}\,e^{-at}$	$\sqrt{\frac{2}{\pi}}\,\tan^{-1}\left(\frac{\xi}{a}\right)$
$t^{p-1}\,,\ 0 < p < 1$	$\sqrt{\frac{2}{\pi}}\,\xi^{-p}\,\Gamma(p)\,\sin\left(\frac{\pi p}{2}\right)$
$t\,e^{-a^2 t^2}$	$2^{\frac{-3}{2}}\,a^{-3}\,\xi\,e^{\left(-\frac{1}{4}\,\frac{\xi^2}{a^2}\right)}$
$\begin{cases} 0 & ,\quad 0 < t < a \\ (t^2 - a^2)^{-\frac{1}{2}} & ,\quad t > a \end{cases}$	$J_0(a\xi)$
$\frac{1}{t(t^2+a^2)}$	$\sqrt{\frac{\pi}{2}}\,\frac{1}{a^2}(1 - e^{-|\xi a|})\,sgn\,\xi$
$H(a - t)\,,\ a > 0$	$\sqrt{\frac{2}{\pi}}\,\frac{1-\cos\,a\xi}{\xi}$
$\operatorname{cosech} t$	$\sqrt{\frac{\pi}{2}}\tanh(\frac{1}{2}\pi\xi)$
$\frac{\operatorname{cosech}(\alpha t)}{\sinh(\beta t)}, 0 < \alpha < \beta$	$\sqrt{\frac{\pi}{2}}\,\frac{1}{\beta}\,\dfrac{\sinh\left(\frac{\pi\xi}{\beta}\right)}{\cosh\left(\frac{\pi\xi}{\beta}\right) + \cos\left(\frac{\pi\alpha}{\beta}\right)}$
$\frac{\sinh(\alpha t)}{\cosh(\beta t)}, 0 < \alpha < \beta$	$\frac{\sqrt{2\pi}}{\beta}\,\dfrac{\sin\left(\frac{\pi\alpha}{2\beta}\right)\,\sinh\left(\frac{\pi\xi}{2\beta}\right)}{\cos\left(\frac{\pi\alpha}{\beta}\right) + \cosh\left(\frac{\pi\xi}{\beta}\right)}$
$\frac{J_0(at)}{t}$	$\begin{cases} \sqrt{\frac{\pi}{2}} & ,\quad a < \xi < \infty \\ \sqrt{\frac{2}{\pi}}\sin^{-1}\left(\frac{\xi}{a}\right) & ,\quad 0 < \xi < a \end{cases}$
$\frac{t\,J_0(at)}{k^2+t^2}$	$\sqrt{\frac{\pi}{2}}\,e^{-k\xi}\,I_0\,(ak)\,,\ \xi > a$

A4 : Laplace Transforms.

$f(t)$	$L\,[f(t)\,;\,t \to p]$
1	$\frac{1}{p}$, $p > 0$
t^ν , $\nu > -1$	$p^{-\nu-1}\,\Gamma\,(\nu+1)$
e^{at}	$\frac{1}{p-a}$, $Re\ p > a$
$\cosh\,(at)$	$\frac{p}{(p^2-a^2)}$, $Re\ p > a$
$\sin\,(at)$	$\frac{a}{(p^2-a^2)}$, $Re\ p > a$
$t\cosh\,(at)$	$(p^2+a^2)(p^2-a^2)^{-2}$, $Re\ p > a$
$t\,\sinh\,(at)$	$2ap((p^2-a^2)^{-2}$, $Re\ p > a$
$\cos\,(at)$	$p\,(p^2+a^2)^{-1}$
$\sin\,(at)$	$a\,(p^2+a^2)^{-1}$
$t\cos\,(at)$	$(p^2-a^2)\,(p^2+a^2)^{-2}$
$t\sin\,(at)$	$2ap\,(p^2+a^2)^{-2}$
$t^{-1}\sin\,(at)$	$\cot^{-1}\,\left(\frac{p}{a}\right)$
$e^{-\frac{1}{4}b^2t^2}$	$\sqrt{\pi}\,b^{-1}e^{\frac{p^2}{b^2}}\,Erfc\,\left(\frac{p}{b}\right)$, $b > 0$
$Erf(\sqrt{t})$	$p^{-1}(p+1)^{-\frac{1}{2}}$
$Erf(\frac{a}{\sqrt{t}})$	$p^{-1}e^{-2a\sqrt{p}}$
$t^\nu J_\nu(at)$	$\frac{(2a)^\nu\,\Gamma\,(\nu+\frac{1}{2})}{\sqrt{\pi}(p^2+a^2)^{\nu+\frac{1}{2}}}$, $\nu > -\frac{1}{2}$
$t^{\nu+1} J_\nu(at)$	$\frac{(2a)^{\nu+1}a^\nu\,p\,\Gamma\,(\nu+\frac{3}{2})}{\sqrt{\pi}(p^2+a^2)^{\nu+\frac{3}{2}}}$, $R\,ep > a > 0$
$t^\nu I_0(at)$	$p\,(p^2-a^2)^{-\frac{3}{2}}$, $R\,ep > a > 0$
$\frac{e^{at}-1}{a}$	$\frac{1}{p(p-a)}$

A4 : Laplace Transforms. (Continued)

$f(t)$	$L\left[f(t)\; ;\; t \to p\right]$
$\sin(a\sqrt{t})$	$\frac{1}{2}\, a\, \sqrt{\frac{\pi}{p^3}}\, e^{-\frac{a^2}{4p}}\; ,\quad Re\; p > a > 0$
$\frac{\sin(a\sqrt{t})}{t}$	$Erf(\frac{1}{2}\frac{a}{\sqrt{p}})\; ,\quad Re\; p > a > 0$
$\begin{cases} 0 & ,\quad 0 < t < b \\ t^n & ,\quad t > b \end{cases}$	$e^{-bp} \sum_{m=0}^{n} \frac{n!\; b^m}{m!\, p^{n-m+1}}\; ,\quad Re\; p > 0$
$\begin{cases} 0 & ,\quad 0 < t < b \\ t/\sqrt{t^2 - b^2} & ,\quad t > b \end{cases}$	$b K_1(bp)\; ,\quad Re\; p > 0.$
$e^{-\frac{1}{4}\frac{a}{t}}$	$\sqrt{a}\, p^{-\frac{1}{2}} K_1(\sqrt{pa})\; ,\quad Re\; a \geqslant 0.$
$\log t$	$-p^{-1} \log(\gamma p)$
$\frac{2}{t} \sinh(at)$	$\log \frac{p+a}{p-a}\; ,\quad Re\; p > |Re\; a|$
$t \cos at$	$\frac{(p^2 - a^2)}{(p^2 + a^2)^2}$
$t \cosh at$	$(p^2 + a^2)/(p^2 - a^2)^{-2}$
$\frac{\sin(at)}{t}$	$tan^{-1}\left(\frac{a}{p}\right)$
$J_0\,(at)$	$(p^2 + a^2)^{-\frac{1}{2}}$
$I_0(at)$	$(p^2 - a^2)^{-\frac{1}{2}}$
$\delta(t - a)$	$exp\,(-ap)\; ,\quad a > 0$
$|\sin at|\; ,\quad a > 0$	$\frac{a}{(p^2 + a^2)} \coth\left(\frac{\pi p}{2a}\right)$

A5 : Hilbert Transforms.

$f(t)$	$H\,[f(t)\,;\,t \to x]$
$\dfrac{t}{t^2+a^2}$	$\dfrac{a}{x^2+a^2}$
$H(t-a)-H(t-b)$	$\dfrac{1}{\pi}\,\log\left|\dfrac{b-x}{a-x}\right|\,,\ b > a > 0$
$\dfrac{H(t-a)}{t}$	$\dfrac{1}{\pi x}\,\log\dfrac{a}{|a-x|}\,,\ a > 0$
$(at+b)^{-1}\,H(t)$	$\dfrac{1}{\pi(ax+b)}\,\log\dfrac{b}{a|x|},\ a,b > 0,\ x \neq \dfrac{b}{a}$
$|t|^{\nu-1}\,,\ 0 < Re\,\nu < 1$	$-ctn\left(\dfrac{\nu\pi}{2}\right)|x|^{\nu-1}\,sgn\,x$
e^{iat}	$i\,e^{iax}$
$\sin(at)/t$	$\dfrac{\{\cos(ax)-1\}}{x}$
$\sin at\,J_1(at)$	$\cos(ax)\,J_1\,(a\,x)$
$\cos at\,J_1(at)$	$-\sin(ax)\,J_1\,(a\,x)$

A6 : Stieltjes Transforms.

$f(x)$	$S\,[\,f(x)\,;\,x \to y\,] = \int_0^\infty \dfrac{f(x)}{x+y}\,dx$
$\dfrac{1}{x+a}$	$\dfrac{1}{a-y}\,\log\left|\dfrac{a}{y}\right|$
$\dfrac{e^{-ax}}{\sqrt{x}}$	$\dfrac{\pi}{\sqrt{y}}\,e^{ay}\,Erfc\,(\sqrt{ay}).$
$\dfrac{1}{a^2+x^2}$	$\dfrac{1}{a^2+y^2}\left[\dfrac{\pi y}{2a}-\log\dfrac{y}{a}\right]$
$x^\nu\,,\ -1 < Re\,\nu < 0$	$-\pi y^\nu\,cosec\,(\pi\nu)$
$\dfrac{x^\nu}{a+x}\,,\ -1 < Re\,\nu < 1$	$\dfrac{\pi(a^\nu-y^\nu)}{(a-y)\sin\nu\,\pi}$
$x^{-\nu}e^{-ax}\,,\ Re\,\alpha > 0,\ Re\,\nu < 1$	$\Gamma(1-\nu)\,y^{-\nu}\,e^{ay}\,\Gamma\,(\nu,\,ay)$

A6 : Stieltjes Transforms. (Continued)

$f(x)$	$S\ [\ f(x)\ ;\ x \to y\ ] = \int_0^\infty \frac{f(x)}{x+y}\ dx$
$\frac{\log x}{(a+x)}$, $\lvert arg\ a\rvert < \pi$	$\frac{1}{2}\ (y-a)^{-1}\left[(\log y)^2 - (\log a)^2\right]$
$\sin(a\sqrt{x})$	$\pi exp(-a\sqrt{y})$
$x^{-1}\sin(a\sqrt{x})$	$\pi y^{-1}\left[1 - exp\ (-a\sqrt{y})\right]$
$\frac{1}{\sqrt{x}}\ \cos(a\sqrt{x})$	$\frac{\pi}{\sqrt{y}}\ exp\ (-a\sqrt{y}\ )$

A7 : Hankel Transforms.

$f(x)$	$H_\nu\ [\ f(x)\ ;\ x \to \xi\ ]$
$\begin{cases} x^\nu & ,\ \ 0 < x < a \\ 0 & ,\ \ x > a \end{cases}\ (\nu > -1)$	$\frac{a^{\nu+1}}{\xi}\ J_{\nu+1}\ (\xi a)$
$(a^2 - x^2)\ H\ (a - x)$	$\frac{4a}{\xi^3}J_1(\xi a) - \frac{2a^2}{\xi^2}J_0(\xi a)$
e^{-px} , $(\nu = 0)$	$p\ (\xi^2 + p^2)^{-\frac{1}{2}}$
$\frac{e^{-px}}{x}$, $(\nu = 0)$	$(\xi^2 + p^2)^{-\frac{1}{2}}$
e^{-px} , $(\nu = 1)$	$\xi\ (\xi^2 + p^2)^{-\frac{1}{2}}$
$\frac{e^{-px}}{x}$, $(\nu = 1)$	$\frac{1}{\xi} - \frac{p}{\xi\sqrt{\xi^2+p^2}}$
$\frac{a}{\sqrt{a^2+x^2}}$, $(\nu = 0)$	$e^{-a\xi}$
$\frac{\sin(ax)}{x}$, $(\nu = 0)$	$(a^2 - \xi^2)^{-\frac{1}{2}}\ H(a - \xi)$
$\frac{\sin(ax)}{x}$, $(\nu = 1)$	$\frac{a}{\xi\sqrt{\xi^2-a^2}}\ H\ (\xi - a)$

A7: Hankel Transforms. (Continued)

$f(x)$	$H_\nu \ [\ f(x)\ ;\ x \to \xi\ ]$
$x^\nu e^{-\frac{x^2}{a^2}}$	$\left(\frac{1}{2}a^2\right)^{\nu+1} \xi^\nu\ e^{-\frac{\xi^2 a^2}{4}}$
$x\,e^{-ax}\ ,\ (\nu = 1)$	$\xi \left(a^2 + \xi^2\right)^{-\frac{3}{2}}$
$\frac{1}{x}\ ,\ (\nu = 0)$	$\frac{1}{\xi}$
$\sqrt{x}\,\left(a^2 + x^2\right)^{-\frac{1}{2}}\ ,\ (\nu = 0)$	$\frac{1}{\sqrt{\xi}}\,e^{-a\xi}$
$\sqrt{x}\,\left(a^2 - x^2\right)^{-\frac{1}{2}}\,H(a - x)\ ,\ (\nu = 0)$	$\frac{1}{\sqrt{\xi}}\,\sin\,(-a\xi)$
$\sqrt{x}\,\left(x^2 - a^2\right)^{-\frac{1}{2}}\,H\,(x - a)\ ,\ (\nu = 0)$	$\frac{1}{\sqrt{\xi}}\,\cos\,(-a\xi)$
$\frac{1}{\sqrt{x}}\,\log x\ ,\ (\nu = 0)$	$-\frac{1}{\sqrt{\xi}}\,\log\,(2\xi)$
$\frac{1}{\sqrt{x}}\,\sin\,(ax)\ ,\ (\nu = 0)$	$\xi^{-\frac{1}{2}}\,\left(a^2 - \xi^2\right)^{-\frac{1}{2}}\,H\,(a - \xi)$
$\frac{1}{\sqrt{x}}\,\cos\,(ax)\ ,\ (\nu = 0)$	$\xi^{\frac{1}{2}}\sqrt{(\xi^2 - a^2)}\,H\,(\xi - a)$
$\sqrt{x}\,\cos\left(\frac{a^2 x^2}{2}\right)\ ,\ (\nu = 0)$	$\frac{\sqrt{\xi}}{2a^2}$
$H(a - x)\ ,\ (\nu = 0)$	$\frac{a}{\xi}\,J_1(a\xi)$
$\frac{1}{x^2}\,(1 - \cos ax)\ \ ,\ (\nu = 0)$	$\cosh^{-1}\left(\frac{a}{\xi}\right)\,H\,(a - \xi)$
$\left(x^2 + a^2\right)^{-\frac{1}{2}}\ ,\ (\nu = 0)$	$\frac{1}{\xi}\,e^{-a\xi}\left\{\sqrt{\xi^2 + a^2} - a\right\}^2$
$\frac{e^{-ax}}{x}\ ,\ (\nu = 1)$	$\frac{1}{\xi}\left[1 - \frac{a}{(\xi^2 + a^2)^{\frac{1}{2}}}\right]$

A8 : Finite Hankel Transforms.

$f(r)$	$\int_0^a f(r)\, r\, J_n\, (r\xi) dr = H_\nu\, [\, f(r)\, ;\, r \to \xi\,]$
$c,\quad (n{=}0)$	$c\frac{a}{\xi} J_1(\xi a)$
$(a^2 - r^2)\ ,\ (n = 0)$	$\frac{4a}{\xi^3}\, J_1(\xi a) - \frac{2a^2}{\xi^2}\, J_0(\xi a)$
$r^n\ ,\ (n > -1)$	$\frac{a^{n+1}}{\xi}\, J_{n+1}\, (\xi a)$
$\frac{J_0(ar)}{J_0\,(a)} - 1\ ,\ (n = 0)$	$\dfrac{-J_1(\xi)}{\xi\left(1 - \frac{\xi^2}{a^2}\right)}$
$\frac{1}{r}\ ,\ (n = 1)$	$\frac{1}{\xi}\{1 - J_0(a\xi)\}$
$(a^2 - r^2)^{-\frac{1}{2}}\ ,\ (n = 0)$	$\frac{1}{\xi}\, \sin a\xi$
$\frac{1}{r}(a^2 - r^2)^{-\frac{1}{2}}\ ,\ (n = 0)$	$\frac{(1-\cos a\xi)}{(a\xi)}$

A9 : Mellin Transforms.

$f(x)$	$M\,[f(x)\, ;\, x \to s] = \int_0^\infty f(x)\, x^{s-1} dx$
e^{-px}	$p^{-s}\, \Gamma\,(s)\ ,\ \ Re\ s > 0$
e^{-x^2}	$\frac{1}{2}\, \Gamma\,\left(\frac{s}{2}\right)$
$(1 + x)^{-a}$	$\Gamma\,(s)\, \Gamma(a - s)/\Gamma(a)\ ,\ 0 < Re\ s < Re\ a$
$(1 + x^a)^{-1}$	$\frac{\pi}{a}\, \mathrm{cosec}\, \frac{\pi s}{a}\ ,\ 0 < Re\ s < Re\ a$
$\sin x$	$\Gamma\,(s)\, \sin\left(\frac{s\pi}{2}\right),\ 0 < Re\ s < 1$
$\cos x$	$\Gamma\,(s)\, \cos\left(\frac{s\pi}{2}\right),\ 0 < Re\ s < 1$
$H(a - x)$	$\frac{a^s}{s}$

A9 : Mellin Transforms. (Continued)

$f(x)$	$M[f(x) ; x \to s] = \int_0^\infty f(x) x^{s-1} dx$
$(1-x)^{p-1} H(1-x)$	$\frac{\Gamma(s)\Gamma(p)}{\Gamma(s+p)}$, $Re\ p > 0$
$\log(1+x)$	$\frac{\pi}{s} \cosec s\pi$
$(1-x)^{-1}$	$\pi \cot(s\pi)$, $0 < Re\ s < 1$
$x^\nu H(1-x)$	$(s+\nu)^{-1}$, $Re\ s > -Re\ \nu$
$\log x . H(a-x)$	$\frac{1}{sa^s}\left(\log a - \frac{1}{s}\right)$, $Re\ s > 0$
$\frac{\{\log x\}}{(a+x)}$, $|arg\ a| < \pi$	$\pi a^{s-1} \cosec \pi s\ [\log a - \pi \cot \pi s]$, $0 < Re\ s < 1$
$x^\nu \log x\ H(1-x)$	$-(s+\nu)^{-2}$, $Re\ s > -Re\ \nu$
$e^{-x}(\log x)^n$	$\frac{d^n}{ds^n}[\Gamma(s)]$, $Re\ s > 0$
$\log|1-x|$	$\pi s^{-1} \cot(\pi s)$, $-1 < Re\ s < 0$
$\log\left|\frac{1+x}{1-x}\right|$	$\frac{\pi}{s} \tan \frac{\pi s}{2}$, $-1 < Re\ s < 1$
$J_\nu(ax)$, $(a > 0)$	$2^{s-1} \Gamma\left(\frac{s}{2} + \frac{\nu}{2}\right)a^s \Gamma\left(\frac{\nu}{2} - \frac{s}{2} + 1\right)$, $Re\ \nu < Re\ s < \frac{3}{2}$
$\log\frac{a}{x}H(a-x)$	$\frac{a^s}{s^2}$
$x^{-1} \log(1+x)$	$\pi(1-s)^{-1} \cosec \pi s$

A9 : Mellin Transforms. (Continued)

$f(x)$	$M[f(x) ; x \to s] = \int_0^\infty f(x) x^{s-1} dx$
$x^a (1+x)^{-b}$	$\Gamma(a+s) \Gamma(b-a-s)\Gamma(b)$
$(1+ax)^{-n}$	$\frac{\Gamma(s) \Gamma(n-s)}{a^s \Gamma(x)}$, $0 < Re \ s < n$

A10 : Kontorovich - Lebedev Transforms.

$f(x)$	$K[f(x) ; \tau] \equiv \bar{f}(\tau) = \int_0^\infty x^{-1} f(x) K_{i\tau}(x) dx$
1	$\frac{\pi}{2\tau} \ \text{cosech} \left(\frac{\pi\tau}{2}\right)$
x	$\frac{\pi}{2} \ \text{sech} \left(\frac{\pi\tau}{2}\right)$
e^{-x}	$\frac{\pi}{\tau} \ \text{cosech} (\tau\pi)$
e^x	$\frac{\pi}{\tau} \ \text{coth} (\tau\pi)$
xe^{-x}	$\pi\tau \ \text{cosech} (\tau\pi)$
$e^{-x \cosh t}$	$\frac{\pi \cos(\tau t)}{\tau \sinh(\pi\tau)}$
$x \ e^{-x \cos t}$	$\frac{\pi \sinh(\tau t) \text{cosec } t}{\sinh(\pi\tau)}$
$x \ e^{-x \cosh t}$	$\frac{\pi \sin(\tau t)}{\sinh(\pi\tau) \sinh t}$

A11 : Mehler - Fock Transforms (of zero order)

$f(\alpha)$	$f_0^*(\tau) = \Phi[f(\alpha) \ ; \ \alpha \to \tau]$
$H(t - \alpha) \ (ch \ t - ch \ \alpha)^{-\frac{1}{2}}$	$\sqrt{2} \ \tau^{-1} \sin \ t\tau$
$H(\alpha - t) \ (ch \ \alpha - ch \ t)^{-\frac{1}{2}}$	$\sqrt{2} \ \tau^{-1} \ \text{cth} \ (\pi\tau) \ \cos(\tau \ t)$
$(ch \ \alpha + \cos \ \beta)^{-1}$	$\pi \ \text{sech} \ (\tau\pi) \ P_{-\frac{1}{2}+i\tau} \ (\cos\beta), \ -\pi < \beta < \pi$
$(ch \ \alpha + \cos \ \beta)^{-\frac{1}{2}}$	$\frac{\sqrt{2}}{\tau} \frac{ch \ (\beta\tau)}{sh \ (\pi\tau)} , \pi < \beta < \pi$
$\text{sech} \ (\frac{\alpha}{2})$	$2 \ \tau^{-1} \ \text{cosech} \ (\pi\tau)$
$\sqrt{\text{sech} \ (\alpha)}$	$\frac{1}{\sqrt{2\tau}} \ \text{cosech} \ (\frac{\pi}{2}\tau)$

A11 : Another form of the Mehler-Fock Transforms.

$f(x)$	$\bar{f}_0(\tau) = \Phi_0[f(x) \ ; \ x \to \tau] = \int_1^\infty \ f(x) \ P_{-\frac{1}{2}+i\tau} (x) \ dx$
e^{-ax}	$\sqrt{\frac{2}{\pi a}} \ K_{i\tau} \ (a) \ , \ \ |arg \ a| < \frac{\pi}{2}$
$x^{-\frac{1}{2}}$	$\sqrt{2} \ \text{sech} \ (\frac{\pi}{2}\tau)$
$x^{-\frac{3}{2}}$	$\frac{2\sqrt{2}}{3} \ \tau \ \text{cosech}(\frac{\pi\tau}{2})$
$\frac{1}{x+t}$	$\pi\text{sech} \ (\pi\tau)P_{-\frac{1}{2}+i\tau} \ (t), \ \ |t| < 1$
$\frac{1}{\sqrt{x+t}}$	$\frac{\sqrt{2} \ ch(\tau \cos^{-1} t)}{\tau \ sh \ (\pi\tau)}$

A12 : The Z - Transforms

y_n	$\sum y_n z^n$
$\begin{cases} 1 & , \quad n = k \\ 0 & , \quad n \neq k \end{cases}$	z^k
c	$\frac{c}{1-z}$
n	$\frac{z}{(1-z)^2}$
n^2	$\frac{z(1+z)}{(1-z)^3}$
n^3	$\frac{z(z^2+4z+1)}{(1-z)^4}$
c^n	$\frac{1}{(1-cz)}$
nc^n	$\frac{cz}{(1-cz)}$
$\begin{pmatrix} n+k \\ k \end{pmatrix} c^n$	$\frac{1}{(1-cz)^{k+1}}$
$\begin{pmatrix} b \\ n \end{pmatrix} c^n \, a^{b-n}$	$(a+cz)^b$
$\frac{c^n}{n}$, $(n = 1, 2, 3, \cdots)$	$-ln(1-cz)$
$\frac{c^n}{n}$, $(n = 1, 3, 5, \cdots)$	$\frac{1}{2} \, ln \left(\frac{1+cz}{1-cz} \right) = \tanh^{-1}(cz)$
$\frac{c^n}{n}$, $(n = 2, 4, 6, \cdots)$	$-\frac{1}{2} \, ln \left(1 - c^2 z^2 \right)$
$c^n / n!$	e^{cz}
$c^n / n!$, $(n = 1, 3, 5, \cdots)$	$\sinh(cz)$
$c^n / n!$, $(n = 0, 2, 4, \cdots)$	$\cosh(cz)$
$(lnc)^n / n!$	c^n

A12 : The Z - Transforms (Continued)

y_n	$\sum y_n z^n$
$\sin(cn)$	$\dfrac{z \sin c}{z^2 + 1 - 2z \ \cos c}$
$\cos(cn)$	$\dfrac{1 - z \cos c}{z^2 + 1 - 2z \ \cos c}$
$b^{-an} \sin(cn)$	$\dfrac{z \sin c}{b^a + b^{-a} z^2 - 2z \cos c}$
$b^{-an} \cos(cn)$	$\dfrac{b^a - z \ \cos \ c}{b^{-a} z^2 + b^a - 2z \ \cos c}$

Bibliography

The following is a sketch of the bibliography which can not be considered as comprehensive and exhaustive. It consists of books and research papers to which reference is made in the text or otherwise. Many other related materials from other books or papers are also included in the text. Acknowledgement is due to all the authors.

(1) Bateman, Harry (1954). Tables of Integral Transforms, Mc Graw-Hill Book Company Inc.

(2) Chakraborty, A and Williams, W.E. (1980). A note on a singular Integral Equations, J. Inst. Maths. Applics, 26, 321-323.

(3) Churchill, R. V (1951). The Operational calculus of Legendre Transforms, J. Math. and Phy, 33, 165-178.

(4) Churchill, R.V. and Dolph, C.L. (1954). Inverse Transform of product of Legendre Transforms, Proc. Amer. Math. Soc., 5, 93-100.

(5) Comninou, M (1977). The interface Crack, J. Appl. Mech., 44,631-636.

(6) Das, S, Patra, B and Debnath, L (2004). On elasto-dynamical problem of interfacial Griffith cracks in composite media, Int. J. Engng. Sci., 42, 735-752.

(7) Debnath, L (1995). Integral Transforms and their applications, C R C Press Inc.

(8) Debnath, L and Thomas, J (1976). On finite Laplace Transformation with Applications, ZAMM, 56, 559-563

(9) Dunn, H.S. (1967). A generalisation of Laplace Transform, Proc. Camb. Phil. Soc., 63, 155-161.

(10) Gakhov, F. D. (1966). Boundary Value Problems, Pergamon Press, London.

(11) Kober, H. (1967). A modification of Hilbert Transform, the Weyl integral and functional equations, J. Lond. Math. Soc., 42, 42-50.

(12) Muskhelishvili, N.I. (1953). Singular Integral Equations, Noord-hoff, Groningen

(13) Magnus, W and Oberhettinger, F. (1943). Formulas and Theorems for the functions of Mathematical Physics. Chelsea Publishing Co., NY

(14) Naylor, D. (1963). On a Mellin Type Integral Transform, J. Math. Mech., 12, 265-274

(15) Okikiolu, G. O. (1965). A generalisation of Hilbert Transform.

(16) Peters, A.S. (1972). Pair of Cauchy Singular Integral Equations, Comm. Pure. Appl. Maths, 25, 369-402

(17) Riemann, B. (1976). Über die Anzahi der Primzahlen under eine Gegebenen Grösse, Gesammelte Math. Werde, 136-144.

(18) Sneddon, I. N. (1946). Finite Hankel Transform, Phil Mag., 37, 17.

(19) Sneddon, I. N. (1974). The use of Integral Transform, Tata Mc Graw-Hill Publishing Co. Ltd.

(20) Senddon, I.N. (1951). Fourier Transforms, Mc Graw-Hill Book Company, NY.

(21) Tricomi, F.G. (1951). On the finite Hilbert Transformation, Q.J. Maths., Oxford, 2, 199-211

(22) Tricomi, F.G. (1955). Integral Equations, Dover Publications, Inc, NY.

(23) Titchmarsh, E.C. (1959). An Introduction to the Theory of Fourier Integrals, Second Edition, Oxford Univ. Press, Oxford.

(24) Ursell, F. (1983). Integrals with large Parameters : Hilbert Transforms, Math. Proc. Comb. Phil. Soc., 93, 141-149.

(25) Gradshteyn, I.S. and Ryzhik, I.M. (1980). Table of Integrals, Series and Products, Academic Press, NY.

(26) Oberhettinger, F(1974). Tables of Mellin Transforms and Application, Springer-Verlag, N.Y.

(27) Watson, E. J. (1981). Laplace Transforms and Applications, Van Nostrand Reinhold, N.Y.

Index

Printed and bound by CPI Group (UK) Ltd, Croydon, CR0 4YY

17/10/2024

01775682-0018